Sage Classics 1

SOCIAL EXPERIMENTATION

Donald T. Campbell
M. Jean Russo

SAGE Publications
International Educational and Professional Publisher
Thousand Oaks London New Delhi

For information:

SAGE Publications, Inc.
2455 Teller Road
Thousand Oaks, California 91320
E-mail: order@sagepub.com

SAGE Publications Ltd.
6 Bonhill Street
London EC2A 4PU
United Kingdom

SAGE Publications India Pvt. Ltd.
M-32 Market
Greater Kailash I
New Delhi 110 048 India

Printed in the United States of America

Library of Congress Cataloging-in-Publication Data

Campbell, Donald Thomas, 1916-
Social experimentation / by Donald T. Campbell and M. Jean Russo.
p. cm.
Includes bibliographical references and index.
ISBN 0-7619-0404-2 (cloth: acid-free paper)
ISBN 0-7619-0405-0 (pbk.: acid-free paper)
1. Sociology—Methodology 2. Social sciences—Methodology. 3. Experimental design. I. Russo, M. Jean. II. Title.
HM24.C2294 1998
301'.01—ddc21 98-25447

This book is printed on acid-free paper.

99 00 01 02 03 10 9 8 7 6 5 4 3 2 1

Acquiring Editor: C. Deborah Laughton
Editorial Assistant: Eileen Carr
Production Editor: Wendy Westgate
Editorial Assistant: Denise Santoyo
Designer/Typesetter: Janelle LeMaster
Cover Designer: Ravi Balasuriya

SOCIAL EXPERIMENTATION

SAGE CLASSICS

The goal of the **Sage Classics Series** is to help bring new generations of social scientists and their students into a deeper, richer understanding of the roots of social science thinking by making important scholars' classic works available for today's readers. The Series will focus on those social science works that have the most relevance for impacting contemporary thought, issues, and policy—including increasing our understanding of the techniques, methods, and theories that have shaped the evolution of the social sciences to date.

The works chosen for the Series are cornerstones upon which modern social science has been built. Each volume includes an introduction by a preeminent leader in the discipline, and provides an historical context for the work while bringing it into today's world as it relates to issues of modern life and behavior. Moving into the future, opportunities for application of the findings are identified and references for further reading are provided.

It is our hope at Sage that the reissuance of these classics will preserve their place in history as having shaped the field of social science. And it is our wish that the Series, in its reader-friendly format, will be accessible to all in order to stimulate ongoing public interest in the social sciences as well as research and analysis by tomorrow's leaders.

CONTENTS

PREFACE

Donald T. Campbell was a giant in the post-World War II social science community. One can travel the world over and, when conversing with social scientists, always receive at least a glimmer of recognition at the mention of Dr. Campbell's name. He remained active in academe until well into his 70s, and as the 1990s progressed, he believed that a budding generation of social science practitioners might benefit from a selected compendium of his writings. This collection of essays by Dr. Campbell provides such a compendium.

Certain debates over methodological issues will always rage in the social scientific community, and Dr. Campbell had unfailingly been at the forefront of those debates. Times will change, scientists' goals and ideas will evolve, but Dr. Campbell's writings will always serve as a resource to steer those in the field away from the dangerous pitfalls inherent in social research.

This book is a presentation of Dr. Campbell's ideas in a format that is intended to guide the reader through a selection of his collected teachings. It is not a critique. Rather, it is edited with the needs of the social science reader in mind. The special challenge was to clarify the essence of Dr. Campbell's work without watering it down or diverting the reader from his special themes. Readers should note that many of the original articles from which these chapters were derived were written before much attention was paid in social science—or any other discipline, for that matter—to the issue of gender-neutral language. Therefore, in the majority of chapters there is a nearly exclusive use of the pronoun *he* to refer to all persons. Likewise, *man* is used when today *humanity* would likely prevail.

It is my hope that this book will serve to challenge, to intrigue, and ultimately to assist and guide the social science reader. Dr. Campbell's writings invite the reader to think about the challenges, the problems, the ultimate purpose, and the significance of social research.

I do not intend to argue Dr. Campbell's case to those who have been his antagonists. On the contrary, it was Dr. Campbell's wish, and my goal, to show others the basic building blocks, the tools, and the philosophical underpinnings that formed the basis of his thinking over the years. If this can be accomplished, then others will gain the benefit of his brilliance, so that they can themselves carry on the great, ongoing debate over the focus and direction of social research. We lost Dr. Campbell in 1996, but his legacy lives on to enhance our thinking and to embolden us to take on new challenges.

I will always remember that summer day in 1995 when Dr. Campbell called to ask me if I would assist him with the project. I studied under him at Lehigh University where I did my graduate work. I had stayed on at Lehigh after my postdoctoral work to pursue several research projects. My schedule left me with precious little free time, but when he asked me to work on this with him, my response was a resounding "yes." I would gladly labor evenings and weekends to have the opportunity to delve into the treasure trove of material this icon of social science had put together over the course of his career.

We met several times to lay out the framework for this book. Dr. Campbell chose the topics himself, as well as the order in which they were to appear. His vision, and that of his friend, Sara Miller McCune of Sage Publications, was to make his work more "accessible" to the student. Dr. Campbell had a limitless vocabulary and a consuming passion for words. If a word did not exist to express an idea he had, he was known to create one. For example, when we met a colleague from his early years at the University of Chicago, Dr. Campbell described himself as "postcocious" (as opposed to precocious) because he had been late in completing his PhD because of his service in the U.S. Naval Reserves during World War II.

If anything critical can be said of Dr. Campbell's prose, it is perhaps that it extended beyond the grasp of many of his readers. While working on this book, I attempted to tone down some of the more esoteric facets of his writing. That said, it is not always appropriate to eliminate technical terms, because to do so could mislead or confuse the reader. In order to derive the maximum benefit from this book, however, it is recommended that the reader possess a working knowledge of elementary statistics and research basics. Also, some of the terms in the theory of science part may be unfamiliar to those who do not possess a background in philosophy. One might seek working definitions for these terms in a reference book, such as the *Cambridge Dictionary of Philosophy;* however, the ideal way to master the themes espoused by Dr. Campbell is to experience them within the rich flow of his writings.

In my own readings of the articles, I found that I often stopped to reread passages to be certain I understood what was being said. When I finished an article, I was a bit overwhelmed by the amount of information that had been presented. There was always the fear that I overlooked an important point or overstressed a minor one. It took some time for me to synthesize the ideas and summarize the central points of the article. To save the student from this uncertainty, I offer my own chapter overview preceding each chapter.

These overviews are intended to serve as a road map: They tell you what you are about to read. But do not conclude that the introductions convey the full meaning of the chapter, any more than a road map conveys the beauty of a landscape. As a suggestion, perhaps it would be helpful to read an introduction prior to delving into a chapter, and then reread the introduction after completing the chapter. The first reading draws your attention to the salient points, and the second reading tends to clarify and summarize the chapter. In the final analysis, though, the ultimate reward is derived by reading the thoughts of Dr. Campbell, a gifted scholar who will always be remembered as one of the most creative and ingenious minds in the social sciences.

It was of vital importance to Dr. Campbell to complete this book as well as a second volume on social measurement. As we worked together on the initial stages, he remarked to me on several occasions, "I've set this up in a way that you could continue on this book even if I should die." I joked to him that he would not be allowed to die, because frankly, I just had too many questions to ask. Unfortunately, his words were prophetic. I was left to continue this work on my own. It is my heartfelt wish to have adequately presented the collected works of this great man in a way that will fulfill the faith that he had in my abilities.

To *Donald T. Campbell* and *Donald P. Russo,*
who both encouraged me to grow.

ACKNOWLEDGMENTS

s with any project of this magnitude, there are many individuals who offered their assistance and to whom I am deeply grateful. First, of course, is Don Campbell, whose fertile and expansive mind produced this great body of work. His generosity to his students was well-known—he referred to us as his "colleagues for the next steps." I am honored that he entrusted this former student with the important task of disseminating his seminal ideas to a new generation of social scientists and for his confidence in my abilities to bring this to fruition.

Several of Dr. Campbell's colleagues and former students kindly agreed to review my introductions to ensure that my interpretations were true reflections of Dr. Campbell's ideas. These noted social scientists also offered suggestions for contextual background information to help the student appreciate the importance of the works. My sincerest thanks to Roy Herrenkohl and Robert Perloff, who pored over all of the introductions and overviews, and to William Shadish, Paul Wortman, Thomas Cook, Robert Boruch, and Mark Bickhard, who read selected portions. Others spent time discussing the theory of science issues that were so important to Dr. Campbell in the latter part of his career. I am grateful to Robert Rosenwein, who helped me arrive at a deeper understanding of Dr. Campbell's beliefs, and to Gordon Bearn, who met with me in the beginning to elucidate some of the philosophical terms.

Throughout this process, I received a great deal of encouragement as well as useful suggestions from Brenda Egolf from the Center for Social Research at Lehigh University. The Media Production Department at Lehigh was also extremely helpful in reproducing the many charts used so effectively by Dr. Campbell to demonstrate his points.

I must include C. Deborah Laughton, senior editor at Sage, who was a sounding board for my ideas and frustrations, for the fine direction she gave to this neophyte in the publishing world.

Finally, I could not have completed this project without the loving support of my husband, Donald P. Russo. His editing tips enhanced the readability of the text, and his encouragement kept me focused on the task ahead. For his good advice, patience, and unwavering confidence in me, I will be forever grateful.

—M. Jean Russo

PART I

THE CONCEPT OF AN EXPERIMENTING SOCIETY

In the 1960s, Presidents Kennedy, Johnson, and Nixon focused special attention on social issues. The government developed a number of social programs aimed at improving the lot of America's poor. The obvious underlying assumption was that these social reforms and interventions would provide increasingly beneficial services to American citizens. These programs were very costly, and the social scientists of the day were urged to ascertain whether the new strategies were successful at solving the social problems they were intended to address. Within this social context, Campbell struggled to discern the best way to conduct valid and hardheaded evaluation.

Campbell concluded that social experimentation provided the best method for reaching valid conclusions about the success of a planned intervention. So committed was he to this idea that he envisioned an idealistic society based on the rigorous scientific principles embodied in experimentation. Chapter 1 in this part provides his in-depth exploration of such a futuristic society. He outlines the ideals that formed the cornerstone of The Experimenting Society, and at the same time considers the standards dictated by the scientific method. Campbell critically examines the consequences of his proposed society and, in so doing, intuits that the practices that would ensure valid experimental results are not always compatible with his utopian ideals. For example, the internally valid randomized treatment design, if enforced, violates the egalitarian and volunteeristic ideals of the proposed society. Issues such as these prevented him from granting full advocacy, and he urged his readers to decide whether these dilemmas were insurmountable.

The essay contains two exhortations to social scientists. First, Campbell encourages those in the field to envision alternative future societies and to examine their every aspect to assess their viability as well as to uncover unintended—and presumably negative—results. This chapter was his own attempt to do just that. Second, if social scientists believe that The Experimenting Society has merit, they are encouraged to address the methodological problems that would prevent the evolution of such a society.

There is no evidence that the idea of an experimenting society is any closer to becoming a reality than when this article was first published. Campbell himself was ambivalent, and perhaps his legacy is an experimenting attitude rather than a successful experimenting society. Throughout his life, however, Campbell maintained that knowledge is accumulated incrementally by testing many options and retaining those that are successful. This is the basic premise behind evolutionary epistemology, which encourages "blind variation and selective retention." This concept will be discussed in greater detail in Part III.

As mentioned earlier, the core assumption of social reforms is that a successful intervention will enhance the lives of the individuals affected by it. If social policies are aimed at improvement, then one would expect that the incremental improvements documented through hardheaded evaluation, and the eventual dissemination of the responsible programs, would result in more and more satisfaction among members of the society. If increasing contentment of the individuals is a central issue in social planning, then the individual's subjective experience becomes a relevant concern. Yet in 1971, when the article in Chapter 2 was first published, few policy planners or program evaluators considered how psychological mechanisms through which an individual experiences pleasure and pain could impact social planning.

Although today a discussion of adaptation-level theory might seem a bit outdated, it serves as an important reminder to the present-day reader to consider the psychological processes in play when assessing the consequences of a social program. For example, in Chapter 2, Brickman and Campbell ponder whether the phenomenon described by adaptation-level theory might foil any attempt to create a society in which its members become increasingly satisfied with the services provided. The theory suggests that because individuals adapt to levels of pleasure, it would take constant inputs and an inexhaustible supply of resources to keep them in a pleasurable state. The authors considered the pessimistic view that all efforts to improve the society would fail simply because of our human nature; however, they settled on the more optimistic view that society can be improved by distributing goods and services in such a way as to establish the greatest good for the greatest number of people.

OVERVIEW OF CHAPTER 1

The question of how a society might provide better services to its people is one that continues to intrigue and plague social scientists. Campbell pondered this question and posed one possible alternative—that of building a society rooted in the scientific principle of experimentation and evaluation. Campbell exhorted those in the field to approach this question by envisioning alternative future societies. In doing so, all aspects of the proposed alternative should be examined carefully, because although the ideals may be admirable, they may be accompanied by unforeseen consequences with negative impacts.

Campbell gives the reader his own vision of an alternative future, one that is rooted in scientific inquiry. He labeled this "The Experimenting Society" (1988d). He set the stage by reviewing the norms of scientific inquiry that are applied to social science when evaluating governmental programs aimed at solving society's problems. Campbell pointed out that social scientists can recommend ideal ways to implement programs so that their effectiveness can be determined, yet he recognized that no political system exists for doing such comprehensive evaluation. In preparing this book, Campbell recognized that Communism, which was such a dominant political force when this article was first written, was no longer a threat in Eastern Europe. Still, he felt that the values of his proposed society should be viewed in the context of different political systems. Also, his admiration for the rebels in Hungary and Dubcek in Czechoslovakia, who sought a more responsive and honest society in the face of Soviet oppression, inclined him to retain these references in this version of The Experimenting Society.

The society envisioned by Campbell should not have a rigid structure, but should be fluid, making changes based on evaluation results. It would be an active society, preferring the untidiness of trial and error to the sterility of inaction. Honesty along with reality testing would be the most cherished characteristic of this society. The Experimenting Society would be nondogmatic in that it would not defend its policies blindly without regard to evidence that points to ineffectiveness. This society would concern itself with implementing a scientific mode of inquiry. As now, available science would be used in social planning, but more important, scientific methods would be employed to establish whether or not the implemented programs have met their goals. All individuals in this society should have access to public records and have input into the policies that will be implemented and the methods used to evaluate them. Society would be decentralized so that the ameliorative efforts of different departments could be replicated and verified. The society would be concerned with means as well as ends. Because perfect endstates are unattainable and hence not the goal, overall, the means should result in incremental improvements. The society would be popularly responsive, volunteeristic, equalitarian, and amenable to continuous change.

The role of the social scientist in The Experimenting Society is to use research to ascertain the effectiveness of innovative policies and programs. This is in contrast to the social scientist's role as advisor, in which the scientist can indulge in overadvocacy while neglecting the essential step of determining whether or not the innovation was truly effective. Because the state of theory development in the social sciences lags behind the physical sciences, and because of the complexities involved in chronic social problems, the social scientist must admit that the best advice, oftentimes, is only conjecture. Changes based on that advice must be evaluated for their success in ameliorating the problems. Evaluation should also identify unanticipated side effects resulting from those changes.

One problem in attempting to evaluate social programs is resistance by administrators to evaluation. Their fears may be based on negative experiences with evaluations in the past, and suggestions are given on how to allay those fears. Campbell then pointed out the shortcomings of the methodologies available for evaluating social policies. In fact, conflicts would arise between some of The Experimenting Society's ideals and the current methods available to study the effects of the policies to implement those ideals. For example, random assignment and forced participation, important factors in enhancing valid interpretation of evaluation results, would come in conflict with the ideals of equalitarianism and volunteerism. It is the responsibility of the social scientist to resolve the problems of invalidity and lack of adherence to the ideals of The Experimenting Society. Campbell stressed the importance of cross-validation of research results and suggested the cross-validation model for local programs or competitive replication of research results.

In the society proposed by Campbell, all individuals would participate in selecting programs and the methods to evaluate them. If opinion surveys are used to influence policy decision making (much like voting), respondents would have to be apprised of who is funding the survey and which policy decisions will be affected by their answers. This may cause individuals to answer in a way that they believe will afford them the greatest benefit. This illustrates how quantitative indicators are subject to corrupting influences as they become the basis by which policy decisions are made. Campbell gave several other illustrations to stress the importance he ascribed to this problem. To circumvent the problem, he suggested using multiple indicators. Because this might result in information overload, a panel of social scientists might be established whose responsibility would be to sift through the data and disseminate the information to policy makers.

Another problem with a society that is responsive to evaluation results is that the effects of social experiments often occur years or even decades after the intervention. There are further complications caused by snowballing or dissipating effects of ameliorative interventions. In the former case, this means that effects may not be apparent in the short-term but are effective over the long run. Dissipating effects may show immediate results that fade as time passes. Longitudinal designs are preferred because they can track effects over time. There are problems inherent in longitudinal studies—for example, losing track of research participants, diminished interest of the evaluator, and lack of consistent funding. Campbell offered some suggestions to make longitudinal designs more feasible for the researcher and more successful when undertaken.

Political reality dictates that administrators and legislators need to receive prompt credit for solutions for difficult problems. Because time is so important, it becomes difficult for them to support programs that may take years to provide any demonstrable effects. By educating them regarding the importance of such programs and the resistance they might encounter, the problem may be alleviated. They must be encouraged to emphasize the importance of the problem and the need to try multiple innovative approaches to solving it. In this way, they can avoid the overadvocacy trap in which too much must be promised in order for a program to be tried.

Campbell's vision of The Experimenting Society represents Kuhn's depiction of normal science. This is characterized by incremental changes in which the merits of the changes can be assessed. Campbell believed that gradual change with demonstrated outcomes can result in an optimal society.

Campbell withheld full advocacy of The Experimenting Society because of the many unresolved problems. He urged researchers to become methodologists to anticipate and address these problems.

ONE

THE EXPERIMENTING SOCIETY

SEVERAL INTRODUCTIONS

Some Ideology of Science

Science requires a disputatious community of "truth seekers." The ideology of the scientific revolution agrees with Popper's (1952, pp. 216–222) epistemological sociology of science. The norms of science are explicitly anti-authoritarian, antitraditional, antirevelational, and proindividualistic. Truth is yet to be discovered. Old beliefs are to be systematically doubted until they have been reconfirmed by the methods of the new science. Persuasion is to be limited to equalitarian means, potentially accessible to all: visual demonstrations and

Campbell, D. T. (1988d). The experimenting society. In E. S. Overman (Ed.), *Methodology and epistemology for social science: Selected papers* (pp. 290–314). Chicago: University of Chicago Press.

Campbell, D. T. (1979a). Assessing the impact of planned social change. *Evaluation and Program Planning, 2,* 67–90.

Campbell, D. T. (1994). How individual and face-to-face-group selection undermine firm selection in organizational evolution. In J. A. C. Baum & J. V. Singh (Eds.), *Evolutionary dynamics of organizations.* New York: Oxford University Press.

logical demonstrations. The community of scientists is to stay together in focused disputation, attending to each others' arguments and illustrations, mutually monitoring and "keeping each other honest," until some working consensus emerges (but mutual conformity in belief per se is rejected as an acceptable goal). Note how the ideology explicitly rejects the normal social tendency to split up into like-minded groups on specific scientific beliefs, but at the same time it requires a like-mindedness on the social norms of the shared inquiry. This is a difficult ideology to put into practice from the perspective of sociology. Merton (1973) has phrased the requirement as *organized skepticism.* Both features are required, yet "organized" and "skepticism" are inherently at odds. Societal and institutional settings in which it can be approximated are rare and unstable. Nonetheless, it may be regarded as a viable sociological thesis about a system of belief change that might plausibly improve beliefs about the physical world, including especially the not directly observable physical world.

Such a social system could not have emerged in just any problem area. It required collective success experiences, built around novel theory related to visually compelling demonstrations that could be independently replicated time and time again. It no doubt helped that many of the problems (as in static and magnetic electricity) were ones on which political and religious powers held no important fixed beliefs. The social sciences are not an arena in which the social system of successful science could have first emerged. If pure or applied social studies are to merit the term scientific, their problem areas will have to be "colonized" from the successful sciences. Such colonization will be dependent on a valid theory of the social system of validity-enhancing belief change of the successful sciences (Campbell, 1986c; Ravetz, 1971).

The Applied Social Science of Program Evaluation

My central concerns over the past 25 years (my concerns, at least, as known to my fellow social scientists) have been as a methodologist trying to extend the epistemology of the experimental method into nonlaboratory social science. They know me for my lists of threats to validity in quasi-experimental research and for my list of design alternatives that will help render these threats less plausible (Campbell, 1957, 1979a; Campbell & Stanley, 1966; Cook &

Campbell, 1979). Since 1969 (Campbell, 1969c) at least, this concern has been focused on applied social science, on treating the ameliorative efforts of government as field experiments. In the American social science community, this falls within the designation of *policy research.* My aspect of it is known as *program evaluation.* I am reporting here on our experience in developing methods for program evaluation and their implications for the question, "Can the open society be an experimenting society?"

Two aspects of program evaluation research push us toward speculating about an experimenting society. On the one hand, as we try to implement high-quality program evaluations, we meet with continual frustration from the existing political system. It seems at times set up to prevent social reality testing. This leads us to think about alternative political systems better designed for evaluating new programs.

On the other hand, if we look at our own recommendations to government regarding how to implement programs so that their impact can be evaluated, one can see that we evaluation methodologists are, in fact, often proposing novel procedures for political decision making. We are designing alternative political systems. If we were self-consciously aware of this, we would, I believe, often make different recommendations.

Out of these influences comes the imagery of an experimenting society, one that would vigorously try out possible solutions to recurrent problems and would make hard-headed, multidimensional evaluations of outcomes, and when the evaluation of one reform showed it to have been ineffective or harmful, would move on to try other alternatives. There is no such society anywhere today. Although all nations are engaged in trying out innovative reforms, none of them are yet organized to adequately evaluate the outcomes of these innovations.

The Experimenting Society and Utopian Thought

Benefiting, I hope, from years of self-conscious American discussion and practice, this is nonetheless a speculative exercise in utopian thought. We should bear in mind, as we consider it, all of the problems of utopian thought that Marx (even though in the end he exemplified these problems rather than avoided them) and Popper have noted. Several features of this model seek to avoid those problems—how well they do so, we must discuss.

First, a critical utopianism is intended. I call on a number of you to make a career commitment to being theorists and methodologists of the experimenting society. I charge you *not only to be ready* with appropriate scientific methods for evaluating new programs when the political will for an experimenting society appears, *but also to contribute to the best possible exploration-in-advance of what such a society would be like.* Though this most thorough exploration *may lead you and I in the end to oppose the experimenting society, this outcome too would thoroughly justify our career commitment,* as a morally responsible and socially necessary function. My own explorations have raised many problems that (even at the a priori utopian level) we have not solved. Until we have more plausible solutions, I myself withhold full advocacy. This is, I argue, very much in Popper's spirit of "letting our ideas die instead of ourselves." We must engage in conjectures and vigorous mutual efforts at refuting, selecting, revising, and purifying our ideas in advance of trying them out. This is particularly important where those practical trials are bound to be costly and socially disruptive in ways that might in part have been anticipated.

Although my own work is focused on this one utopia (The Experimenting Society), I recommend that it be seen as part of that important perspective of modern future studies, namely the strategy of *alternative futures.* We are to elaborate and conceptually explore, with mutual criticism, a variety of alternative utopias. In guiding our political action, we should choose the one that compares best with the others based on as complex and wise a comparison as our detailed elaboration and mutual criticism can provide. (How tragic that the journal *Alternative Futures* died after only 5 years of publication. Fortunately, we still have other "futures studies" journals.) In the alternative futures strategy for guiding current political thought, validity cannot be achieved in the form of true theories established at one time as absolutely true and superior to all potential alternatives. Rather, we must live by theories fallibly selected from among the few alternatives explicitly available in the contemporary competition among alternatives. Let us make this choice as competently as possible through the most thorough, critical conceptual exploration of the widest range of alternatives.

Second, the experimenting society is a *process utopia,* not a utopian social structure per se (Haworth, 1960; Schwarzlander, 1978). It seeks to implement that recommendation of Popper's, "A social technology is needed whose results can be tested by piecemeal social engineering" (1945/1952, p. 222). The implication of this feature in avoiding the errors of most utopian thought will be explored in more detail in what follows.

The Ideology of the Experimenting Society

Before getting into the methodological details, a little more needs to be said about the nature of the experimenting society as an ideology for a political system.

It would be an *active society* (Etzioni, 1968) preferring exploratory innovation to inaction. It would be a society that experiments, tries things out, explores possibilities in action (as well as, or even instead of, in thought and simulation). It would borrow from epistemology and the history of science the truism that one cannot know for certain in advance that a certain amount of trial and error is essential. Faced with a choice between innovating a new program or commissioning a thorough study of the problem as a prelude to action, the bias would be toward innovating. It will be committed to *action research,* to action as research rather than research as a postponement of action (Lewin, 1946, 1948; Sanford, 1970). It will be an *evolutionary, learning society* (E. S. Dunn, 1971).

It will be an *honest society,* committed to *reality testing,* to self-criticism, to avoiding self-deception. It will say it like it is, face up to the facts, be undefensive and open in self-presentation. Gone will be the institutionalized bureaucratic tendency to present only a favorable picture in government reports. This freedom to be honest will be one of the strongest attractions of The Experimenting Society. The motive of honesty in political reform, revolution, and personal heroism has been generally neglected until recently. It is of course a dominant theme among young idealists, showing up in their standards for their own interpersonal relations and in their criticism of the cowardly hypocrisy, double-talk, and dishonesty of their elders. It also emerged as a major political force within the communist countries, as Polanyi (1966a) argued in the case of the Hungarian uprisings of 1956. Although that revolution no doubt had complex roots, many of which might accurately be called reactionary or fascist, Polanyi persuaded me in his analysis of the motives of the "Petofi Circle." These elite communist journalists, well rewarded by their establishment with power and wealth, were motivated by the pain of continually having to write lies and by the promise of a society in which they could write the truth as they saw it. [For students who have come of age knowing only an Eastern Europe with democratic reforms and free market economic development, the significance of Hungary's October 1956 revolution may perhaps be lost. The Hungarian people at that time revolted against the Communist government, which had been installed by the Soviet Union. In November of 1956, however, ten divisions of Soviet Red Army troops poured into Hungary to crush the revolt. The 1950s were indeed the darkest period of the Cold War, but for a few brief days in October of 1956, the world hoped that Eastern Europe had begun to break free from Soviet domination. Despite these short-lived expectations, the world would have to wait until 1989 for these changes to occur.]

Honesty was also among the prime motives of the Czechoslovak leadership and followership of 1968, accounting for both its great popularity and its dysfunctionally provocative excesses in exposing past lies. This sounds more unsympathetic than I feel. Although I share the belief that more moderation in press statements, in the program of retrials, and in other corrections of past falsehoods probably would have avoided the external interventions while allowing the continuation of the more tangible innovations, I also sympathize with the spirit that rejected such compromise. In October 1968, I, along with fifty others, attended the International Conference on Social Psychology at Prague. The numerous Czechoslovak people we met, although profoundly distressed by the occupation, were still glowing with the excitement of their January to August experiment and were eager to talk. They freely conceded that any economic reforms that might have been initiated could not have had any effect in that short time and perhaps would not have even in the long run. But the honesty reforms had immediate effects and were profoundly enjoyed. These were most obvious, perhaps, in newspapers, television, and radio, which became exciting as specific reporters were really allowed to describe things to the best of their personal beliefs and knowledge. It is hard to exaggerate the contrast this represented over the prior times in which carefully worded, multiply vetoed statements said only what was wanted said and even disguised policy changes under ritual jargon. As it affected personal lives, the honesty reforms took the shape of dismantling the thought-police apparatus, cessation of persecutions because of beliefs of statements, reestablishment of falsely maligned reputations, and full exposure of past lies of the state. Honesty was a great part of Dubcek's amazing personal popularity. Here was a weak compromise candidate whose past history gave no more promise of greatness than did Harry Truman's in the United States, for example. He had been great in how he carried out a role, rather than being one who obtained the role through greatness. Prior to Dubcek, every official appearing on television read carefully from a multiply censored, cautiously expressed, and often dishonest text. Dubcek instead spoke freely without notes, describing things as he saw them, naively expressing honest emotions in words, facial expressions, and tone of voice.

As one came to know how decisively (and flagrantly) they had been challenging, changing, and criticizing the past regime and exposing its cruelty and dishonesty, one naturally wondered if they would not have been wiser to have moved more slowly and less provocatively and thus have avoided the occupation and retained moderate gains. But this was our question as outsiders. The issue was never raised by them in anything like these terms, even though they were uniformly deeply pessimistic over what was in store for them. When one raised the question, they said, "Perhaps so, but Dubcek had no choice. He was the most conservative and moderate of the democratic majority." The real answer, one felt, was that their joy and pride in their brief, outspoken, free, and optimistic period made it something they would not want to have missed, that it had a great value that compromise would have spoiled, that it was an experience worth the price of the risk they ran, even when viewed after the gamble had been lost.

Their reform as they saw it was within Communism or Socialism, not at all changing the ownership of the means of production, and quite interpretable as compatible with Marx and Lenin. (A quotation from Marx was widely displayed that damned large states for using their power to impose their will on small states.) Their slogan was, "Communism with a human face," a humane Communism, a democratic Communism. Their main target was inhumane bureaucracy. Their reforms were primarily in the location of authority: workers' councils for factories, local authorities as decision makers and censors, and so on. Coming as it did from within the party apparatus as a majority position, and inspired by their most gifted Marxist writers, the movement served to create a new enthusiasm for worn-out ideology. A revitalized exportable evangelical communism might well have resulted, appealing to the disenchanted in both capitalist and communist countries. [The significance of the Dubcek government's reforms may be lost on the modern day observer of world events. In 1968, however, to Campbell and other intellectuals around the globe, it was heartening to view a national Communist government with a "human face." The Dubcek experiment, for a brief moment, offered the world an opportunity to observe a Marxist government that was free of the tyranny and repression endemic to the old style, Soviet-bloc Cold War Communism.]

The Experimenting Society will be a *nondogmatic society.* Although it will state ideal goals and propose wise methods for reaching them, it will not dogmatically defend the value and truth of these goals and methods against disconfirming evidence or criticism.

It will be a *scientific society* in the fullest sense of the word "scientific." The scientific values of honesty, open criticism, experimentation, willingness to change once-advocated theories in the face of experimental and other evidence will be exemplified. This usage should be distinguished from an earlier use of the term "scientific" in social planning. In this older usage, one scientific theory of society is judged to be established as true. On the basis of this scientific theory, extrapolations are made to the design of an optimal social organization. This program is then put into effect, but without explicit mechanisms for testing the validity of the theory through the results of implementing it. Such social planning becomes dogmatic, nonexperimental, and is not scientific in the sense used here even though the grounds of its dogma are the product of previous science. Such dogmatism has been an obvious danger in implementations of Marxist socialism. It is also a common bias on the part of governmental and industrial planners everywhere. In such planning, there is detailed use of available science but no use of the implemented program as a check on the validity of the plans or of the scientific theories on which they were based. Thus economists, operations researchers, and mathematical decision theorists trustingly extrapolate from past science and conjecture, but in general fail to use the implemented decisions to correct or expand that knowledge.

It will be an *accountable, challengeable, due-process society.* There will be public access to the records on which social decisions are made. Recounts, audits, reanalyses, reinterpretations of results will be possible. Just as in science objectivity is achieved by the competitive criticism of independent scientists, so too The Experimenting Society will provide social organizational features making competitive criticism possible at the level of social experimentation. There will be sufficient separation of governmental powers so that meaningful legal suits against the government are possible. Citizens not a part of the governmental bureaucracy will have the means to communicate with their fellow citizens disagreements with official analyses and to propose alternative experiments. It will be an *open society* (Popper, 1945/1952).

It will be a *decentralized society* in all feasible aspects. Either through autonomy or deliberate diversification, different administrative units will try out different ameliorative innovations and will cross-validate those discoveries they borrow from others. The social system independence will provide something of the replication and verification of successful experiments found in science. Semiautonomy will provide some of the competitive criticism that promotes scientific objectivity.

It will be a society committed to *means–idealism* as well as *ends–idealism.* As in modern views of science, the process of experimenting and improving will be expected to continue indefinitely without reaching the asymptote of perfection. In this sense, all future periods will be mediational and transitional, rather than perfect-goal states. Ends cannot be used to justify means, for all we can look forward to are means. The means, the transitional steps, must in themselves be improvements.

It will be a *popularly responsive society* whose goals and means are determined by collective good and popular preference. Within the limits determined by the common good, it will be a *voluntaristic society,* providing for individual participation and consent at all decision levels possible. It will be an *equalitarian society,* valuing the well-being and the preferences of each individual equally.

This too brief sketch has glossed over many problems. A few of these will be considered later on. A few need to be considered now. Although The Experimenting Society described thus far has many attractive features, it obviously has many dangers and costs. We methodologists for The Experimenting

Society should be sensitive to our own and others' ambivalences about it. We should anticipate its dangers and misuses as well as its promises. We should, of course, design ways of obviating these where possible. But we should keep open the possibility that we will end up opposing it. We who now specialize in thinking about The Experimenting Society should be the first to decide that it is unworkable or in the net undesirable, if indeed it is and if this can be ascertained in advance of trying it.

The ideals described as characterizing The Experimenting Society are for the most part ideals that all of today's major ideologies claim as their own. They are endorsed in both communist and capitalist countries, most clearly in each one's criticisms of the other. Although neither type of society is as yet achieving these ends, The Experimenting Society could grow out of either, or out of both. A competition to see which could best and soonest implement it might even be envisaged. In any event, we methodologists for The Experimenting Society should consider the problems of implementing it under all available forms of political–economic organization and should regard designing all such routes as a part of our methodological challenge. Such universalism is made easier both by the common ideals and by the fact that the ideology of The Experimenting Society is a *method* ideology, not a content ideology. That is, it proposes ways of testing and revising theories of optimal political–economic–social organization rather than proposing a specific political and economic system. (Of course, this is too simple, for the requirements of The Experimenting Society exclude many forms of political–economic–social organization.)

Once implemented in both capitalist and Communist countries, one might expect that social experimentation would tend to produce increased similarity of social organization, much as industrialization has tended to do. This expectation is based on the assumption of some universals in human preferences for the good life and the good society. Once these preferences become the selective criteria for choosing among alterations of existing forms, these universals will presumably shape societies toward a common optimum. (This needs qualifying in terms of the theory of the conditions of convergence and nonconvergence in iterative processes.)

There are, of course, forces operating against the development of The Experimenting Society in both capitalist and communist countries, both in ideology as well as in current practice. Thus the Marxist–Leninist commitment to the necessity of a dictatorship during the transitional phase has served in practice to justify a dogmatism and intolerance of criticism that are inconsistent with The Experimenting Society. Yet in the total body of Marxist theory, there are stated perspectives and ideals quite sufficient to justify a truly experimental Socialism.

Within western democratic Capitalism, there are a number of favorable features. These include the legal tradition, the successful achievement of changes in government through elections, and the genuine pluralism of decision-making units. The so-called "market mechanisms" of capitalist economic theory can be regarded in ideal form as self-regulatory cybernetic feedback systems implementing the collective aspects of the preferences of individual decision makers. But the ideological justification and effective practice of the accumulation of great inequalities in individual and corporate wealth, and the role of wealth in providing grossly uneven weightings of some persons' preferences over those of others, provide great obstacles that may effectively sabotage program decisions genuinely based on the public good.

Within both Capitalist and Communist countries there are shared aspects of political processes that work against the emergence of The Experimenting Society and that may, in the long run, preclude it. We will examine a number of such problems, each of which needs technical solutions before we can advocate an experimenting society.

The Social Scientist as Servant of The Experimenting Society

Societies will continue to use preponderantly unscientific political processes to decide on ameliorative program innovations. Whether it would be good to increase the role of social science in deciding on the content of the programs tried out is not at issue. The emphasis is, rather, on the more passive role for the social scientist as an aid in helping society decide whether or not its innovations have achieved desired goals without damaging side effects. The job of the methodologist for The Experimenting Society is not to say *what is to be done,* but rather to say *what has been done.* The aspect of social science that is being applied is primarily its research methodology rather than its descriptive theory, with the goal of learning more than we do now from the innovations decided on by the political process. As will be elaborated later, even the conclusion drawing and the relative weighting of conflicting indicators must be left up to the political process.

This emphasis seems to me to be quite different from the present role as government advisors of most economists, international relations professors, foreign area experts, political scientists, sociologists of poverty and race relations, psychologists of child development and learning, and so forth. Government asks what to do, and scholars answer with assurance quite out of keeping with the scientific status of their fields. In the process, the scholar–advisors too fall into the overadvocacy trap and fail to be interested in finding out what happens when their advice is followed. Certainly the idea that one already knows precludes finding out how valid one's theories are. We social scientists could afford more of the modesty of the physical sciences, should more often say that we cannot know until we have tried. For the great bulk of social science where we have no possibility of experimentally probing our theories, we should be particularly modest. Although the experiments of The Experimenting Society will never be ideal for testing theory, they will probably be the best we have, and we should be willing to learn from them even when we have not designed them. More important, measuring the effects of a complex, politically designed ameliorative program involves all of the problems of inference found in measuring the effects of a conceptually pure treatment variable—all and more. The scientific methods developed for the latter are needed for ameliorative program evaluation.

The distinction is overdrawn. It reflects my own judgment that in the social sciences, including economics, we are scientific by intention and effort, but not yet by achievement. We have no elegantly successful theories that predict precisely in widely different settings. Nor do we have the capacity to make definite choices among competing theories. Even if we had, the social settings of ameliorative programs involve so many complexities that the guesses of the experienced administrator and politician are apt to be on the average as wise as those of social scientists. But whatever the source of the implemented guess, we learn only by checking it out. Certainly in The Experimenting Society, social scientists will continue to be called on to help design solutions to social problems, and this is as it should be. Perhaps all I am advocating in emphasizing the role of servant rather than leader is that social scientists avoid cloaking their recommendations in a specious pseudoscientific certainty and, instead, acknowledge their advice as consisting of wise conjectures that need to be tested in implementation.

The servant–leader contrast is overdrawn in other senses also. The truism that measurement itself is an agent of change is particularly applicable to The Experimenting Society. Advocating hard-headed evaluation of social programs is a recommendation for certain kinds of political institutions. In considering the methodological challenges of The Experimenting Society in what follows, appeals to the theory and content of the social sciences will be made, as well as to their methodology.

METHODOLOGICAL PROBLEMS

Resistance to Assessing the Outcome of Praxis and the Overadvocacy Trap

In the United States one of the pervasive reasons why interpretable program evaluations are so rare is the widespread resistance of institutions and administrators to having their programs evaluated. The methodology of evaluation research should include the reasons for this resistance and ways of overcoming it.

A major source of this resistance in the United States is the identification of the administrator and the administrative unit with the program. An evaluation of a program under our political climate becomes an evaluation of the agency and its directors. In addition, the machinery for evaluating programs can be used deliberately to evaluate administrators. Combined with this, there are a number of factors that lead administrators to correctly anticipate a disappointing outcome. As Rossi (1969) has pointed out, the special programs that are the focus of evaluation interests have usually been assigned the chronically unsolvable problems—those on which the usually successful standard institutions have failed. This in itself provides a pessimistic prognosis. Furthermore, the funding is usually inadequate, both through the inevitable competition of many worthy causes for limited funds and because of a tendency on the part of our legislatures and executives to generate token or cosmetic efforts designed more to convince the public that action is being taken than to solve the problem. Even for genuinely valuable programs, the great effort required to overcome institutional inertia in establishing any new program leads to grossly exaggerated claims. This produces the "overadvocacy trap" (Campbell, 1969c), so that even good and effective programs fall short of what has been promised, which intensifies fear of evaluation.

The seriousness of these and related problems can hardly be exaggerated. As methodologists, we in the United States are called on to participate in the political process in efforts to remedy this situation. But before we do so, we should sit back in our armchairs in our ivory towers and invent political–organizational alternatives that would avoid the problem. This task we have hardly begun, and it is one in which we may not succeed. Two minor suggestions will illustrate. I recommend that we evaluation–research–methodologists should refuse to use our skills in ad hominem research. Although the expensive machinery of social experimentation can be used to evaluate persons, it should not be. Such results are of very limited generalizability. Our skills should be reserved for the evaluation of policies and programs that can be applied in more than one setting and that any well-intentioned administrator with proper funding could adopt. We should meticulously edit our opinion surveys so that only attitudes toward program alternatives are collected and such topics as supervisory efficiency excluded. This prohibition on ad hominem research should also be extended to program clients. We should be evaluating not students or welfare recipients but alternative policies for dealing with their problems. It is clear that I felt such a prohibition is morally justified. But I should also confess that in our U.S. settings it is also recommended out of cowardice. Program administrators and clients have it in their power to sabotage our evaluation efforts, and they will attempt to do so if their own careers and interests are at stake. Although such a policy on our part will not entirely placate administrators' fears, I do believe that if we conscientiously lived up to it, it would initiate a change toward a less self-defeating political climate.

A second recommendation is for advocates to justify new programs on the basis of the seriousness of the problem rather than on the certainty of any one answer and to combine this with an emphasis on the need to go on to other attempts at finding a solution should the first one fail (Campbell, 1969c). Shaver and Staines (1971) have challenged this suggestion, arguing that if an administrator takes this attitude of scientific tentativeness, it constitutes a default of leadership. Conviction, zeal, enthusiasm, and faith are required for any effective effort to change traditional institutional practice. To acknowledge only a tentative faith in the new program is to guarantee a halfhearted implementation of it. But the problem remains; the overadvocacy trap continues to sabotage program evaluation. Clearly, we should address our social–psychological and organizational–theoretical skills to this problem.

The Use of Experimentation in The Experimenting Society

Social scientists often argue that they, like the astronomers, can build a science on correlational evidence alone and can do without the outmoded concept of cause. For a while, major philosophers of science agreed. The concepts of cause and effect are being readmitted to the most sophisticated philosophy of science (see Cook & Campbell, 1979, chap. 1, for a review). Whatever stand one may take on this, it must be admitted that intentional projects to improve society (praxis, if you will) accept both the concept of cause and the concept of learning from praxis. Implementing social change efforts in natural settings is a praxis akin to experimentation, differing only in the equivocality of interpreting the outcome. The big question is whether or not we can improve the interpretability of intentional praxis in our political life without destroying the open society in the process.

This is not to deny that the concept of cause may be a logical hodgepodge, involving a number of analytical criteria of no logical relationship to each other and of no entailed status beyond observation of past correlation. Let us accept the fact that humanity's deeply ingrained concept of cause is a product of biology, psychology, and evolution (Campbell, 1974b) rather than a pure analytic concept. If so, it reflects the adaptive advantage of being able to intervene in the world to deliberately change the relationship of objects. From among all the observable correlations in the environment, humans and their predecessors focused on those few that were manipulable correlations. From this emerged the human predilection for discovering "causes" rather than mere correlations. In laboratory science, this search is represented in the experiment, with its willful, deliberate intrusion into ongoing processes. Similarly for the ameliorative social scientist: Of all of the correlations observable in the social environment, we are interested in those few that represent manipulable relationships, in which—by intervening and changing one variable—we can affect another. No amount of passive description of the correlations in the social environment can sort out which are"causal" in this sense. To learn about the manipulability of relationships one must try out manipulation. The scientific, problem-solving, self-healing society must be an experimenting society. (The causal modeling of current sociology can only generate focal hypotheses still needing cross-validation in praxis [Cook & Campbell, 1979, chap. 7]).

Accepting the important role of social experimentation at once puts clarity of scientific inference into conflict with means–idealism and acceptable political processes. Experiments involving randomized assignment to treatments are undoubtedly the most efficient and valid. Although I am a strong advocate of good quasi-experimental designs in which randomization is not possible, detailed consideration of specific cases again and again reinforces my belief in the scientific superiority of randomized-assignment experiments. As advisors to The Experimenting Society we will often recommend such research designs. Yet they present special moral problems that we will have to consider. First, such experiments are best done when those designing and directing the study have most complete and arbitrary control over the people participating in the study—that is, in total institutions such as prisons and armies. One needs optimally to be able to randomly assign persons to experimental treatments and to enforce 100% participation in these treatments. Moreover, to avoid reactive arrangements, the participants should be unaware of the experiment, unaware that other people are deliberately being given different treatments (Campbell, 1969c, pp. 367–377). Speaking from the point of view of humanistic existential socialism, Janousek (1970) has criticized "Reforms as Experiments" (Campbell, 1969c) on these grounds. He has argued that the whole orientation of assigning persons to treatments by randomization betokens an authoritarian, paternalistic imposition, treating citizens as passive recipients rather than as coagents directing their own society, treating fellow citizens as "subjects" in the psychologist's and monarchist's sense, as "victims" of the experiment rather than as collegial agents of the experiment.

The enforcement of assigned treatments also violates the egalitarian and voluntaristic ideals of The Experimenting Society. The disguised experiment violates these too and, in addition, the values of openness, honesty, and accountability. However much we may weight the value of scientific information in deciding on the ethics of deceit and lack of informed consent in harmless experimentation done to test scientific theories (Campbell, 1969c, pp. 270–277), social experimentation for policy decisions must adhere meticulously to means–idealism on these issues and should include no research procedures that would be excluded as a part of regular governmental procedures. Participation in policy experiments is more akin to participating in democratic political decision making than to participating in the psychology laboratory. These restrictions all have costs in the validity of experimental inference. They are costs that we must live with and try to compensate for in other ways.

Thus the task of first priority for the methodologists of The Experimenting Society is to design experimental arrangements that obviate these difficulties. Janousek (1970) has suggested that some system of rotation between the roles of experimenter and participant be designed. Although this is not obviously feasible, it is worth more detailed consideration. The following suggestions are likewise only initial fumblings toward possible solutions. For example, experiments using volunteers who are informed of the treatment and control-group alternatives and agree to accept whatever random assignment they draw—such experiments seem to me well worth doing. On some problems, such as public housing versus rental vouchers, they might be as informative as disguised experiments; but even if not, The Experimenting Society may have to make do with them. In the New Jersey Negative Income Tax Experiment (Kershaw & Fair, 1976a; Watts & Rees, 1977a, 1977b), one of the most famous social experiments, participants were told about the particular treatment they were asked to volunteer for but were not informed of what others would be getting. Should the control group have been told of the $1000 supplemental income it was missing? Perhaps The Experimenting Society of the future will decide it should have been. Today's methodologists, however, regard the envy and resentment that might thus be generated as too great a threat to experimental validity to be tolerated. Here is a tangle of problems we should be working on right now.

It is my tentative judgment that a variety of random-assignment experiments will be acceptable (see Cook & Campbell, 1979, chap. 8). But even more important, we must be elaborating high-quality quasi-experimental approaches that, although more ambiguous in terms of clarity of scientific inference, are usually more acceptable as processes we will be willing to make a permanent part of our political system. Note, however, that the most common of these quasi-experiments (multivariate regression approaches, covariance in the absence of randomization, path analysis, and causal modeling) have had the bias of making those special programs given to the neediest look mistakenly harmful (Campbell & Boruch, 1975; Cook & Campbell, 1979, chaps. 3, 4, and 7).

Getting Mutual Criticism and Competitive Replication Into Social Experimentation and Program Evaluation

As my introductory citation of Karl Popper made clear, the objectivity of physical science does not come from the fact that single experiments are done by reputable scientists according to scientific standards. It comes instead from a social process that can be called competitive cross-validation (Campbell, 1986c) and from the fact that there are many independent decision makers capable of rerunning an experiment, at least in a theoretically essential form. The resulting dependability of reports (such as it is, and I judge it to usually be high in the physical sciences) comes from a social process rather than from dependence on the honesty and competence of any single experimenter. Somehow in the social system of science a systematic norm of distrust (Merton's [1973] "organized skepticism") combined with ambitiousness leads people to monitor each other for improved validity. Organized distrust produces trustworthy reports.

This competitive cross-validation will be very hard to achieve, and specific plans must be made to get some semblance of it. To judge from the American experience, government prefers to turn a major pilot study over to a single investigative team. The U.S. Congress is apt to mandate an immediate, nationwide evaluation of a new program to be done by a single evaluator, once and for all, subsequent implementations to go without evaluation. Here are a variety of recommendations to correct this, none of which are practiced in the United States.

The Contagious Cross-Validation Model for Local Programs

A generous and concerned government provides funds for developing local programs addressed to a chronic problem. This local program funding includes funds for whatever evaluation the program designers want, including funds for academic consultants. As a result, there are lots of local programs. When any one of them, after a year or so of debugging, is deemed to be a program others would consider worth borrowing, only at that time would there be program evaluation in a serious sense. Our slogan would be, "Evaluate only proud programs!" (Contrast this with our present ideology, in which Washington planners in Congress and the executive branches design a new program and command immediate nationwide implementation, with no debugging, plus an immediate nationwide evaluation.)

When the high-morale programs and program results were disseminated, there would, no doubt, emerge a group of willing adopters. At this stage, our national funding would support adoptions that included locally designed cross-validating evaluations, including funds for appropriate comparison groups not receiving the treatment. (We might at this or the next stage have large-scale "external" evaluations, as long as these did not preclude interpretable comparisons at each site not depending on full national implementation.) After 5 years we might have 100 locally interpretable experiments. We would also have a community of applied social scientists familiar with them all who had cross-examined each others' data, suggested and done reanalyses, performed bias-sensitive meta-analyses, and so forth. Many of these scholars would be tenured university or public school faculty whose job security did not depend on the outcome. From the consensus of this mutually monitoring research community we would advise government and potential adopters.

The contagious cross-validation model is much closer to the model of the physical sciences. Applied social sciences has more, rather than less, need for mutual criticism, argumentative reanalyses, and cross-validation than does physical science. This is so for several reasons. We lack the possibility of experimental isolation. Our data have to be generated through the cooperation of persons with strong stakes in the outcome. Finally, applied science (either physical or social) is done in an arena in which the rival interests in what the outcome is are so powerful that objective description can become a minor motive.

Getting Competitive Replication Into National Policy Pilot Studies

The contagious cross-validation model is only appropriate in instances in which the program under study can be implemented autonomously by a local unit (be it school, classroom, city, retail store, factory, and so on). In the case in which the program being piloted has to be eventually implemented nationally, different sources of competitive cross-validation must be sought.

A. Rather than awarding a single contract, each should be *split into two or more independent experiments,* so that all of the hundreds of discretionary decisions about how to present the experimental treatment and design the questionnaires and interviews would be made and implemented by at least two independent research teams. Such heteromethod replication (Campbell, 1969c; Cook & Campbell, 1979) is needed for interpretative validity. It would also provide a small group of informed scientists for competitive cross-examination.

B. Involve *adversarial stakeholder* participation in the design of each pilot experiment or program evaluation and again in the interpretation of results (Bryk, 1983; Krause & Howard, 1976). We should be consulting with the legislative and administrative opponents of the program as well as the advocates, generating measures of feared undesirable outcomes as well as promised benefits.

C. Encourage and fund *competitive reanalysis* of data from the big studies. The Office of Economic Opportunity created a great precedent we have inadequately lived up to. Through this office, the Institute for Research on Poverty at the University of Wisconsin is a repository for data sets on related topics. Available for reanalysis are the data tapes for the New Jersey Negative Income

Tax Experiment, and proper scientific disagreements are emerging, for example, about how they handled the attribution problem (Boeckmann, 1981). Another such data set is the first big Head Start evaluation, a data set with a fine record for productive second-guessing (Barnow, 1973; Bentler & Woodward, 1978; Magidson, 1977; Smith & Bissell, 1970). Major classics in this area come from my Northwestern University colleagues (Boruch, 1978a; Boruch et al., 1981; Cook et al., 1975; Trochim, 1982). Also available is the original Coleman report (Coleman et al., 1966) on educational desegregation, which has been thoroughly reanalyzed. As a result, we could assemble a half dozen volumes the size of Mosteller and Moynihan's (1972), and from a modern postpositivist theory of science, we can recognize that only now do we have a competent applied social science community ready to use the Coleman report in conjunction with all related research prior and subsequent to guide governmental policy.

Although these secondary analyses are of great value, and should become obligatory for all expensive data collections, we should remember that they cannot fully correct for the hundreds of idiosyncratic discretionary judgments involved in the initial data collection.

D. *Legitimating dissenting-opinion research reports* from members of the research team. The Freedom of Information Act of the late 1960s was one of the great social inventions increasing the possibility of a valid, policy-relevant, applied social science. Although Rights of Subjects legislation (another great innovation) has often been used to greatly curtail its practical implementation (needlessly so—see Campbell, Boruch, Schwartz, & Steinberg, 1977; also Boruch & Cecil, 1979, 1982; Campbell & Cecil, 1982), the legitimating value is still there. What I propose is that we use such freedom of information and right of reanalysis to give to every research assistant on any social research the right to publish independently on the data collected.

A background for my argument is the great value that "whistle-blowing" has had for the validity of physical and biological research results when these have been done under conditions of extreme policy relevance. (I am thinking of research on the dangers of chemicals to manufacturing workers and food consumers, the dangers to and effects on humans and sheep of irradiation from nuclear experiments and power generators.) Although such whistle-blowing occurs, it is still experienced as a guilt-producing team disloyalty, both by the whistle-blower and coworkers, who may react with ostracism. It would improve the scientific and political validity of applied physics, chemistry, and biology if whistle-blowing were legitimated by reconceptualizing it as the right and duty to generate *dissenting-opinion research reports,* and if all laboratory staff were provided official access to all data for this purpose. Insofar as our research results are inherently more ambiguous, we are more in need of this in applied social science.

Opinion Surveys as Voting Opportunities

Opinion surveys would be of central importance in The Experimenting Society, as "social indicators" of the effects of new programs. Treating opinion surveys as an ideal political decision-making process akin to voting would require great changes. Here are a few: Interviewees would be told who had paid for the questions and how their answers would be used. They would know what programs were being evaluated by their answers (Campbell & Cecil, 1982). They would be given the results of the survey, just as they are given voting results. They would be allowed to use these results in political debates. They would be "co-owners" of the opinions they had created. From the point of view of present-day social science methodology, these changes would make opinion surveys less valid. Respondents would distort their opinions in deliberate efforts to influence governmental decisions about which programs should be continued or which regions were most in need of more resources. There would be political campaigns to get respondents to reply in the particular ways that the local political organizations see as desirable, just as there are campaigns to influence the vote. There would be efforts comparable to ballot-box stuffing. Interviewer bias would become even more a problem. Bandwagon effects—in other words, conformity influence from the published results or prior surveys —must be anticipated. New biases, like exaggerated complaint, would likely emerge.

These costs in "validity" are probably unavoidable. Opinion surveys, however, would still be useful and informative, just as are votes now, once we got used to these new conditions of meaning. The methodological problems involved are ones we should be working on and indeed, are ones best researched in the transition period.

The Corrupting Effect of Using Social Science Indicators

The social indicators that are now being used include public records as well as opinion surveys—records of deaths, diseases, crimes, accidents, incomes, and school-achievement test scores. In The Experimenting Society of the future, these would probably be used even more as indicators of how programs are doing. It might be thought that such records would be more resistant to bias than are interview data. More resistant, yes, but still subject to a discouraging law that seems to be emerging: *The more any quantitative social indicator is used for social decision making, the more subject it will be to corruption pressures and the more apt it will be to distort and corrupt the social processes it is intended to monitor.* Let me illustrate these two laws with some evidence that I take seriously, although it is predominantly anecdotal.

Take, for example, a comparison between voting statistics and census data in the city of Chicago: Surrounding the voting process, there are elaborate precautionary devices designed to ensure its honesty; surrounding the census-taking process, there are few, and these could be so easily evaded. Yet, the voting statistics are regarded with suspicion whereas the census statistics are widely trusted (despite underenumeration of young adult, Black males). I believe this order of relative trust to be justified. The best explanation for it is that votes have continually been *used*—have had real implications as far as jobs, money, and power are concerned—and have therefore been under great pressure from efforts to corrupt. On the other hand, until recently our census data were unused for political decision making. (Even the constitutional requirement that electoral districts be changed to match population distribution after every census was neglected for decades.)

Another example: In the spirit of scientific management and accountability, police departments in some jurisdictions have been evaluated by "clearance rates." This refers to the *proportion* of crimes solved, and considerable administrative and public pressure is generated when that rate is low. Skolnick (1966) provided illustrations of how this pressure has produced both corruption of the indicator itself and a corruption of the criminal justice administered. Failure to record all citizens' complaints, or to postpone recording them unless solved, are simple evasions that are hard to check, because there is no independent record of the complaints. A more complicated corruption emerges in combination with plea bargaining. *Plea bargaining* is a process whereby the prosecutor and court bargain with the prisoner and agree on a crime and a punishment to which the prisoner is willing to plead guilty, thus saving the costs and delays of a trial. Although this is only a semilegal custom, it is probably not undesirable in most instances. However, combined with the clearance rates, Skolnick finds the following miscarriage of justice. A burglar who is caught in the act can end up getting a lighter sentence the more prior unsolved burglaries he is willing to confess to. In the bargaining, he is doing the police a great favor by improving the clearance rate, and in return, they provide reduced punishment. Skolnick has argued that in many cases the burglar is confessing to crimes he did not in fact commit. Crime rates are in general very corruptible indicators. For many crimes, changes in rates are a reflection of changes in the activity of the police rather than changes in the number of criminal acts (Gardiner, 1969; Zeisel, 1971). It seems to be well-documented that a well-publicized, deliberate effort at social change—Nixon's crackdown on crime—had as its main effect the corruption of crime-rate indicators (Morrissey, 1972; Seidman & Couzens, 1974; Twigg, 1972), achieved through underrecording and by downgrading the crimes to less serious classifications.

For other types of administrative records, similar use-related distortions are reported (Garfinkel, 1967; Kitsuse & Cicourel, 1963). Blau (1963) provided a variety of examples of how productivity standards set for workers in government offices distort their efforts in ways deleterious to program effectiveness. In an employment office, evaluating staff members by the number of cases handled led to quick, ineffective interviews and placements. Rating the staff by the number of persons placed led to concentration of efforts on the easiest of cases, neglecting those most needing the service, in a tactic known as "creaming" (Miller, Roby, & Steenwijk, 1970). Ridgeway's pessimistic essay on the dysfunctional effects of performance measures (1956) provides still other examples.

From an experimental program in compensatory education comes a clearcut illustration of the principle. In the Texarkana "performance contracting" experiment (Stake, 1971), supplementary teaching for undereducated children was provided by "contractors" who came to the schools with special teaching machines and individualized instruction. The corruption pressures were high because the contractors were to be paid on the basis of the achievement-test score gains of individual pupils. It turned out that the contractors were teaching the answers to specific test items that were to be used on the final pay-off testing. Although they defended themselves with a logical–positivist operational–definitionalist argument—that their agreed on goal was defined as improving scores on that one test—this was generally regarded as scandalous. However, the acceptability of tutoring the students on similar items from other tests is still being debated. From my own point of view, achievement tests may well be valuable indicators of general school achievement under conditions of normal teaching aimed at general competence. But when test scores become the goal of the teaching process, they both lose their values as indicators of educational status and distort the educational process in undesirable ways. (Similar biases of course surround the use of objective tests in courses or as entrance examinations.) In compensatory education in general there are rumors of other subversions of the measurement process, such as administering pretests in a way designed to make scores as low as possible so that larger gains will be shown on the post test, or limiting treatment to those scoring lowest on the pretest so that regression to the mean will provide apparent gains. Stake (1971) lists still other problems. Achievement tests are, in fact, highly corruptible indicators.

That this serious methodological problem may be a universal one is demonstrated by the extensive USSR literature (reviewed in Berliner, 1957, and Granick, 1954) on the harmful effects of setting quantitative industrial production goals. Prior to the use of such goals, several indexes were useful in summarizing factor productivity—for example, monetary value of total product, total weight of all products produced, or number of items produced. Each of these, however, created dysfunctional distortions of production when used as the official goal in terms of which factory production was evaluated. If monetary value, then factories would tool up for and produce only one product to avoid the production interruptions in retooling. If weight, then factories would produce only their heaviest item (e.g., the largest nails in a nail factory). If number of items, then only their easiest item to produce (e.g., the smallest nails). All these distortions led to overproduction of unneeded items and underproduction of much needed ones.

We return to the U.S. experience for another example. During the first period of U.S. involvement in Vietnam, the estimates of enemy casualties put out by both the South Vietnamese and our own military were both unverifiable and unbelievably large. Through the influence of the Secretary of Defense, Robert McNamara, an effort was instituted to substitute a more conservative and verifiable form of reporting, even if it underestimated total enemy casualties. Thus the "body count" was introduced, an enumeration of only those bodies left by the enemy on the battlefield. This became used not only for overall reflection of the tides of war, but also for evaluating the effectiveness of specific battalions and other military units. There was thus created a new military goal, that of having bodies to count, a goal that came to function instead of more traditional goals, such as gaining control over territory. Pressure to score well in this regard was passed down from higher officers to field commanders. The realities of guerrilla warfare participation by persons of both genders and of a variety of ages added a permissive ambiguity to the situation. Thus poor Lieutenant Calley was merely engaged in getting bodies to count for the weekly effectiveness report when he massacred the civilians in the village of My Lai. His goals had been corrupted by the worship of a quantitative indicator, leading both to a reduction in the validity of that indicator for its original military purposes and a corruption of the social processes it was designed to reflect.

I am convinced that this problem must be solved if we are to achieve meaningful evaluation of our efforts at planned social change. It is a problem that will get worse, the more common quantitative evaluations of social programs become. We must develop ways of avoiding this problem if we are to move ahead. We should study the social processes through which corruption is being uncovered and try to design social systems that incorporate these features. In the Texarkana performance-contracting study, it was an "outside evaluator" who uncovered the problem. In a later U.S. performance-contracting study, the Seattle Teachers' Union provided the watchdog role. We must seek out and institutionalize such objectivity-preserving features. We should also study the institutional form of those indicator systems such as the census or the cost of living index in the United States, which seem relatively immune to distortion.

A final example of how the validity of the quantitative indicators can be corrupted is found by examining the evolution of management information systems (MIS). This concept of the "scientific management" of programs was first introduced into government by McNamara in the 1940s and 1950s. Much of the information used to manage programs involve internal reporting systems, self-reports, evaluation of others, and evaluation of subgroup status.

The standardized reporting forms used at all levels of government and business greatly increase the red-tape paperwork. Those who regularly receive and complete them suspect that they are invalid and irrelevant. On the other hand, those who design the forms (perhaps experimental psychologists rather than street-wisdom sociologists) seem to have assumed that the only motive in filling them out is to report accurately; however, evolutionary analysis (Campbell, 1994, p. 36) has led to the suspicion that any of the following motives may come into play:

1. To describe accurately. (There is widespread evidence of a general preference for not lying if nothing else is at stake.)
2. To influence the decisions one anticipates will be based on the reporting form in one's personally preferred direction.
   - 2a. For the good of the organization.
   - 2b. For the good of one's own face-to-face group within the organization or the individuals within it.
   - 2c. For one's own well-being and that of one's family.

Once the use of such reporting forms becomes routinized, it is likely that motives 2b and especially 2c dominate.

Further, such management information systems produce proxy variables that are complex in that they reflect several components. Not all the components are relevant to the outcome that the indicator is intended to monitor. When used in managerial control, "irrelevant" components that will produce the desired score come to dominate. In this way, the validity of the proxy variable is undermined. There is longstanding literature on the harmful effects of setting quantitative goals in USSR Five-Year Plans, in U.S. employment agencies, and so on. The classics are Ridgeway (1956) and Blau (1963). The literature is reviewed by Campbell (1979b, pp. 83–86) and Ginsberg (1984).

Recognizing the distortions and biases in quantitative indicators and developing appropriate safeguards is necessary if the information they supply is to be useful. In The Experimenting Society, social indicators will be used more than they are at present, and the corruption pressures will thus be greater. This problem seems so serious that it provides one more reason that we should not rush forward into The Experimenting Society until it is solved. My tentative solution to this problem is to distinguish carefully between two movements in "scientific government." I end up opposing the use of quantitative indicators for achieving managerial control—in other words, the *accountability* movement. The regularized use of such measures and the focus on evaluating specific social units and their administrators seems to me to create more evils than it cures. The other movement is *program evaluation.* As I have expressed under "Resistance to Assessing Outcomes," the temporary use of quantitative measures in evaluating alternative programs, which existing staffs could implement, seems to me still beneficial to The Experimenting Society.

Integrating the Evidence Into Political Decision

One solution to bias is to use multiple indicators of the same problem, each of the indicators being recognized as imperfect, but so chosen as to have different imperfections, different susceptibilities to distortion. This produces a variety of estimates of program effectiveness, benefits, and problems. The judgmental task of pooling all of these indicators is once more appropriate for an elected legislature than for a committee of social scientists. Furthermore, in an experimenting society we would be doing scientific evaluations on many more programs than we are doing now. With *multiple measures on multiple programs* we will have created a monster of measurement, a formidable information overload. How to reconcile our need for facts with democratic decision making is another problem we must solve before we welcome The Experimenting Society. New institutions will be needed, such as an auxiliary legislature of quantitative social scientists, each appointed by one real legislator. Such an auxiliary legislature could process advisory decisions, with full attention to the scientific evidence. The real legislator could then guide his or her own decisive vote by the auxiliary legislature's actions and the issues raised. This awkward and expensive procedure seems to me better than delegating this process to an appointed scientific elite.

I offer this unwieldy suggestion to testify to my estimate of the seriousness of the problem and to emphasize our need for creative speculation about how it might be solved. Such conjectures can right now be exposed to vigorous criticism. The ones we judge most plausible we will want to try out in practice. After implementing them—or even before—we may decide that they remain so unsolved as to be a reason to withhold our advocacy of The Experimenting Society.

Legitimating and Facilitating Evaluation by Nonprofessional Participants and Observers

We applied social scientists, methodological servants of The Experimenting Society, are like any profession (see Ivan Illich) in danger of becoming a self-serving elite. It becomes in our best professional interest to make program evaluation an esoteric art requiring our services, computers, and complex statistical adjustments that make our conclusions immune from criticism, even from well-placed, competent observers who saw the program in action. We are also apt to become unwittingly coopted into a pervasive bias in favor of the already-established governmental and extragovernmental powers, who, after all, will usually be the sources of our past and future salaries.

To avoid such biases, we must devise ways that are readily comprehensible to the participating staffs, recipients, and other well-placed observers for them to collect, formulate, and summarize their estimations of program effectiveness (Campbell, 1978). We must recognize that such summaries may have a validity comparable to the statistical analysis of more formal measures. Usually these perspectives will agree, but where they do not, we should remember that the statistical analyses involve simplifying assumptions that may be seriously in error (Campbell & Boruch, 1975; Campbell & Erlebacher, 1970).

We must also remember that in social experimentation, the lack of controlled conditions makes necessary the technique of explicitly developing and evaluating "threats to validity" and "plausible rival hypotheses" (Campbell & Stanley, 1966; Cook & Campbell, 1979). For this purpose, it is those who have situation-specific information who make the best critics, and the best judges, of the plausibility of most of the rival hypotheses in their specific setting. We must develop procedures for eliciting such criticisms and judgments.

In this process, we must provide these nonprofessional observers with the self-confidence and opportunity to publicly disagree with the conclusions of the professional applied social scientists.

Long-Term Follow-Up

The relevant efficacy of social experiments is long term, the effect in subsequent decades (rather than, or in addition to, months or years). For experiments involving children (compensatory education, preventive interventions, kibbutz childrearing), the effects on their adult lives—15 to 20 years after the experiment—are most relevant. Such long-delayed evaluations are jeopardized by several different classes of problems.

1. Locating and assessing experimental participants 20 years later presents formidable problems. (Some U.S. compensatory-education efforts have had 50% failures in efforts to recontact and remeasure, after only 1 year.)
2. The scientists doing research on the initial intervention are unlikely to still have that research focus 10 to 20 years later. Their focal scientific community for the experimental problem will also have dispersed.
3. The governmental initiative funding the research will have dissipated; new focal concerns will have higher priority.

Thus it is essential to have immediate outcome measures, focused on proximal indicators of those traits that the theory involved specifies as mediating the long-term benefits. But because the theories involved are not themselves well tested, and indeed, can only be tested by long-term follow-ups, it is a major responsibility to prepare the basis for later follow-ups.

Snowballing Versus Dissipating Effects

One should distinguish between two contrasting forms of impact for interventions delivered at a specific time: snowballing versus dissipating effects. For most interventions, there is an unknown threshold point between them. Consider two randomized experiments using samples of 100 college seniors. For one experiment, members of the treatment group are each given $5000, their controls nothing. It might be anticipated that in each subsequent year the difference between the net worth of the experimentals and controls would diminish until 5 years later no significant difference would remain. On the other hand, were the experimental treatment to be $50,000, a snowballing effect might be found, in which the differences between the experimental and control group steadily increased in subsequent years.

For many interventions, for example those that leave children in the same environment and involve small percentages of their waking hours, we must expect dissipating effects, unless a particularly sensitive "imprinting" time has been hit. The laws of learning, interference, forgetting, and spontaneous recovery provide theoretical support for this prediction (Campbell & Frey, 1970). Think of the multiplicity of change agents, *other* than the intervention, that impinge on the outcome variable in one direction or another. These collectively produce a random scatter of effects, more numerous the more time has elapsed. Relative to the sum total of these effects, the intervention is but a drop in the bucket. Moreover, as time goes on, weights of preintervention influences increase relative to that of the intervention when it first occurred. (Such a theory argues for small dosage, short-term experiments to winnow out a few promising treatments. In later experiments these would be applied intensively over many years, looking for a long-term effect.)

But such theory can lead to undue pessimism. Several reports on long-term follow-ups of preschool compensatory education show lasting effects (Consortium for Longitudinal Studies, 1983; Crain, Hawes, Miller, & Peichert, 1984; Lazar & Darlington, 1982; Weikart, Berrueta-Clement, Schweinhart, Barnett, & Epstein, 1984). For cognitive effects as measured by achievement tests, the dissipating pattern has been found (considering these as akin to vocabulary tests makes the learning-theory expectations more obvious). But for academic self-confidence, persistence in seeking higher education, and avoidance of delinquency, impressive impacts have been found 15 years later. Note that on these variables no immediate proximal mediator was effectively measured. Making decisions on long-term follow-ups based on test scores received 5 years later would have precluded discovery of the important long-term effects. Many exploratory studies will validly be deemed unworthy of long-term follow-ups, based on the experimental pilot studies employing short-term outcome measures. But if the costs of archiving the records that would make long-term follow-ups possible is not too great, I recommend routine recording of such data, because later theoretical reconsiderations may lead to changes in such decisions.

Another classic in long-term follow-up on a deliberate experimental intervention is McCord's (1978, 1981) 30-year follow-up on the Cambridge–Somerville delinquency prevention project. Here again, very significant long-term impact was documented, but in this case, unanticipated harmful effects the opposite of those intended. Caplan (1968), for the Chicago Boys Club study, has likewise found what appear to be harmful effects from an extensive gang-worker program in a 5-year follow-up study. The methodological difficulties of long-term follow-ups are such that one might well be skeptical of the outcomes claimed, whether beneficial or harmful. So far as we are aware, a thorough scrutiny of such problems and the plausible alternative explanations of the outcomes they provide has not yet been undertaken. Differential attrition for experimentals versus controls during treatment and differential rates of locating cases at follow-up time are just two of the possibilities. We have major responsibility to make a profound cross-examination of these studies and others, both for the intrinsic interest in their outcomes and, in general, for their methodological implications in preparing plans for possible long-term follow-ups.

A Survey of Methodological Problems and Solutions in Long-Term Follow-Up Studies

We propose a methodological review in this area. As a beginning, one might make extended visits to scholars and centers having experience with long-term follow-up, such as those cited previously. Along with deliberate experimental interventions, correlational or impact studies of exogenous "treatments" should be examined. Examples of such studies are the impact of the great depression of the 1930s, parental employment, parental death, or mode of childrearing, and so on. A major center for such studies is in the Institute for Human Development at the University of California, where follow-ups are still being conducted on longitudinal study cases started in the early 1930s by the Institute of Child Welfare under grants from the Laura Spellman Rockefeller Foundation. At least three studies are involved; two beginning at birth or before (The Guidance Study, initiated by Walker Macfarlane, and The Growth Study, under Nancy Bailey) and one (The Adolescent Study, by Harold E. Jones, Mary Cover Jones, et al.) beginning with 5th-grade students. The Fels

Institute of Yellow Springs, Ohio, also started longitudinal studies of infants and families in the 1930s with some follow-ups into the late 1940s at least.

This methodological review should include not only the experimental threats to validity, including the massive "pretest" effects, but also should pay attention to sociology-of-science issues. For instance, we should examine what kinds of institutional arrangements or accidents of careers made it possible to carry out repeated follow-ups (as in the University of California Berkeley group) or follow-ups after long periods of no contact with the participants (as in the McCord [1978] classic, the Crain [Crain et al., 1984] study, and some of those in the cooperating group of Lazar and Darlington [1982]). Even such issues as who provided the documentation, storage space, and retrieval competence in the archiving are important. Analysis of whether current legislation and custom on privacy and other rights of research participants would have permitted such follow-up would also be germane. Other stages of this survey might be achieved by contracted papers and conferences, bringing together those who have done such follow-ups or have planned them.

Guidelines for Preparing for Long-Term Follow-Up

Subject to revision after the study described previously, the following temporary recommendations are offered:

1. Names, addresses, parents' names, place of birth, and social security numbers should be recorded and archived under confidentiality-preserving conditions for all designated experimental and control participants.

2. Informed consent should include mention of the possibility of long-term follow-up. Institutional Review Board approval need not be obtained for the follow-up because that will not be part of the research currently funded. However, the importance of possible later long-term follow-ups should be used as justification to the internal review board for the retention of identifying information facilitating such follow-up.

3. Identifying information and reason for loss of each case should be retained for all types of sample attrition from the very first designation of target populations. (Because many of the potential long-term outcomes will have no

appropriate "pretest," outcomes for all those randomly assigned, even if lost before any pretesting or treatment, will have to be analyzed to estimate the degree to which selection bias through differential attrition explains the results [Cook & Campbell, 1979, chap. 8]).

4. Feasibility of long-term follow-up should be one criterion in selecting among research projects to be supported. (Particularly attractive are "encompassing measurement frameworks" that will provide later indicators for those who fall away as a result of attrition as well as those who remain within the experimental and control samples. School records, earnings subject to withholding tax, intake records from mental hospitals, and the criminal justice bureaucracy provide such.) Sometimes a sample will be chosen just because the experimental and control samples can be from a relevant measurement framework. Thus a study might be centered on life insurance salespersons just because sales volume was a relevant measure of effective adjustment. (A little thought should provide better examples than this.)

Strategies for Achieving Continuity Necessary for Long-Term Follow-Up

We have already noted that new members of Congress and new administrators have a need to introduce innovations that can be credited to them. This works against project longevity because there is a need for news of prompt "breakthroughs." Those legislators and administrators who take political risks to establish and maintain a given experiment are reluctant to put emphasis on pilot studies that will not pay off until 15 years later. An adequate, hypothetically normative sociology of scientific validity would address this problem and provide theoretical understandings that could suggest remedies. We are not at that stage yet, but we should begin brainstorming on the problem.

Educating our sponsors to the problem might be tried. They need to be apprised of the predicament, the understandable reactions to this predicament, and the destructive effects those reactions will have. Once pointed out, and with the compensatory education results as illustrations, the necessity for long-term follow-up should be obvious.

In our dealings with our patrons (and in their dealings with their constituencies), it may help to emphasize the importance of the problem and how that importance justifies our trying out many approaches to see what works. This avoids falling into the tempting "overadvocacy trap" (Campbell, 1969c) in which we (and our patrons) promise too much for specific new "solutions" just to get them tried out at all. This strategy has been criticized (Campbell, 1971a; Shaver & Staines, 1971) at the level of treatment program leadership, but the disadvantages alleged would not seem to hold for many experimental interventions.

The Experimenting Society as Normal Rather Than Extraordinary or Revolutionary Science

As a final issue, let me confess or claim that in terms of Thomas Kuhn's (1970) still influential dichotomy, The Experimenting Society represents normal rather than revolutionary science. Popper can be quoted two ways on this. As a depiction of science (Popper, 1970), he has deprecated normal science and called for the ideal of continual revolution. But in his model for The Experimenting Society, he has asserted, "A social technology is needed whose results can be tested by piecemeal social engineering." This is what I am classifying as a normal science model for an experimenting society, in Kuhn's category system.

According to Kuhn, there are normal periods of scientific growth during which there is general consensus on the rules for deciding which theory is more valid. In contrast, there are extraordinary or revolutionary periods in science in which the choices facing scientists have to be made on the basis of decision rules that are not a part of the old paradigm. Initially, the choice of the new dominant theory after such a scientific revolution is unjustified in terms of the decision rules of the prior period of normal science.

For social experimentation, the Kuhnian metaphor of revolution can be returned to the political scene. Evaluation research is clearly something done by, or at least tolerated by, a government in power. It presumes a stable social system generating social indicators that remain relatively constant in meaning so that they can be used to measure the program's impact. The programs that are implemented must be small enough not to seriously disturb the encompassing social measurement system. Thus the technology I have been discussing is not available to measure the social impact of a revolution. Even within a stable political continuity, it may be limited to relatively minor innovations. It presupposes a stable society with governmental stability both in general and particularly in record-keeping. It presupposes the meaningful comparison of present, past, and future. The experimenting proposed here is within the framework of such a society, any given experiment being of such a small magnitude as to not fundamentally change that society. For any given step, it is limited to small changes rather than fundamental framework changes.

We gradualists will argue that fundamental changes can be made in this way, that by small steps, each validated as improvements, we can move toward any optimal society. The revolutionist will agree with Kuhn that the framework is the problem, not the details, that radical changes going beyond the framework are needed and that these are not likely to be judged scientifically by the criteria of the preexisting framework.

For intentional social changes at any level, I would argue that we should not hold up implementing them just because their impacts cannot be measured. A trivial example that comes to mind is the shift in the U.S. public schools some years ago from traditional teaching of arithmetic to the "new math" based on set theory. Because the "new math" made the available achievement tests and records based on old math inappropriate, the superiority of "new math" had to be taken on the basis of consensus of expert judgment, rather than pilot studies demonstrating its superiority. (Twenty years after the implementation of "new math," we could have done rather elegant quasi-experimental evaluations of the impact of new math on mathematics courses taken at the university level and the quality of performance in them. This was possible because the innovation was decided on by local school districts, at a variety of dates over several years, rather than being implemented simultaneously.)

Extending this reasoning, I will concede that it is not a crucial argument against political revolution to say that it may destroy the measurement series by which the revolution's impact might be precisely measured. If there be a net argument against revolutions over cumulative, gradual, tested-out, step-wise change, it has to be made on the basis of other side effects. After Popper's *The Open Society and Its Enemies* (1945/1952) and his *Poverty of Historicism* (1944), we can no longer credit revolutionary programs with superior scientific status over reformist ones. But although revolution may not provide in itself an evaluable social experiment, it could be followed by an experimental approach to fine tuning its praxis in order to achieve the optimal implementation of the ideals and idealism that brought it into being.

CONCLUDING COMMENTS

We have gone over a number of the problems that have to be solved before we can wholeheartedly advocate an experimenting society. Although it is clear on many counts that a totalitarian society is not likely to be an experimenting society, it is also clear that it holds many problems for an open society.

For now, let a number of us become methodologists for The Experimenting Society, remembering that as we develop in detail the procedures, possibilities, and problems of The Experimenting Society, we will be acquainting ourselves with what it would be like as well as this can be done in advance. As this portrait emerges in greater clarity, it will be our duty to continually ask ourselves if we really want to advocate this monster of measurement and experimentation. We must share the developing picture with the most articulate and hostile critics of such a society and consider in detail their warnings.

If it is not a future we want, who should know better or sooner than we, the ambivalent methodologists of The Experimenting Society.

OVERVIEW OF CHAPTER 2

If social scientists undertake the task of envisioning new societies as proposed in Chapter 1, they will be required to examine both the positive and negative impacts of the new world orders they envision. Armed with such models about the immediate and long-term impacts of future societies, they will be in a better position to offer guidance regarding policy decisions. In planning a better society in which the individuals are happier and more content, it is important to consider the principles of the adaptation level (AL) theory discussed by Brickman and Campbell in Chapter 2.

AL maintains that pleasure is subjective and transient. Whereas one's feelings of pleasure increase initially, the individual eventually reaches a point where the previously pleasurable experience becomes the status quo. Because the status quo is at a higher level of pleasure, it now it takes more input to experience pleasure. This is referred to as *habituation.* One can make the assumption that this same phenomenon occurs with regard to other societal rewards, such as money and prestige, and personal rewards, such as food and sex. Thus, in planning a society in which all members are happy (known as a utopia), this AL theory must be considered; happiness for this society is fleeting.

Brickman and Campbell attempt to synthesize what is known about adaptation levels and examine the effect on social planning. There are two themes, a pessimistic one which suggests that individuals will always be in a state of seeking further pleasures, and an optimistic one, which suggests that there are ways in which goods and services can be distributed to provide the maximum good for the maximum number of people. To do this, the pessimistic viewpoint must be weighed, taking into consideration habituation due to adaptation levels.

The authors do not suggest who should make the decisions regarding the distribution of goods and services. In fact, the level of satisfaction with the rewards are related as much to the level of participation in the decision making as to the rewards themselves. Ideally, all members of a society should participate in the decision-making process. There remains the question of whether people would prefer a society in which all are equal, one that rewards people according to their merit, or one in which chance or randomness determines who should be rewarded.

The problem in a society that seeks to increase the pleasure experiences of its people is that of keeping stimulus levels above the adaptation level, so that the pleasurable experience continues. One solution is to keep trying to increase the pleasure stimulus, but this is not a feasible solution. The other solution is to prevent adaptation levels from rising continually. Much of this chapter deals with how this might be accomplished.

Brickman and Campbell discuss three mechanisms for establishing an individual's adaptation level. The first mechanism for fixing on an AL is in relation to one's own history of rewards. Another mechanism is to view one's level of reward on one dimension relative to the level received in another dimension of an individual's life. The third mechanism for establishing the AL is to compare one's own reward against the experiences of another relevant person or persons.

The first mechanism is labeled the temporal comparison in which the individual compares the present reward with a past reward or some composite of past rewards. Reaching some level of aspiration (LA) produces a pleasurable experience. Since this is so, a society can satisfy its members by providing the minimum of reward above the AL that satisfies the LA. As ALs rise, though, the problem becomes one of continually increasing the inputs for everyone. For a society to even attempt to satisfy the LAs of its members, the LA of each would have to be known, and it becomes necessary to devise a system where people would benefit by stating honestly their AL and LAs. Another issue involves whether it would be preferable to distribute rewards gradually or all at once. While sizable jumps provide the greatest pleasure, there may be times when absence of rewards followed by a larger reward may be desirable.

The second mechanism that shapes an individual's AL is referred to as spatial comparison. This involves comparing the level of reward on one dimension of an individual's life (perhaps income) to the level of reward on another dimension (perhaps job prestige). If there is a discrepancy in one dimension, for example if a person's income level is not in accord with his or her job prestige, the individual may become discontented with that income. Societies can deal with this in different ways. One is by keeping all the individuals' dimensions at the same level as would be found in a caste society. Another is by allowing the discrepancies to be resolved fully through social mobility processes. The intermediate course is to allow the inconsistencies to exist but to mitigate discontent by providing greater satisfactions in certain dimensions that may be more available and hence of less importance.

The third mechanism used by individuals in setting their ALs is comparing oneself to another relevant person or group of relevant persons. Yet, individuals who are relatively better off than others often find themselves to be more discontented. Rather than compare themselves to those who are worse off, they compare themselves to others who are similar but have greater rewards. Their comparison AL rises faster than it would if they were still in an inferior position. This is the concept of relative deprivation. Relative enrichment, where individuals receive more than they expect, does not appear to generate the same kind of discontent. There is some evidence that overreward encourages people to work harder to justify the reward; however, when overreward cannot be justified, it too may cause discontent. The first implication is that a good society should allow individuals to maintain favorable comparisons and protect themselves from unfavorable comparisons. The second implication is that rewards should be perceived as fair and equitable, that is, the rewards are in proportion to merits or inputs. These perceptions of similarity and merit have been used by traditional societies to help people "know their place."

Brickman and Campbell then ask: "How might people react to rewards given by the principle of equity versus rewards given by the principle of equality?" The principle of equity, reward based on merit, is in opposition to the principle of equality, where distribution takes place according to need. When the principle of equity is used as a method of distribution, it results in further handicapping the less talented. If equality is used to distribute rewards, would the overall group be more or less satisfied? The good society must determine the extent to which its people will accept the compensatory equity principle (compensating for past inadequacies) or the equal equity principle (compensating all equally).

The ideal for social welfare programs is to establish the greatest good for the greatest number of people. Campbell and Brickman suggest that evaluation is the best weapon for estimating satisfaction over the various distributions of goods over times, modes of competence, and persons. This could answer many questions, including which method of comparison is most influential in determining the adaption level. If this is known, evaluations of social programs must consider which method of comparison is being evoked by the questionnaires and other tools used in the evaluation. The evaluations themselves can add to our knowledge of how ALs affect judgments about the goodness of life.

While the pessimistic theme suggests that given rising ALs, there may be no solution to the problem of happiness, the optimistic theme suggests that various of distributions of goods and rewards be attempted and evaluated. Further, we must search for new humane, and behaviorally realistic, means of controlling rising ALs.

TWO

PLEASURE/PAIN RELATIVISM AND PLANNING THE GOOD SOCIETY

In specifying only the pursuit of happiness as an inalienable right, the writers of the Declaration of Independence may well have expressed an intuitive understanding of adaptation-level (AL) theory (Helson, 1964), as indeed have certain philosophers since the time of the Stoics and the Epicureans. Although happiness, as a state of subjective pleasure, may be the highest good, it seems to be distressingly transient. Even as we contemplate our satisfaction with a given accomplishment, the satisfaction fades, to be replaced finally by a new indifference and a new level of striving. This is, of course, a derivation from the fundamental postulate of AL theory—namely, that the subjective experience of stimulus input is a function not of the absolute level of that input but of the discrepancy between the input

Brickman, P., & Campbell, D. T. (1971). Hedonic relativism and planning the good society. In M. H. Appley (Ed.), *Adaptation-level theory: A symposium* (pp. 287–302). New York: Academic Press.

and past levels. As the environment becomes more pleasurable, subjective standards for gauging pleasurableness will rise, centering the neutral point of the pleasure–pain, success–failure continuum at a new level such that once again as many inputs are experienced as painful as are pleasurable (e.g., Beebe-Center, 1932).

In making this derivation, one must assume that the same laws governing the experience of specific sensations will also govern the experience of generalized symbolic rewards (namely, money, prestige), and also even the experience of qualitative pleasures that lack clear external or extrinsic reference scales. Not all these assumptions have been subject to test. It should be noted, however, that this derivation does not rest on the assumption that an increase in the input level for a given good must lead to a further increase in the level of aspiration (LA) for that good, as in achievement tasks (Lewin, Dembo, Festinger, & Sears, 1944). Even if LA does not increase, the important AL principle remains: Habituation will produce a decline in the subjective pleasurableness of the input.

If the planning of a society in which people are happy is a task of Sisyphus, it would seem no wonder that the only societies that have set themselves this goal are the fictional creations called (sometimes scornfully) utopias. Interestingly enough, one of the main criticisms raised about proposals for ideal societies is that the utopias themselves are too static, too unchanging, too unchallenging, and that the inhabitants would soon grow bored and restless with the "ideal" conditions the authors find so desirable (see Dahrendorf, 1958). We may take it as given that if a society in which people are happy is possible at all, it will be possible only if it incorporates an understanding of AL phenomena, an understanding that past writers appear to have lacked.

And yet there is some sense in which society, without any such understanding of AL effects, has already begun this task of Sisyphus. Society has committed itself to providing for its members certain minimal levels of goods and services, for example, minimal levels of income and education. These levels may not be explicitly justified in terms of providing for happiness. Instead, mention may be made of fairness or equity, which, as we shall see, are related concepts. Where hedonic value is invoked to justify social programs, the value is much more likely to be one of minimizing discontent, which may be quite different from maximizing happiness. Thibaut and Kelley (1959) drew a distinction between comparison level for outcomes and comparison level for alternatives. A person will be satisfied if his outcomes are above his comparison level for outcomes (which may also be his general AL); however, even if his outcomes sink below this level and he becomes unhappy, he will not necessarily become violently discontent. That will occur only if his outcomes also sink below his comparison level for an alternative society, or a violent revolution. Thus, society's present commitment seems only to be keeping people above their comparison level for violent (or nonviolent) revolution, and not keeping them above their AL for hedonic satisfaction. However, it is still true that the satisfaction with the level of goods society seeks as a minimal goal for all its members will decrease as a result of habituation, even if this subjective decrease does not bring it below people's comparison level for alternative societies. Thus, the long-term effects of any societal program to ameliorate the general condition or the specific condition of particular groups cannot be understood without reference to the temporal course of subjective AL, even if the specific programs limit themselves to talking about objective quantities like tons of food or number of available jobs.

Economists have long been very gloomy about measuring and comparing the subjective satisfactions that people derive from goods (see Mishan, 1960, for a review of the relevant area of welfare economics), because such measurement would involve assumptions that they feel are untenable (such as that people have equal capacity for satisfaction). But social psychologists have long been committed to trying to assess such satisfactions in laboratory settings and have shown interest in trying to assess more general satisfactions with the "quality of life" (e.g., Bradburn & Caplovitz, 1965). One of the purposes of this chapter will be to demonstrate that social psychology has accumulated enough knowledge in this regard to at least ask some of the right questions. This chapter will attempt to organize what we already know about the determinants of hedonic ALs to assess the implications for social planning, and to point out new areas of relevant research.

Throughout this article there will run two themes: a pessimistic one and an optimistic, or at least an ameliorative, one. The pessimistic theme is that the nature of AL phenomena condemns men to live on a hedonic treadmill, to seek new levels of stimulation merely to maintain old levels of subjective pleasure, to never achieve any kind of permanent happiness or satisfaction. The optimistic theme is that regardless of this ultimate impossibility, there are still wise and foolish ways to pursue happiness, both for societies and for individuals, and, from a planner's point of view, there are certain distributions of goods over time, persons, and modalities that will result in greater happiness than others. The pessimistic theme is that subjective pleasure is, as a state, by its very nature transient, and, as a goal, an ever-receding illusion. The optimistic theme is that society is overdue to undertake an explicit, experimental commitment to maximize the greatest subjective good for the greatest number. Whether the reader resonates more to the pessimistic theme or the optimistic theme is probably in large part a matter of his own philosophical bent. This chapter does not reconcile these two conflicting themes, except to the extent that it argues that no planning for the good society (a task for optimists) can be successful unless it is done by people who thoroughly understand the relativistic and elusive character of subjective pleasure (an understanding of pessimism).

It is important here to enter a disclaimer on the question of who should make decisions about the distribution of rewards. This article limits itself to a treatment of the probable consequences of different distributions. We do not mean to imply that social scientists should make the decisions as to which of these are preferable. Indeed, it may well be that the question of how decisions about the distribution of rewards are made is more important in determining satisfaction with outcomes than the nature of the particular decisions—that is, people will be satisfied with allocations of resources to the extent that they participate in the allocation decisions. Even here, however, it should be noted that how satisfied people are with their participation in decision making will itself be a function of an AL for participation. It may also be that certain distributions of rewards that satisfy our demands for maximizing general hedonic satisfaction will not work because they will self-destruct—in other words, a particularly ill-favored group will seek violent redress, or a particularly well-favored group will move to seek still further reward. This alerts us to a further issue not treated: that differences in rewards are not merely a matter of hedonic satisfaction with present states, but also a serious issue of social power to affect future states. A man's wealth or his prestige affects not only his satisfaction but also his ability to influence others, and this too must be considered.

The present authors believe that all members of society should participate in making these decisions about the distribution of rewards, both as regards their own individual states and as regards the interrelationships among everyone. What this would lead to can only be speculated about at the present time. As will be discussed, there are some grounds for believing that people will prefer a system in which everyone is equal, but there are also grounds for believing that they will prefer a system in which people are rewarded differentially according to merit. And indeed, they might even prefer a system in which people are rewarded differentially by chance, for example, in which monthly or yearly lotteries decide who lives extravagantly and who lives in genteel poverty.

There are in general two possible solutions to the problem of keeping stimulus levels above ALs so that the stimulus levels continue to be experienced as pleasurable. The first is to continually increase stimulus levels in some fashion so as to keep them above the hedonic AL. For a variety of reasons—limited resources, declining marginal utilities for goods, and social comparison complexities (to be discussed)—this seems an impractical solution in the general case. This article will concentrate on the second possible solution, preventing hedonic ALs from continuously rising, or specifying situations in which AL phenomena either do not hold or can be modified. Let us first consider the various factors that determine hedonic ALs and how these might be manipulated.

TEMPORAL COMPARISON: DISCREPANCY BETWEEN PRESENT AND PAST REWARDS

The first mechanism for establishing AL (Helson, 1964) is temporal pooling of stimuli, or, for hedonic AL, the individual history of reward. Temporal comparison may be considered to involve a person's comparing his level of reward at the present time t_n with his level of reward at time t_{n-1}, or with some temporally weighted average of his rewards over all times up to t_n. It has been demonstrated that past experiences of rewarding or punishing sensation are pooled and that stimuli are most effective as reinforcers when they depart most strongly from the level of past experience (Bevan, 1963; McHose, 1970; Quinsey, 1970). It has also been shown that people typically set their LA slightly above their past performance level (Lewin et al., 1944).

Although a strict AL theory would predict that the larger the positive discrepancy between reward and AL, the greater the pleasurable sensation, there is considerable evidence from a variety of domains that outcomes very much in excess of AL may not be positively reinforcing (Aronson & Carlsmith, 1962; Lewin et al., 1944; McClelland, Atkinson, Clark, & Lowell, 1953). Although this is not fully settled, because there is also evidence that outcomes far above AL are primary sources of elation and pleasure (Dittes, 1959; Verinis, Brandsma, & Cofer, 1968), we are probably safe in following the lead of Siegel (1957), who showed that outcomes above the LA might produce further increments in satisfaction, but none so great as that produced by reaching the LA.

This leads to the first important implication of hedonic AL theory: Namely that we can most efficiently allocate goods to produce satisfaction by first trying to give everyone the minimum above his AL that brings him to his LA. Indeed, this principle has been suggested as the basis for proportional or progressive taxation (Duesenberry, 1949; Vertinsky, 1969). As ALs continue to rise, however, society is then faced with the problem of trying to simultaneously increase its inputs to everyone. Doing this—that is, expanding production, rather than redistributing goods—has been the historic secret of the success of capitalism (Galbraith, 1958), although there are limits on the expansion of production such as the accumulation of hidden costs in the form of pollution and the finite nature of natural resources. Society is given a break in this process, however, by the fact that members with high consumption functions or high ALs are continually dropped out of the picture (through death), and new members starting with low ALs are added (through birth). Thus, society may manage something very much like the continual pacing of stimulus inputs ahead of AL across the individual life cycle by increasing the monetary rewards, freedom of choice, and social status that accrue to the individual, as he gets older just fast enough to keep his inputs above his AL. Indeed, apart from the frustrations of living, one of the reasons that people may not be much attracted by the idea of living their lives over is that such a renewal would mean starting with rewards that are much below the standards to which they have been accustomed. And one of the basic, though often unrecognized, factors in today's "generation gap" may be that the continuity of the incremental reward function with age has been disrupted, so that the younger generation is seen as getting "too much, too soon," and are themselves strongly challenged by having to cope with their disproportionate affluence.

There are, of course, some difficulties with taking reward according to AL as an explicit principle of distribution. A teacher who was limited in the number of high grades she could give out might know that she could maximize satisfaction in the class by giving each person just what the person expected and no more (and it might be hypothesized that teachers seek to learn student expectancies for precisely this reason). However, the moment the students learned that this was her grading principle, they would take great pains to state as high expectancies as possible, and to escalate their demands even beyond normal AL effects. It would seem difficult to recommend as a basis for social organization a principle that could succeed only if members of the social system remained in ignorance of it. The only alternative, however, would be to devise a system in which it was truly to people's advantage over the long run to state only their "honest" ALs and LAs.

Next we might consider the related question of how shifts in level of reward should be managed. Should an upward shift be made gradually or suddenly in order to maximize the satisfaction gained from the new level of reward? Adaptation-level theory and evidence (e.g., Quinsey, 1970) are unequivocal in predicting that the affective value of the final level of reinforcement will be reduced if the transition is gradual rather than sudden. This by itself would argue that we should try to avoid too much "smoothing out" of positive transitions, while conversely making negative transitions as smooth as possible. Dissonance theory, on the other hand, would suggest that individuals are much more likely to accept new levels of reward without dissonance if these new levels are not sharply discrepant with their expectancy levels—that is, if the transition to a higher level of reward is gradual (Aronson & Linder, 1965) versus if the transition is sudden (Aronson & Carlsmith, 1962). It should be noted that this question cannot be answered by considering the affective value of the reward level alone, but must also take into account the satisfaction experienced by the person over the whole series of rewards, where the sum total of rewards received over all points in time is equal in all conditions. The evidence reviewed earlier would seem to suggest that a series of gradual increments would be preferable to an alternative series, equal in total reward, that delayed the person longer at his initial level and then jumped him all at once to his final level.

If overall shifts are to be gradual, is there any way we can preserve the virtues of sudden discrepancies, without at the same time drastically affecting the gradual course of the AL? We may now mention a second implication of AL theory for the societal scheduling of rewards that society has already discovered: the use of periods of abstinence and periods of special indulgence. Adaptation-level theory would suggest that periods of abstinence could serve the very useful function of lowering AL in certain areas and thus permitting future rewards to be experienced with greater satisfaction. These periods of abstinence must be clearly connected to the ordinary course of events. On the other hand, it would be equally useful to allow periods of special indulgence that are specifically defined as special, and hence not allowed to affect the ordinary AL, much as the tray lifted in the middle of a series of weights does not affect the AL for judging the heaviness of the weights (Brown, 1953). Both these practices are features of traditional religion that have become less common as traditional religion has lost its central place in society. It might be well both to gear some of our future research to understanding the effects of these practices and to consider whether they might be reintroduced into social custom with a new rationale.

Finally, besides manipulating AL through programming variations in current experience—for example, introducing recurrent periods of abstinence—we might consider achieving comparable effects by playing with memory functions. Adaptation-level theory would suggest that there is an optimal degree of recollection for past sadness. It would not recommend that people dwell on past unhappy states, for that would probably cause them to assimilate current events to past levels. On the other hand, AL theory would not recommend that people forget or suppress entirely these sad times, for the measure of their present happiness is the contrast with the less pleasurable past. Thus, perhaps the happiest adult is one who had a moderately unhappy childhood. There is good evidence that people ordinarily forget unhappy memories faster than they do happy ones (Holmes, 1970). AL theory would suggest that we employ occasional reminders of such events.

SPATIAL COMPARISON: DISCREPANCY BETWEEN AREAS OF COMPETENCE

The second mechanism for establishing AL in perceptual experiments is the spatial pooling of stimuli, or the averaging of the stimuli received from the synchronous background. With regard to hedonic AL, *spatial comparison* may be somewhat loosely taken to refer to a person comparing his level of reward on dimension A with his simultaneous level of reward on dimension B, or with some weighted average of his rewards on all other dimensions. The literature that has been most nearly devoted to this topic is the literature on status incongruence or status inconsistency (see Sampson, 1963, for a review). Status incongruence exists when a man's standing on one dimension is discrepant with his standing on another dimension—for instance, when a man's occupational prestige is higher or lower than his income level. Though here again there are conflicting findings (Lenski, 1967; Segal, Segal, & Knoke, 1970), much of the work in this area has demonstrated that status incongruence is a source of activation, strain, and discontent. Thus, Lenski (1967) reported that people with discrepant status inputs are more likely to support radical political parties that promise to help secure their status at a favorable level. In part, then, it seems that stimulus inputs on the more favorable dimension may come to serve as AL for the inputs on other dimensions, thereby greatly increasing discontent, not unlike the golfer who allows a once-in-a-blue-moon score in the 1980s to make him thereafter unhappy with his usual 110s.

Societies traditionally have coped with the potential problems of discrepant status inputs for individual ALs in several different ways. One solution refuses to allow status inconsistencies to arise in the first place—the definition of a caste society, in which an individual's placement in one status category automatically determines his location on all other dimensions. The other logical solution permits status inconsistencies to be fully resolved, in whatever fashion, by social mobility processes. Of most interest for our purposes, however, is an intermediate course in which such inconsistencies are allowed to arise and to persist without resolution; yet comparison among these dimensions is mitigated or prevented, for example, by role segregation. In this manner society may even be able to turn the multidimensional nature of status and prestige to its advantage. For instance, lower participants may be "cooled out"

(Goffman, 1952) by having their attention diverted to relatively satisfactory returns they may be receiving on unimportant dimensions on which higher status persons are much more likely to make concessions (Jones, Gergen, & Jones, 1963).

Research in the area of status incongruence has not been couched in an AL framework, and it is not fully possible to assess the hedonic implications of the research at this point. Instead, it may be more appropriate simply to phrase some of the relevant questions. It would seem rare for an individual to be outstanding on all dimensions, so as to receive top rewards in all areas. The question then arises as to whether it would be more satisfying for a person to be outstanding on dimension A and relatively poor on dimension B, or to be located at an intermediate (fairly good) position on both A and B. Or in a case involving social comparison, would two people together derive more satisfaction if each were superior on one dimension and inferior on the other, or if both were identical (at some intermediate position) on both dimensions? Is it possible that receiving a varied pattern of rewards—including at least one domain in which the individual is in a steady state of high reward, and one in which he is experiencing sharply rising rewards—can protect a person from dissatisfaction as a result of an unsatisfactory level of reward in other areas—that is, a low level or even a declining level of reward? Or will it be true, as some of the status-incongruence literature implies, that the added dissatisfaction from the lower levels will more than counterbalance the added satisfaction from the higher levels?

SOCIAL COMPARISON: DISCREPANCY BETWEEN SELF AND SIMILAR OTHERS

The third and perhaps most potent mechanism for establishing hedonic AL is social comparison, or the comparing of the rewards accruing to oneself with the rewards accruing to a relevant other person or to some weighted average of the rewards accruing to all relevant others. Social comparison may of course have both temporal and spatial components. The literature on social comparison has established that the reward levels of various others will contribute to the person's hedonic AL for his own rewards to the extent that these others are seen as similar to the person seeking comparison (see Festinger, 1954; Latane,

1966). In one of the best studies on this point, Hoffman, Festinger, and Lawrence (1954) found that individuals were less likely to compete with an opponent who had gained a head start if that opponent were defined as superior to them in ability than if he were defined as similar ability. Much the same findings on this important point have been obtained on a societal level. Stern and Keller (1953) and Runciman (1961) found that it was relatively rare for respondents in France and England, respectively, to take as standards for comparison persons who were perceived as belonging to different social classes than their own.

The concept of relative deprivation (Merton & Kitt, 1950) was evolved to explain why in a number of instances persons who were objectively better off appeared to be more discontent with their lot. Studies of the American soldier in World War II, for instance, found that soldiers with more education were significantly less content with their chances of being promoted, even though they had significantly better chances of being promoted than did the less well-educated soldiers. The better-educated soldiers were not comparing themselves to their less well-educated peers, however, but to others who were similar to them in education but were getting better treatment (others who were officers). Since then, the notion of relative deprivation has been used to explain why people may become less satisfied, rather than more satisfied, as their objective condition improves—because the improvement of their condition raises their comparison AL at an even faster rate. Thus, Pettigrew (1964) discussed the psychological losses that accompany the actual gains achieved by the civil rights movement, and Sears and McConahay (1970) explained the fact that Northern-born Blacks are both more optimistic in their aspirations and also more discontented and more likely to participate in riots than their Southern-born neighbors.

Loose application of the concept of relative deprivation in explaining the dynamics of social change can make those whose lot is improving seem relatively ungrateful and shortsighted. Thus, it may be well to say a word in passing about other possible interpretations of these data. As Sawyer (1971) pointed out, the notion that the improvement has stimulated a sense of relative deprivation, on the basis of present data, cannot be distinguished from the possibility that the improvement has activated a general desire to make all men equal. Thus, it may be unnecessary to assume that the improvement results in a shift in AL (resulting from a shift in social comparison), but merely that the improvement generates well-known goal gradient effects as a long-standing goal comes nearer (see J. S. Brown, 1961). The effects of blocking efforts to reach a goal are greatly enhanced when the goal is close versus when it is far off (Haner & Brown, 1955), a fact that has been incorporated in the theory of revolution (Brinton, 1965; Davies, 1962).

Also in passing, it might be noted that there is no parallel concept of "relative enrichment," or at least no concept that implies that overreward can present a similar danger to a social system. Indeed, giving people more than they deserve may be seen as an effective way of inducing them to increase their commitment to the system, much as Adams (1965) found that individuals worked harder when they were overpaid in order to "justify" their level of reward. But we may suspect that overreward too can result in discontent, especially when it cannot be made up for or justified by extra work. That some members of the privileged classes perceive such inequity and are not satisfied to deal with it by the traditional gestures of charity has meant that a small but historically important group of revolutionaries has always come from the upper classes. The American New Left is the latest case in point (Franks & Powers, 1970).

The first major implication of hedonic AL theory in the area of social comparison is simply that we should recognize the importance of allowing people to maintain favorable comparisons and to protect themselves from unfavorable comparisons. Thus we might note that traditional society encouraged covert comparison with those less fortunate in the form of "counting one's blessings." Charity simultaneously provided an occasion for comparing one's own lot with those worse off and at the same time worked to relieve the guilt that might come from such comparison. And we should consider the possibility that Castro may have raised the psychic well-being of the Cubans simply by getting rid of the rich tourists who came in numbers sufficient to constitute a very unfavorable comparison group. An unrestricted range of social comparison may itself have detrimental effects on hedonic ALs, as in the case of academic conventions in which the vast number of similar others set hedonic ALs very high. Ideally, one principle of freedom that a good society might embody is freedom of comparison, both in the sense of allowing a person to seek comparison where and when he chooses, and in allowing him to protect himself from comparison where and when he chooses.

The principle of equity (Adams, 1965) or of distributive justice (Homans, 1961) specifies the most important dimension people use in comparing their own rewards with those of others: their relative merits, or inputs. The equity principle implies that if people are unequal in their abilities or their past accomplishments or in what they contribute to a job, we would prefer to see them rewarded unequally, and indeed, to see them rewarded in direct proportion to their merits. Thus, Exline and Ziller (1959) found that groups in which members had power or status directly proportionate to their skills were more satisfying than groups in which such resources were allocated in inverse proportion to merit.

From all this we may draw a second major conclusion for the allocation of rewards with regard to social comparison ALs: Rewards should be allocated so as to be congruent with these comparison ALs, and thus to be perceived as fair, equitable, and legitimate. They should neither be allowed to fall below these levels nor to rise above them (in contrast to the situation for temporal ALs) and, if it is important to conserve the supply of available resources, these comparison ALs should be lowered by reducing people's perceptions of what they deserve.

The control of perceptions of similarity and merit has of course been the principal means in traditional society whereby people were induced to accept discrepancies between wealth and poverty even more striking than those in modern industrial society. However familiar a servant might be with the standard of living in the castle, beliefs about differences in birth, breeding, and station protected the servant (and the master) from the servant's taking life in the castle as the basis for his own hedonic AL. No such inhibitions are imposed in the descriptions of how other people live by our own egalitarian media—especially movies and television. By choice or necessity—perhaps the media are the message—these emphasize the common humanity of people rather than their class or regional differences. This has had the momentous consequence of making the range of social comparison wider—rendering similar more and more others who were believed to be dissimilar (and hence perhaps deserving of dissimilar rewards). The erosion of the traditional controls on social comparison is of course the so-called "revolution of rising expectations" that is shaking the world today, a revolution that many of us would judge good, many would judge inevitable, and all would agree is highly explosive.

It is interesting to note that as the perception of similarity has become more important, the desire to control information about rewards as a means of minimizing discontent from social comparison has probably increased. Experienced interviewers know that people are even more reluctant to reveal their incomes than to reveal their sexual habits. A study by Lawler (1965), however, indicates that secrecy may not work to minimize dissatisfaction but just the opposite; Lawler found that managers overestimated the pay of subordinates and peers and thus saw their own pay as too low by comparison. However, abandoning secrecy about rewards in the face of increasing perceived similarity would increase pressure for full equality, perhaps even regardless of equity.

There is reason to hypothesize that the principle of equality will tend to gain strength simply as information about rewards and decisions about rewards become more public. It is interesting to note the contrast between the division of reward on athletic teams in salaries (which each player negotiates individually with management) versus prize money or bonuses for winning (which the players as a whole allocate by vote). Salaries are of course traditionally highly uneven, with players receiving what is believed that their skills deserve. Prize money, on the other hand, is traditionally divided into equal shares and given to all members of the team regardless of their contributions to winning. Recently, as players have begun to hire agents to bargain for them at salary time, one general manager (Editors of *Sport*, 1970) has had the thought (not serious) of telling the agents and the players: "Look here's what the club can afford for salaries. You divide it." If the players were called on to divide salary money as they are to divide prize money, would they preserve the traditional equity principle or would we see a sharp swing toward equality? If there were a shift toward equality, would the overall satisfaction of members (averaging those receiving less than "equity" with those receiving more than "equity") be enhanced or decreased? Although no general manager of a professional team is likely to try this experiment in the near future, there is no reason social psychologists should not pursue this question in their own territory.

Finally, it is most important to note that the distinction between a preference for equality and a preference for equity takes on additional importance when it is realized that assigning rewards on the apparently fair principle of equity strongly biases against the possibility that people will ever achieve equality by handicapping further those who begin with lesser abilities or accomplishments. For example, by giving the most promising and most accomplished new graduate students the most attractive forms of aid (fellowships or prestigious research assistantships), and the less accomplished ones are given less attractive forms of aid (teaching assistantships or perhaps no aid), we thereby help to ensure that those who have done better in the past will continue to do better in the future. (We ignore for the moment the narrow perspective of competing systems, each concerned with attracting the best talent.) Thus, the principle of merit, or to each according to his achievement, directly contradicts the principle of welfare, or to each according to his need—which may be one reason why this country has found it so difficult to design truly effective compensatory education and welfare programs.

We must discover the limits of Homans's distributive justice principle, or the principle of assigning rewards in direct proportion to merit (thus far called "the" equity principle, but there is in fact no reason to give this principle a monopoly on the word "equity"). We must discover when people will accept either the principle of assigning rewards in inverse proportion to accomplishment (the compensatory equity principle) or the principle of assigning rewards equally irrespective of merit (the equal equity principle). Beck (1967) suggested that labeling is the crucial difference, and that welfare will work (in the sense of benefiting or rehabilitating its recipients) only when it is not called *welfare.*

THE GREATEST GOOD FOR THE GREATEST NUMBER

The question of hedonic relativism has been largely—though not entirely—ignored in welfare economics (see Brown, 1952; Duesenberry, 1949; Mishan, 1960). The relativistic perspective considerably complicates the question of whether a given distribution of welfare is optimal or not—that is, maximizes the greatest possible good for the greatest number. The most basic and enduring optimality principle was formulated by Pareto and is called "Pareto optimality." This principle states that a given distribution is optimal if there is no change that will make any person better off without also making somebody worse off—or, conversely, a change is to be preferred if it makes at least one person better off without making anyone worse off, or (in subsequent versions) if the gain to the first party is such that he can compensate the second party for the loss and still have some left over. In any event, we should note that the idea that utilities for goods change over time as AL changes, or the idea that increases for certain parties may automatically produce decrements either for others or for themselves by changing social comparison ALs, materially complicates the task of defining optimal distributions. But welfare economics is already so overburdened by abstract complications (for example, an unwillingness to assume that utility or satisfaction can be compared across persons, or the demand that there exist unequivocal social choice functions for all possible combinations of individual preference orders [Arrow, 1951]) that it seems unfair to add to its theoretical problems. Instead what we recommend, both for welfare economics and for social psychology, are empirical studies of the effects of various distributions of goods over times, modes of competence, and persons.

Even without any theoretical underpinning at all, it would seem most useful simply to map out, empirically, how not only the mean but also the variance and the skewness of a distribution of rewards affects satisfaction. This should be done for each of the distributions discussed (temporal, spatial, social). If we remove all variance from such a distribution, do we thereby destroy that distribution as a basis for satisfaction? If we make all people equal, do we thereby destroy social comparison as a basis for satisfaction? Or do we thereby maximize satisfaction, or minimize dissatisfaction—which may be quite different things? If we insist on an unequal distribution of rewards, should the range be wide or narrow? Should the distribution be negatively skewed (so that most people lie above the mean, but a few lie far below it), positively skewed (so that most people lie below the mean, but a few lie far above it), or bimodal (haves and have-nots)?

It would also seem useful to know whether temporal discrepancies, "spatial" or status discrepancies, or social discrepancies are most influential in determining AL. Given the general importance of social factors in the conception of self (see Jones & Gerard, 1967), we might think that social comparison would be the most important determinant. Yet Fishbein, Raven, and Hunter (1963) found that students preferred to use their own past performances rather than social comparison information in setting their aspirations and expectancies for a future performance. Clearly this too is an area that needs more work.

Finally, it should be noted that AL theory has something very important to say about how the evaluations of the success or failure of social programs should be conducted: It tells us that we must be aware of and indeed explore the subtle effects that may be generated by implicitly evoking one or another AL in our questionnaires. At the same time evaluative research may itself be a very handy arena for building our knowledge about the effects of AL on judgments about the goodness of life.

GETTING OFF THE HEDONIC TREADMILL

In closing, it may be well to return once again to the pessimistic theme, the theme that the relativistic nature of subjective experience means there is no true solution to the problem of happiness. It should be remembered, however, that there are nonrelativistic elements that enter into pleasure, and that there is, in addition, a view that "true happiness" does away with all relativistic perspectives. The nonrelativistic elements are those that come from the satisfaction of recurrently renewed need cycles, like hunger; though we may satiate (and raise our hedonic AL) each time we eat, nature has arranged it so that the effects of this reinforcement dissipate over time and that we will be satisfied with the same level of reinforcement 6 hours later. On this level, we should not ignore the possibility of a purely physical or physiological and nonrelativistic solution to the problem of happiness. There may be a new drug, a method of brain stimulation, or a happiness pill that will move people to any level of good feeling they choose whenever they choose. Short of this, however, there may be no way to permanently increase the total of one's pleasure except by getting off the hedonic treadmill entirely. This is of course the historic teaching of the Stoic and Epicurean philosophers, Buddha, Jesus, Thoreau, and other men of wisdom from all ages. Unfortunately, renouncing the hedonic treadmill is a very difficult thing for men to do, at least until, like St. Augustine, they have traveled the full path from innocence to corruption. Even in renouncing the pleasures of the flesh, however, men may experience AL phenomena in their pursuit of piety or saintliness.

Truly renouncing the hedonic treadmill may mean abandoning all evaluative judgments and even all questions about happiness in pursuit of the notion that happiness is unselfconscious and that when a person is happy he is unaware of it. It may be, however, that evaluative judgments, and AL and LA phenomena, are necessary to the restlessness and the searching that have made human life what it is; perhaps the absence of such phenomena is to be found only in people who are resigned to an oppressive existence or who have surrendered even the basic competence motivation (White, 1959); but if we are to prevent this restlessness from wreaking its most destructive consequences, we must acquire new, humane means of controlling rising ALs.

PART II

THREATS TO THE VALIDITY OF SOCIAL EXPERIMENTS AND HOW THEY CAN BE CONTROLLED

In Chapter 1, Campbell suggested that society could be improved through policy decisions based on hardheaded evaluation. He proposed that the best way to evaluate social reforms was to treat them as experiments through the use of randomization, or by approximating experiments, using what he referred to as *quasi-experiments.* His reason for advocating social experimentation was that he felt this method emphasized internal validity—in other words, it increased confidence that the treatment or reform was indeed the reason for the observed outcome. This is in marked contrast to other scholars of his time, such as Cronbach, who emphasized external validity, or the generalizability of the research results to circumstances not studied.

Campbell argued that if we believe that society can be improved through policy decisions based on social experimentation and evaluation, it is incumbent on the social scientist to ensure the validity of the research results. Billions of dollars could be at stake, not to mention needless human suffering. Yet, an inherent dilemma for the social scientist is that he or she (unlike the physical scientist counterpart) does not work in a laboratory. This makes it more difficult to control factors, other than the phenomenon being studied, that can affect the experimental outcome.

Part II deals with the issue of validity and some common threats to validity. To set the stage for these two chapters, Campbell discussed how the scientific method is applied to the study of social issues in his Forward to Robert Yin's (1984) *Case Study Research.*

Campbell felt that the core of the scientific method is best described by the phrase "plausible rival hypotheses." This strategy involves two processes. First, other implications of the hypothesis are tested to see if they are supported by other available sets of data. Second, the original evidence is examined to see if factors other than the hypothesized cause may be responsible for the result. These other rival hypotheses are rendered implausible by "ramification extinction." This means that the implications of the rivals are tested on other data sets to determine if they fit. These processes are carried out to different degrees, depending on the scientific community and rival hypotheses that have been raised by that community. Although their theories can never be proven, the scientific community can employ this strategy to achieve consensus and cumulative achievements.

There are two ways in which the strategy of "plausible rival hypotheses" can be applied by social scientists. The first method involves randomized assignment to treatments, which renders implausible an infinite number of rival hypotheses without specifying what they are. The other method is characterized as either "experimental isolation" or "laboratory control." For this method, the rival hypotheses must be specified prior to the experiment so that controls can be put in place to render them implausible. They are generally defined by the current scientific community.

Campbell suggested that the quasi-experimental tradition, which characterizes much of applied social research, is closer to the laboratory experiment than to random assignment to treatments. This is because the quasi-experiment and laboratory experiment both require the scientist to specify the threats to validity in advance and design controls specifically for each.

Campbell is probably best known for his seminal work with Julian C. Stanley presented in the book *Experimental and Quasi-Experimental Designs for Research.* In this important work, widely read within the research community, the authors popularized the notions of internal and external validity, and enumerated some of the common threats. They then evaluated some experimental and quasi-experimental designs in terms of their ability to rule out these threats. In a later work, *Quasi-Experimentation: Design and Analysis Issues for Field Settings*, Thomas D. Cook and Campbell further subdivided the types of validity. Because internal validity involves the soundness of the conclusions about a causal hypothesis, statistical conclusion validity is considered a special case of internal validity. This type of validity involves the reasonableness of inferences that are based on statistics. Construct validity pertains to generalizations, a characteristic it shares in common with external validity. It is concerned with the possibility that several underlying factors may be confounded in a single construct; for example, this occurrence in a causal variable might lead different investigators to draw contradictory conclusions regarding which factor was responsible for the observed outcome. It is also concerned with the possibility that the construct as measured might underrepresent the construct as conceptualized.

The concepts in this part perhaps reflect Campbell's greatest legacy. The concepts of internal and external validity are so broadly used that many times the connection to Campbell is not apparent. Yet, some scientists question whether the concept of validity has any meaning for science, and others question whether validity, as characterized by Campbell, is applicable to the science they practice. These issues continue to be debated in the social science community.

The first article in this part, Chapter 3, gives a brief introduction to the concepts of internal and external validity discussed in Campbell and Stanley. It also offers an overview of some of the common threats to validity. The chapter then presents some quasi-experimental designs that have been applied to social reforms and illustrates the weakness in those designs that allowed plausible rival hypotheses to be inferred. Finally, Campbell demonstrated how altering the design can render these threats implausible. Chapter 4 provides a more in-depth discussion of the distinction between various types of experimental validity. This is a very important discussion, because the oft-used concepts of internal and external validity have been misunderstood by many, both students and scholars. Campbell recognized that the labels, internal and external, may have been responsible for some of the confusion surrounding them, so he and his colleague, Tom Cook, attempted to relabel them in order to clarify their meaning and help differentiate between them. Although this chapter goes a long way in elucidating the concepts, the new terminology was never adopted by the research community.

OVERVIEW OF CHAPTER 3

Imagine yourself as an administrator of an important social program. You fought hard to get your program funded, and because you believed your program had the best chance of alleviating the problem, you made bold promises to the legislators responsible for funding it. Now your job depends on showing that your program is successful. How would you feel about an evaluation that might point out the weaknesses of your program? You might be tempted to choose an evaluation design that would ensure illusory positive results, or accept positive results on face value without exploring to see whether factors other than your program might have been responsible for them. These are the issues discussed in this chapter.

Campbell believed that the United States and other modern countries should implement social programs aimed at curing society's ills. These programs should be evaluated to see if they are effective, and those that appear to be effective should be retained, modified, or imitated. In other words, Campbell advocated what is now called *summative evaluation,* which assesses the usefulness of a program in terms of how well it alleviates the problem it was meant to address. Evaluation results, however, are often uninterpretable and can be misleading. Two factors contribute to this problem. First, social programs are not conducted in a laboratory in which outside influences can be controlled. Without careful consideration of these influences, the results of the evaluation may be misinterpreted and perhaps overestimated. Another contributing factor is the political process that must occur for any social program to be implemented. It is politically advantageous to have immediate positive results to report. If the results are less than positive or only trivially positive, or demonstrate that the reform has failed, program administrators rely on ambiguous or falsely positive research results to be able to control the information that is fed to the public, and hence, preserve their reputations. For this reason, they often balk at the hard-headed evaluation procedures that might bring negative results to light.

To address the political issue, Campbell suggested that rather than arguing for the success of a specific program, social reformers should focus on the seriousness of the problem and suggest multiple approaches to solving it. If evaluation results indicate that a program does not live up to expectations, an alternative program should be implemented. Politicians need not fear failure of a specific reform as much as a dearth of ideas to solve serious social problems. Also, avoiding evaluations that focus on an administrator's performance will lessen resistance to evaluation and add an important ally to the evaluator searching for reliable and interpretable results.

Because social reforms cannot be evaluated in a laboratory setting, there are many explanations for the perceived effects beyond the effectiveness of the reform. We refer to these alternative explanations as *rival hypotheses.* True experiments using randomization can rule out some of these rival hypotheses. The social researcher, however, must conduct quasi-experiments, and because of the inherent lack of control in such designs, must take special steps to render rival hypotheses implausible.

Internal validity deals with whether the experimental treatment actually made a difference. Campbell lists nine rival hypotheses, called *threats to internal validity,* that might provide alternative explanations to quasi-experimental results. External validity is the extent to which the effects can be generalized to other populations, settings, and treatment and measurement variables. Six threats to external validity are listed.

As mentioned previously, a number of threats to internal validity are controlled in randomized experiments. Randomized experiments are preferred whenever feasible, because the results are clear and interpretable. Quasi-experimental designs must be evaluated in terms of the threats to validity. Different quasi-experimental designs are known to rule out certain threats. In this article, Campbell discusses some of these quasi-experimental designs, beginning with a very weak *one-group pretest—posttest design* and progressing toward the more interpretable *interrupted time-series design,* and finally to the strongest *control series design.* Using as an example a study in Connecticut to determine whether a crackdown on speeding reduced traffic fatalities, Campbell demonstrates how modifying the design can rule out the threats and improve the interpretability of the results.

Finally, Campbell shows how evaluations can be designed and how ambiguous results can be used by trapped administrators who wish to avoid the negative or threatening findings that might be uncovered by hard-headed evaluation.

THREE

AN INVENTORY OF THREATS TO VALIDITY AND ALTERNATIVE DESIGNS TO CONTROL THEM

The United States and other modern nations should be ready for an experimental approach to social reform. Using this approach, we should try out new programs designed to cure specific social problems, learn whether or not these programs are effective, and retain, imitate, modify, or discard them on the basis of apparent effectiveness on the multiple imperfect criteria available. So long have we had good intentions in this regard that many may feel we are already at this stage. It is a theme of this chapter that this is not at all so, that so many ameliorative programs end up with no interpretable evaluation (Etzioni, 1968; Hyman & Wright, 1967;

Campbell, D. T. (1977b). Reforms as experiments. In F. G. Caro (Ed.), *Readings in evaluation research* (2nd ed.). New York: Russell Sage Foundation.

Campbell, D. T., & Ross, H. L. (1968). The Connecticut crackdown on speeding: Time-series data in quasi-experimental analysis. *Law and Society Review, 3*(1), 33–53.

Schwartz, 1961). We must look hard at the sources of this condition and design ways of overcoming the difficulties. This chapter is a preliminary effort in this regard.

Many of the difficulties lie in the intransigencies of the research setting and in the presence of recurrent seductive pitfalls of interpretation. The bulk of this chapter will be devoted to these problems. But the few available solutions turn out to depend on correct administrative decisions in the initiation and execution of the program. These decisions are made in a political arena, and involve political jeopardies that are often sufficient to explain the lack of hard-headed evaluation of effects. Removing reform administrators from the political spotlight seems both highly unlikely and undesirable, even if it were possible. What is instead essential is that the social scientist—research—advisor understand the political realities of the situation, and that he or she aid by helping create a public demand for hard-headed evaluation by contributing to those political inventions that reduce the liability of honest evaluation and by educating future administrators to the problems and possibilities.

For this reason, there is also an attempt in this chapter to consider the political setting of program evaluation and to offer suggestions as to political postures that might further a truly experimental approach to social reform. Although such considerations will be distributed as a minor theme throughout this chapter, it seems convenient to begin with some general points of this political nature.

POLITICAL VULNERABILITY FROM KNOWING OUTCOMES

It is one of the most characteristic aspects of the present situation that specific reforms are advocated as though they were certain to be successful. For this reason, knowing outcomes has immediate political implications. Given the inherent difficulty of making significant improvements by the means usually provided and given the discrepancy between promise and possibility, most administrators wisely prefer to limit the evaluations to those in which they feel they can control the outcomes, particularly when published outcomes or press releases are concerned. Ambiguity, lack of truly comparable comparison bases, and lack of concrete evidence all work to increase the administrator's control over what gets said, or at least to reduce the bite of criticism in the case of actual failure. There is safety under the cloak of ignorance. Beyond this tangle of advocacy and administration, there is another source of vulnerability in that the facts relevant to experimental program evaluation are also available to argue the general efficiency and honesty of administrators. The public availability of such facts reduces the privacy and security of at least some administrators.

Even where there are ideological commitments to a hard-headed evaluation of organizational efficiency, or to a scientific organization of society, these two jeopardies lead to the failure to evaluate organizational experiments realistically. If the political and administrative system has committed itself in advance to the correctness and efficacy of its reforms, it cannot tolerate learning of failure. To be truly scientific we must be able to advocate without that excess of commitment that blinds us to reality testing.

This predicament, abetted by public apathy and by deliberate corruption, may prove in the long run to permanently preclude a truly experimental approach to social amelioration. But our needs and our hopes for a better society demand we make the effort.

The problem can be reduced by two shifts in attitudes toward ameliorative programs. One involves a political shift from the advocacy of a specific reform to the advocacy of the seriousness of the problem. This avoids the overadvocacy trap discussed in Chapter 1 and stresses the persistence and innovativeness of the advocate rather than the correctness. The administrator could afford honest evaluation of outcomes, because negative results would not jeopardize his job, which would be to keep after the problem until something was found that worked. A second shift involves a general moratorium on ad hominem evaluative research—that is, on research designed to evaluate specific administrators rather than alternative policies. This engenders trust between evaluators and administrators and reduces the temptation by administrators to squelch unwanted research findings.

FIELD EXPERIMENTS AND QUASI-EXPERIMENTAL DESIGNS

In efforts to extend the logic of laboratory experiments into the "field," and into settings not fully experimental, an inventory of threats to experimental validity has been assembled, in terms of which some 15 or 20 experimental and quasi-experimental designs have been evaluated (Campbell, 1957, 1963; Campbell & Stanley, 1966). In this chapter only three or four designs will be examined, and therefore not all of the validity threats will be relevant, but it will provide useful background to look briefly at them all. Following are nine threats to internal validity.

1. *History:* Events, other than the experimental treatment, occurring between pretest and posttest and thus providing alternate explanations of effects.

2. *Maturation:* Processes within the respondents or observed social units producing changes as a function of the passage of time per se, such as growth, fatigue, secular trends, and so on.

3. *Instability:* Unreliability of measures, fluctuations in sampling persons or components, autonomous instability of repeated or "equivalent" measures. (This is the only threat to which statistical tests of significance are relevant.)

4. *Testing:* The effect of taking a test on the scores of a second testing. The effect of publication of a social indicator on subsequent readings of that indicator.

5. *Instrumentation:* Changes in the calibration of a measuring instrument or changes in the observers or scores used may produce changes in the obtained measurements.

6. *Regression artifacts:* Pseudo-shifts occurring when persons or treatment units have been selected on the basis of their extreme scores.

7. *Selection:* Biases resulting from differential recruitment of comparison groups, producing different mean levels on the measure of effects.

8. *Experimental mortality:* The differential loss of respondents from comparison groups.

9. *Selection-maturation interaction:* Selection biases resulting in differential rates of "maturation" or autonomous change.

If a change or difference occurs, there are rival explanations that could be used to explain away an effect and thus to deny that in this specific experiment any genuine effect of the experimental treatment had been demonstrated. These are faults that true experiments avoid, primarily through the use of randomization and control groups. In the approach advocated here, this checklist is used to evaluate specific quasi-experimental designs. This is evaluation, not rejection, for it often turns out that for a specific design in a specific setting the threat is implausible, or that there are supplementary data that can help rule it out even where randomization is impossible. The general ethic, advocated here for public administrators as well as social scientists, is to use the very best method possible, aiming at "true experiments" with random control groups. But where randomized treatments are not possible, a self-critical use of quasi-experimental designs is advocated. We must do the best we can with what is available to us.

Our posture vis-à-vis perfectionist critics from laboratory experimentation is more militant than this: The only threats to validity that we will allow to invalidate an experiment are those that admit of the status of empirical laws more dependable and more plausible than the law involving the treatment. The mere possibility of some alternative explanation is not enough—it is only the plausible rival hypotheses that are invalidating. Vis-à-vis correlational studies, on the other hand, our stance is one of greater conservatism. For example, because of the specific methodological trap of regression artifacts, the sociological tradition of "ex post facto" designs (Chapin, 1947; Greenwood, 1945) is totally rejected (Campbell & Stanley, 1966, pp. 70–71).

Threats to external validity, which follow, cover the validity problems involved in interpreting experimental results, the threats to valid generalization of the results to other settings, to other versions of the treatment, or to other measures of the effect.

1. *Interaction effects of testing:* The effect of a pretest in increasing or decreasing the respondent's sensitivity or responsiveness to the experimental variable. In this situation, the results obtained for a pretested population are unrepresentative of the effects of the experimental variable for the unpretested universe from which the experimental respondents were selected.

2. *Interaction of selection and experimental treatment:* Unrepresentative responsiveness of the treated population.

3. *Reactive effects of experimental arrangements:* "Artificiality"; conditions making the experimental setting atypical of conditions of regular application of the treatment; Hawthorne effects.

4. *Multiple-treatment interference:* Multiple treatments are jointly applied, effects atypical of the separate application of the treatments.

5. *Irrelevant responsiveness of measures:* All measures are complex, and all include irrelevant components that may produce apparent effects.

6. *Irrelevant replicability of treatments:* Treatments are complex, and replications of them may fail to include those components actually responsible for the effects.

These threats apply equally to true experiments and quasi-experiments. They are particularly relevant to applied experimentation. In the cumulative history of our methodology, this class of threats was first noted as a critique of true experiments involving pretests (Schanck & Goodman, 1939; Solomon, 1949). Such experiments provided a sound basis for generalizing to other pretested populations, but the reactions of unpretested populations to the treatment might well be quite different. As a result, there has been an advocacy of true experimental designs obviating the pretest (Campbell, 1957; Schanck & Goodman, 1939; Solomon, 1949) and a search for nonreactive measures (Webb, Campbell, Schwartz, & Sechrest, 1966).

These threats to validity will serve as a background against which we will discuss several research designs particularly appropriate for evaluating specific programs of social amelioration. These are the *one-group, pretest–posttest design,* the *interrupted time-series design,* the *control series design,* and various "true experiments." The order is from a weak but generally available design to stronger ones that require more administrative foresight and determination.

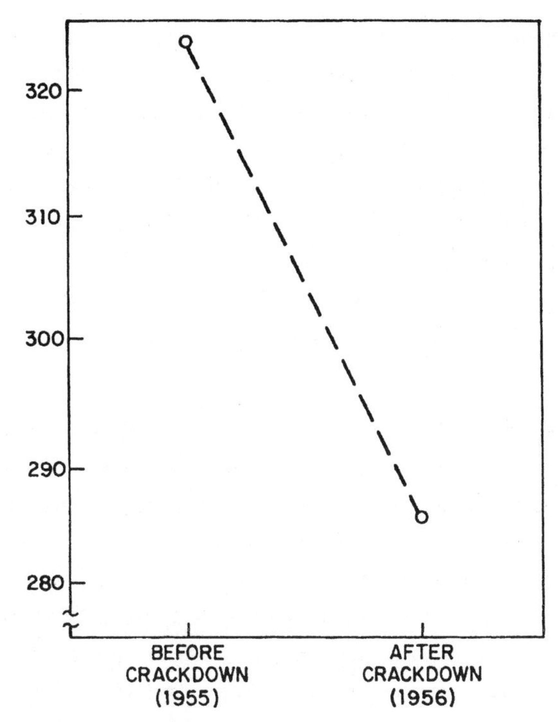

Figure 3.1. Connecticut traffic fatalities: 1955–1956.

ONE-GROUP PRETEST–POSTTEST DESIGN

By and large, when a political unit initiates a reform it is put into effect across the board, with the total unit being affected. In this setting the only comparison base is the record of previous years. The usual mode of utilization is a casual version of a very weak quasi-experimental design, the one-group pretest–posttest design.

A convenient illustration comes from the 1955 Connecticut crackdown on speeding. Traffic fatalities in Connecticut for 1956, compared with 1955, are presented in Figure 3.1. These are the data on which Governor Ribicoff relied in claiming success for the crackdown on speeding. Skillfully presented, such results can look impressive, but can also be fundamentally misleading.

We can speak of the evidence presented in Figure 3.1 as a quasi-experiment: There is a *pretest* (the 1955 figures), an *experimental treatment* (the crackdown), and a *posttest* (the 1956 figures). A substantial change is noted that one would like to ascribe to the "experimental treatment." In quasi-experimental analysis this interpretation is held to be legitimate, provided consideration is given to plausible rival explanations of the differences, with supplementary analyses being added to eliminate these where possible. In the language of quasi-experimental analysis, the data of Figure 3.1 constitute a one-group pretest–posttest design. This design fails to control for the six common threats to the validity of experiments specified next:

1. *History.* This term denotes specific events, other than the experimental treatment, occurring between the pretest and posttest, which might account for the change. It furnishes a "rival hypothesis" to the experimental hypothesis, a competing explanation of the before-to-after change that must be eliminated as implausible before full credence can be given to the experimental hypothesis. For instance, 1956 might have been a particularly dry year, with fewer accidents as a result of rain and snow, or there might have been a dramatic improvement of the safety features on the 1956-model cars. In fact, neither of these is a particularly plausible rival hypothesis in this instance, and we have not encountered more likely ones, so this potential weakness may not be crucial.

2. *Maturation.* This term originates in studies of individuals, where it refers to regular changes correlated with the passage of time, such as growing older, more tired, more sophisticated, and so on. It is distinguished from history in referring to processes, rather than to discrete events. Thus, one could classify the general long-term trend toward a reduction in automobile mileage death rates, presumably as a result of better roads, increased efficacy of medical care, and so on. The better designs discussed subsequently provide evidence concerning this trend in Connecticut in previous years and in other states for the same year.

3. *Testing.* A change may occur as a result of the pretest, even without the experimental treatment. In the present instance, the assessment of the traffic death rate for 1955 constitutes the pretest. In this case it is conceivable that the

measurement and publicizing of the traffic death rate for 1955 could change driver caution in 1956.

4. *Instrumentation.* This term refers to a shifting of the measuring instrument independent of any change in the phenomenon measured. In the use of public records for time-series data, a shift in the government agency recording the fatality statistics could account for such a shift. For example, suicide statistics increased a dramatic 20% in Prussia between 1882 and 1883, when record keeping was transferred from the local police to the national civil service (Selltiz, Jahoda, Deutsch, & Cook, 1959). Orlando Wilson's reforms of the police system in Chicago led to dramatic increases in rates for most crimes, a result presumably of the more complete reporting (Sween & Campbell, 1965). The "new broom" that introduces abrupt changes of policy is apt to reform the record keeping too, and thus confound reform treatments with instrumentation change. The ideal experimental administrator will, if possible, avoid this. He will prefer to keep comparable a partially imperfect measuring system rather than lose comparability altogether.

Another instrumentation threat can be seen in earlier versions of the Connecticut crackdown study. The death rate per 100 million vehicle miles is computed by using the number of gallons of gasoline sold in the state to estimate the number of miles driven. This is done by multiplying the number of gallons of gasoline sold by an empirically derived constant representing the number of miles one could expect to drive per gallon of gasoline. If there was a subsequent decrease in the actual miles obtained per gallon, as through engines of larger horsepower or driving at higher speeds, the death rate per 100 million miles would appear lower because of the inflated estimate of miles driven. Conversely, if the crackdown actually reduced driving speeds, this would increase the miles-per-gallon actually obtained, leading to an underestimate of mileage driven in the post crackdown period, and as a consequence, an overestimate of the fatality rate.

5. *Instability.* A ubiquitous plausible rival hypothesis is that the change observed is a result of the instability of the measures involved. Were Figure 3.1 to show fatality rates for a single township, with the same 12.3% drop, we would be totally unimpressed, so unstable would we expect such rates to be. In general,

as is made explicit in the models for tests of significance, the smaller the population base, the greater the instability. In the uncontrolled field situation, sample size is only one of many sources of instability. Much instability may be a result of large numbers of change-producing events of the type which, taken individually, we have called *history.*

6. *Regression.* Where a group has been selected for treatment just because of its extreme performance on the pretest, and if the pretest and posttest are imperfectly correlated, as they almost always are, it follows that on the average the posttest will be less extreme than the pretest. This regression is a tautological restatement of the imperfect correlation between pretest and posttest, as it relates to pretest scores selected for their extremity. The r of the correlation coefficient actually stands for the percentage of regression toward the mean. An analogous regression problem exists for time-series correlations.

Selection for extremity (and resultant retest regression) can be seen as plausibly operating in two ways: (a) of all states in 1955, this treatment was most likely to be applied to one with an exceptionally high traffic casualty rate; (b) for Connecticut, the most likely time in which a crackdown would be applied would be following a year in which traffic fatalities were exceptionally high.

In the true experiment, the treatment is applied randomly, without relation to the prior state of the dependent variable. In other words, the correlation between pretest scores and exposure to treatment is zero. Likewise, in the most interpretable of quasi-experiments, the treatment is applied without systematic relationship to the prior status of the group. Thus, an analysis of the effects of a tornado or an earthquake can be made with confidence that the pretreatment values did not cause the tornado or the earthquake. Not so here: The high 1955 rates can plausibly be argued to have caused the treatment. That 1956 was less extreme would then be expected because of regression. This occurs because in ordinary correlation, the regression is technically toward the mean of the second variable, not to the mean of the selection variable, if these means differ. In time series, the regression is toward the general trend line, which may of course be upward or downward or unchanging. A more expanded analysis of the regression problem in correlation across persons is contained in Campbell and Clayton (1961) and in Campbell and Stanley (1966).

In giving advice to the experimental administrator, one is also inevitably giving advice to those trapped administrators whose political predicament requires a favorable outcome, whether valid or not. To such trapped administrators the advice is to pick the very worst year and the very worst social unit. If there is inherent instability, there is no where to go but up, for the average case at least.

Advice to administrators who want to do genuine reality testing must include attention to this regression-artifact problem, and it will be a very hard problem to surmount. The most general advice would be to work on chronic problems of a persistent urgency or extremity, rather than reacting to momentary extremes. The administrator should look at the pretreatment time series to judge whether or not instability plus momentary extremity will explain away his program gains. If it will, he should schedule the treatment for a year or two later, so that his decision is more independent of the a year's extremity. (The selection biases remaining under such a procedure need further examination.)

7. *Irrelevant Responsiveness of Measures.* This threat to external validity is most relevant to social experimentation. This seems best discussed in terms of the problem of generalizing from indicator to indicator or in terms of the imperfect validity of all measures that is only to be overcome by the use of multiple measures of independent imperfection (Campbell & Fiske, 1959; Webb et al., 1966).

For treatments on any given problem within any given governmental or business subunit, there will usually be something of a governmental monopoly on reform. Even though different divisions optimally may be trying different reforms, within each division there will usually be only one reform on a given problem going on at a time. But for measures of effect this need not and should not be the case. The administrative machinery should itself make multiple measures of potential benefits and of unwanted side effects. In addition, the loyal opposition should be allowed to add still other indicators, with the political process and adversary argument challenging both validity and relative importance, with social science methodologists testifying for both parties, and with the basic records kept public and under bipartisan audit (as are voting records under optimal conditions). This competitive scrutiny is indeed the

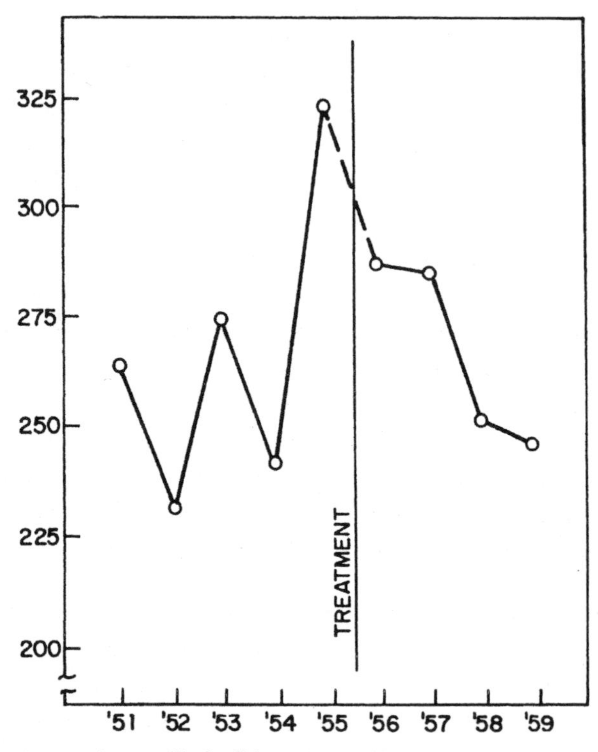

Figure 3.2. Connecticut traffic fatalities: 1951–1959.

main source of objectivity in sciences (Polanyi, 1966b, 1967; Popper, 1963) and epitomizes an ideal of democratic practice in both judicial and legislative procedures.

INTERRUPTED TIME-SERIES DESIGN

Figure 3.2 plots traffic fatalities for 5 years before and 4 years after the crack-down. This mode of quasi-experimental analysis has been labeled *interrupted time-series* design to distinguish it from the time-series analysis of economics. In the latter, the exogenous variable to which cause is imputed is a continuously present variable, occurring in different degrees. In the interrupted time-series design, the *causal* variable is examined as an event or change occurring at a single time, specified independently of inspection of the data. Again a note to experimental administrators: With this weak design, *it is only abrupt and decisive changes that we have any chance of evaluating. A gradually introduced reform will be indistinguishable from the background of secular change, from the net effect of the innumerable change agents continually impinging.*

The interrupted time-series design represents a use of the more extensive data that are often available even when only before-and-after measures are reported. Some potential outcomes of such a time-series analysis greatly reduce the plausibility of certain threats to validity. If the preexposure series shows but minor point-to-point fluctuations and no trend anticipating a big change from the period preceding the treatment to the one following it, then maturation may not be plausible. In most instances the plausible maturation hypothesis would have predicted shifts of the same order as the transtreatment shift in each of the pretreatment stages. Reasonable models of the testing effect would have the same implications. (In our instance, this would be on condition that the annual fatality rates had been given equal publicity.) The outcome in Figure 3.2 is not of this readily interpretable sort, although the trend is perhaps generally upward prior to the treatment, and steadily downward subsequently.

Judgments of the plausibility of instrumentation effects must be based on other than time-series data. However, notice should be taken of a frequent unfortunate confounding: The administrative reform that is meant to produce a social change very frequently is accompanied by a coincident reform of the record keeping, ruling out valid inferences about effects. The Orlando Wilson's Chicago police reform cited earlier is a case in point. In the present instance, we have found no evidence of a change in record keeping or index computing of the type that would produce a pseudo-effect.

The likelihood of regression, or of selection for "treatment" on a basis tending to introduce regression, is supported by inspection of the time-series data. The largest change of any year is not the one after the crackdown, but is instead the upswing in the series occurring in 1954–1955, just prior to the crackdown. In terms of crude fatality rates, 1955 is by far the highest point reached. It thus seems plausible that the high figure of 1955 caused the crackdown, and hence it seems much less likely that the crackdown caused the low figure of 1956, for such a drop would have been predicted on regression grounds in any case.

The graphic presentation of the precrackdown years provides evidence of the general instability of the accidental death rate measure, against which the 1955–1956 shift can be compared. This instability makes the *treatment effect* of Figure 3.1 now look more trivial. Had the drop following the treatment been the largest shift in the time series, the hypothesis of effect would have been much more plausible. Instead, shifts that large are relatively frequent. The 1955–1956 drop is less than half the magnitude of the 1954–1955 gain, and the 1953 gain also exceeds it. It is the largest drop of the series, but it exceeds the drops of 1952, 1954, and 1958 by trivial amounts. Thus the unexplained instabilities of the series are of such a magnitude as to make the 1955–1956 drop understandable as more of the same. On the other hand, it is noteworthy that after the crackdown, there are no year-to-year gains, and in this respect, the character of the time-series design has changed. The plausibility of the hypothesis that instability accounts for the effect can be judged by visual inspection of the graphed figures, or by qualitative discussion, but in addition it is this one threat to validity that can be evaluated by tests of significance.

Our position in regard to tests of significance is an intermediate one. On the one hand, we would agree that they are overly honored and are often mistaken as protecting against all of the threats to valid interpretation, when in fact they are only relevant to instability. On the other hand, they are often useful in ruling out this threat and should be used to do so. These tests are appropriate even where randomization has not been used, because they can rule out the relevant threat to validity that even had these data been assigned at random, differences this large would be frequent (Campbell, 1968).

The simplest tests conceptually are those testing for a difference in slope or intercept between pretreatment and posttreatment observations. As applied in this case, these assume linearity and independence of error. It has been shown that the *proximally autocorrelated* error typical of natural situations (in which adjacent points in time share more error than nonadjacent ones) biases the usual tests in the direction of finding too many significant differences (Sween & Campbell, 1965). Unaffected by this bias is a *t*-test by Mood that compares a single posttreatment point with a value extrapolated from the pretreatment series (Mood, 1950). None of these approached any interesting level of significance for the Connecticut crackdown.

Glass (Glass, 1968; Glass, Tiao, & Maguire, 1971; Maguire & Glass, 1967) introduced into the social sciences a more sophisticated statistical approach based on the work of Box and Tiao (1965). This has the advantages of realistically assuming the interdependence of adjacent points and estimating a weighting parameter thereof, of avoiding the assumption of linearity (at least in a simple or direct manner), and of weighting more heavily the observations closer to the point of treatment. A number of assumptions about the nature of the data must be made, such as the absence of cycles, but these can be examined from the data. Applying this test to monthly data, he finds a drop in fatalities not quite reaching the $p < .10$ level of significance. Using a monthly difference between Connecticut's rate and that of the pool of the four control states, still less of a significant effect is found. In what he regards as the most powerful analysis available, he computes an effect parameter for each of the four comparison states and compares the effect parameter of Connecticut with this. Connecticut shows more effect, with a significance level somewhere between $p < .05$ and $p < .07$, with a one-tailed test.

Thus on the graphic evidence of steadily dropping fatality rates, and on these marginal statistical grounds, there may be an effect. This effect, it must be restated, could be a result of the crackdown, or could be a result of the regression effect. (Regression effects can of course produce "statistically significant" results.)

The next few figures return again to the Connecticut crackdown on speeding and look to some other measures of effect. They are relevant to the confirming that there was indeed a crackdown, and to the issue of side effects. They also provide the methodological comfort of assuring us that in some cases the interrupted time-series design can provide clear-cut evidence of effect. Figure 3.3 shows the jump in suspensions of licenses for speeding—evidence that severe punishment was abruptly instituted.

We would want intermediate evidence that traffic speed was modified. A sampling each year of a few hundred 5-minute films of traffic (random as to location and time) could have provided this at a moderate cost, but they were not collected. Of the public records available, perhaps the data of Figure 3.4 showing a reduction in speeding violations indicate a reduction in traffic speed. But the effects on the legal system were complex, and in part undesirable.

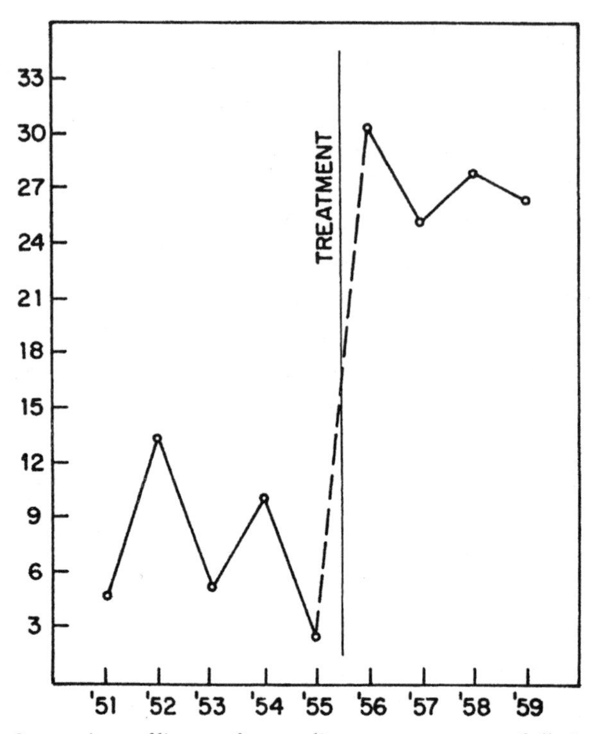

Figure 3.3. Suspensions of licenses for speeding, as a percentage of all suspensions.

Driving with a suspended license markedly increased according to Figure 3.5, at least in the biased sample of those arrested. Presumably because of the harshness of the punishment if guilty, judges may have become more lenient (Figure 3.6), although this effect is of marginal significance.

The relevance of indicators for the social problems we wish to cure must be kept continually in focus. The social indicators approach will tend to make the indicators themselves the goal of social action, rather than the social problems they imperfectly indicate. There are apt to be tendencies to legislate changes in the indicators per se rather than changes in the social problems.

CONTROL SERIES DESIGN

The interrupted time-series design as discussed so far is available for those settings in which no control group is possible, in which the total governmental

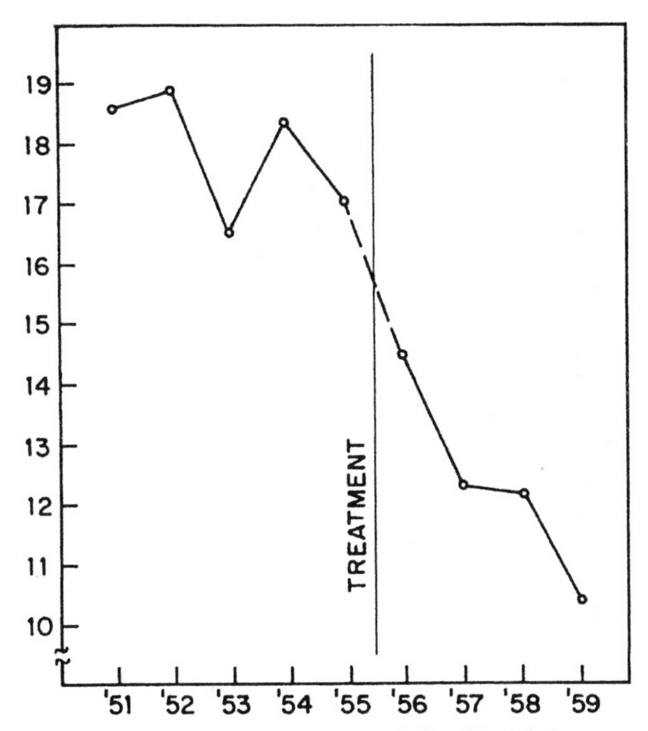

Figure 3.4. Speeding violations, as a percentage of all traffic violations.

unit has received the experimental treatment, the social reform measure. In the general program of quasi-experimental design, we argue the great advantage of untreated comparison groups even where these cannot be assigned at random. The most common of such designs is the nonequivalent control-group pretest–posttest design, in which for each of two natural groups, one of which receives the treatment, a pretest and posttest measure is taken. If the traditional mistaken practice is avoided of matching on pretest scores (with resultant regression artifacts), this design provides a useful control over those aspects of history, maturation, and test–retest effects shared by both groups. But it does not control for the plausible rival hypothesis of selection–maturation interaction—that is, the hypothesis that the selection differences in the natural aggregations involve not only differences in mean level but differences in maturation rate.

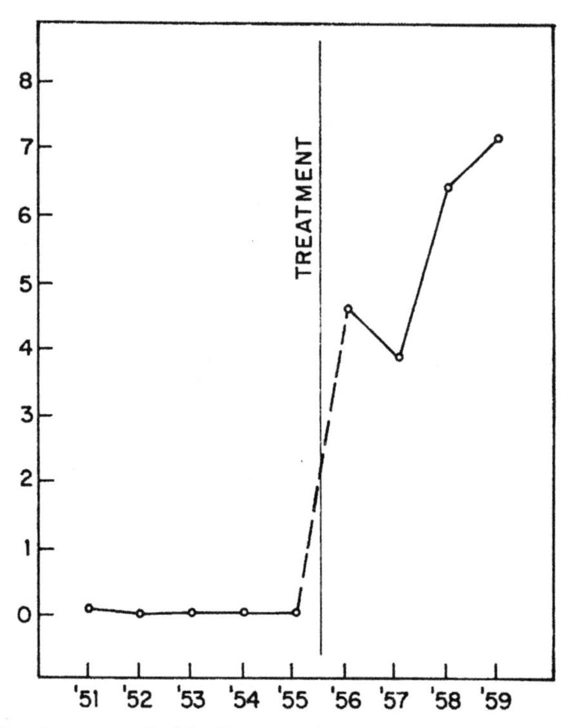

Figure 3.5. Number arrested while driving with a suspended license, as a percentage of suspensions.

This point can be illustrated in terms of the traditional quasi-experimental design problem of the effects of Latin on English vocabulary (Campbell, 1963). In the hypothetical data of Figure 3.7, part B, two alternative interpretations remain open. Latin may have had an effect, for those taking Latin gained more than those not. But, on the other hand, those students taking Latin may have a greater annual rate of vocabulary growth that would manifest itself whether or not they took Latin. Extending this common design into two time-series designs provides relevant evidence, as comparison of the two alternative outcomes shown in Figure 3.7, parts C and D. Thus approaching quasi-experimental design from either improving the nonequivalent control-group design or from improving the interrupted time-series design, we arrive at the control series design.

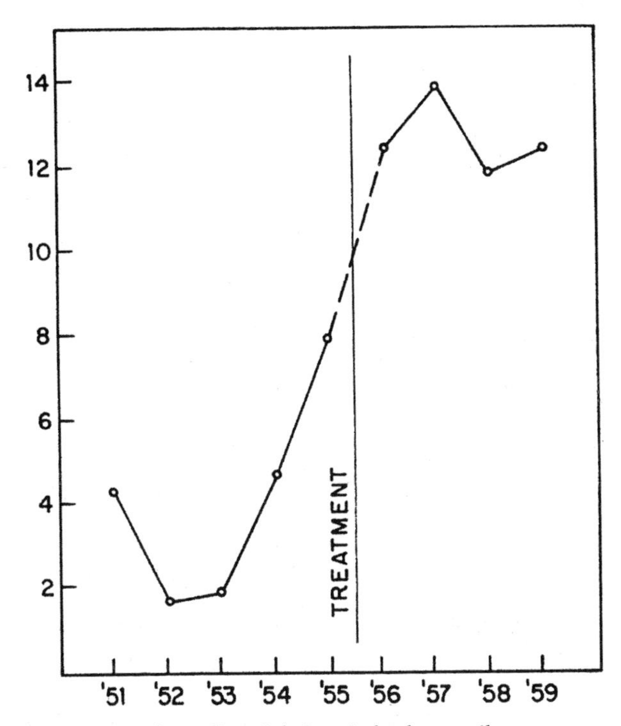

Figure 3.6. Percentage of speeding violations judged not guilty.

Figure 3.8 shows this for the Connecticut speeding crackdown, adding evidence from the fatality rates of neighboring states. For Connecticut, it was judged that a pool of adjacent and similar states—New York, New Jersey, Rhode Island, and Massachusetts—provided a meaningful comparison. Figure 3.8 plots the death rates for the control states alongside Connecticut, all data being expressed on a per-100,000 population base to bring the figures into proximity. The control data are much smoother, because of the much larger base—in other words, the canceling out of chance deviations in the annual figures for particular states.

Although in general these data confirm the single time-series analysis, the differences between Connecticut and the control states show a pattern supporting the hypothesis that the crackdown made a difference. In the pretest years, Connecticut's rate is parallel or rising relative to the control, exceeding it in

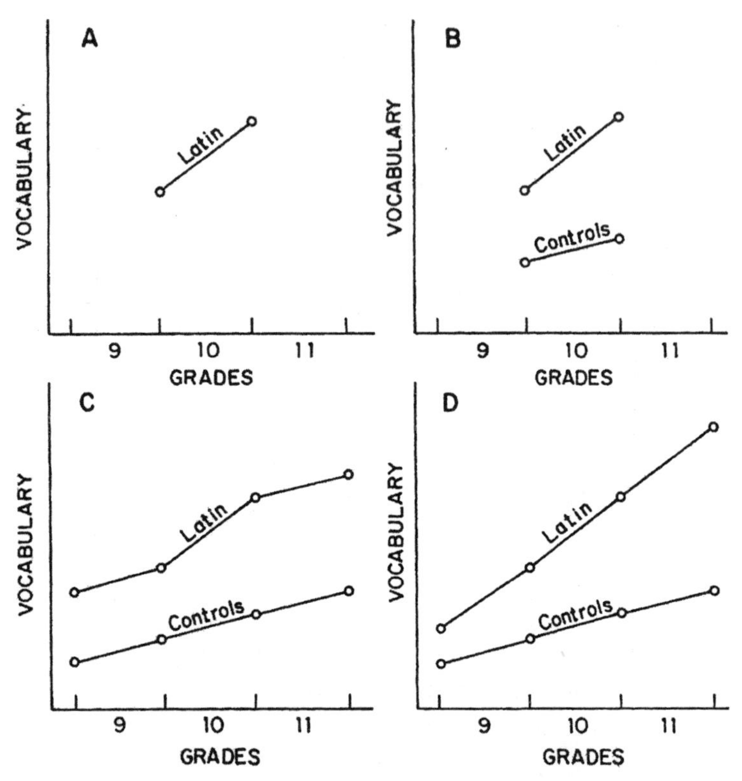

Figure 3.7. Forms of quasi-experimental analysis for the effect of specific course work, including control-series design. Campbell, Donald. "From Description to Experimentation." In Harris, Chester W., ed., *Problems in Measuring Change: Proceedings of a Conference Sponsored by the Committee on Research Council, 1962.* Copyright 1963. Reprinted by permission of The University of Wisconsin Press.

1955. In the posttest years, Connecticut's rate drops faster than does the control, steadily increasing the gap. Although the regression argument applies to the high point of 1955 and to the subsequent departure in 1956, it does not plausibly explain the steadily increasing gap in 1957, 1958, and 1959.

Figure 3.9 shows the comparison states individually. Note that four of the five show an upward swing in 1955, Connecticut having the largest. Note that all five show a downward trend in 1956. Rhode Island is most similar to Connecticut in both the 1955 upswing and 1956 downswing, actually exceeding

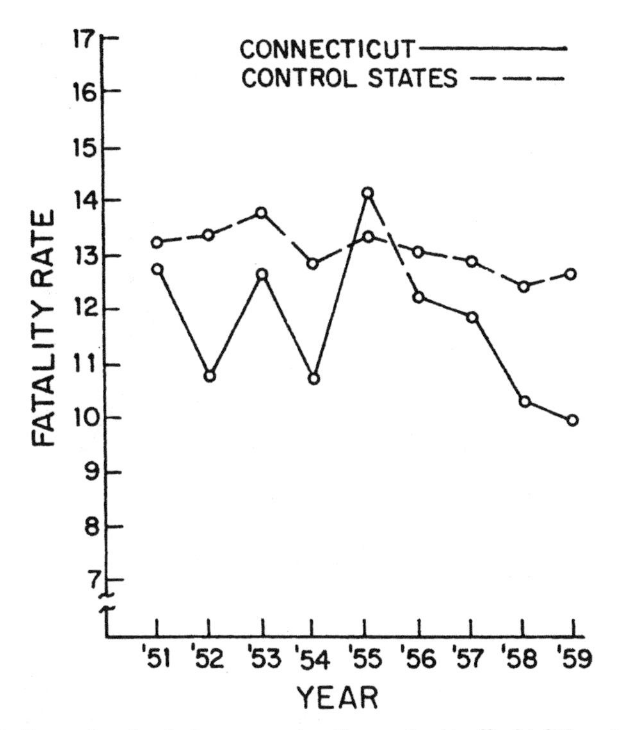

Figure 3.8. Control-series design comparing Connecticut traffic fatalities with those of four comparable states.

Connecticut in the latter—in a striking argument against the hypothesis of a crackdown effect. However, the trend in 1957, 1958, and 1959 is steadily upward in Rhode Island, steadily downward in Connecticut, supporting the concept of effect.

The list of plausible rival hypotheses should include factors disguising experimental effects as well as factors producing pseudo-effects. Thus, to the list should be added *diffusion*, the tendency for the experimental effect to modify not only the experimental group but also the control group. Thus the crackdown on speeding in Connecticut might well have reduced traffic speed and fatalities in neighboring states. The comparison of posttreatment levels of Connecticut and the neighboring states might thus be invalid, or at least underestimate the effects. Conceivably one might for this reason prefer the

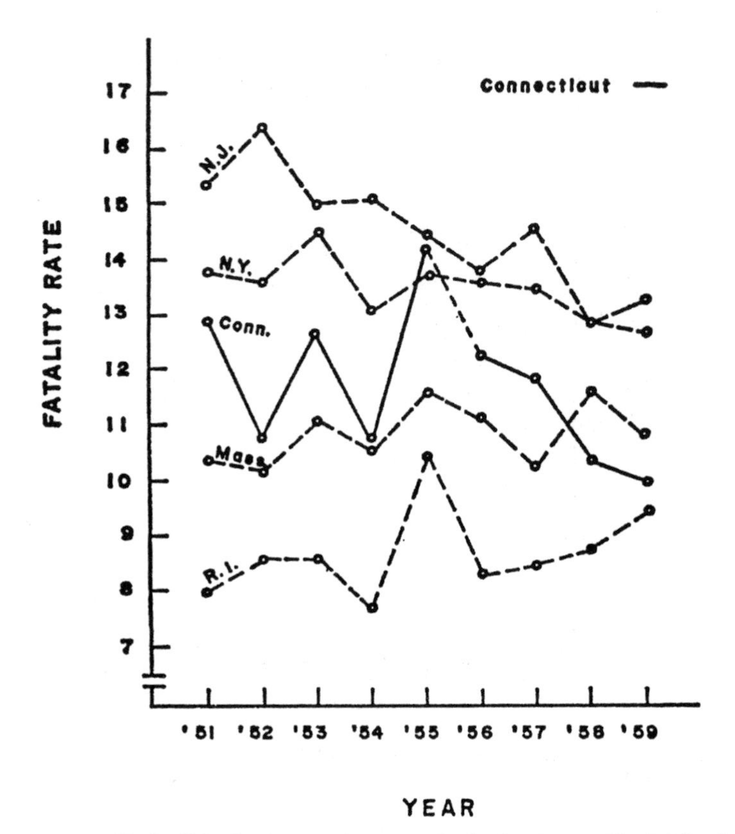

Figure 3.9. Traffic fatalities for Connecticut, New York, New Jersey, Rhode Island, and Massachusetts (per 100,000 persons).

single time-series analysis to the multiple time-series one. If highly similar remote states were available, these would make better controls, but adjacent states are more likely to be similar in weather or culture.

Glass (1968) has used our monthly data for Connecticut and the control states to generate a monthly difference score, and this too shows a significant shift in trend in the Box and Tiao (1965) statistic. Impressed particularly by the 1957, 1958, and 1959 trend, we are willing to conclude that the crackdown had some effect, over and above the undeniable pseudo-effects of regression (Campbell & Ross, 1968).

The advantages of the control series design point to the advantages for social experimentation of a social system allowing subunit diversity. Our ability to estimate the effects of the speeding crackdown, Rose's (1952) and Stieber's (1949) ability to estimate the effects on strikes of compulsory arbitration laws, and Simon's (1966) ability to estimate the price elasticity of liquor were made possible because the changes were not being put into effect in all states simultaneously, because they were matters of state legislation rather than national. I do not want to appear to justify on these grounds the wasteful and unjust diversity of laws and enforcement practices from state to state. But I would strongly advocate that social engineers make use of this diversity while it remains available, and plan cooperatively their changes in administrative policy and in record keeping to provide optimal experimental inference. More important is the recommendation that, for those aspects of social reform handled by the central government, a purposeful diversity of implementation be planned so that experimental and control groups be available for analysis. Properly planned, these can approach true experiments, better than the casual and ad hoc comparison groups now available. But without such fundamental planning, uniform central control can reduce the present possibilities of reality testing—that is, of true social experimentation. In the same spirit, decentralization of decision making, both within large government and within private monopolies, can provide a useful competition for both efficiency and innovation, reflected in a multiplicity of indicators.

One further illustration of the interrupted time series and the control series will be provided. The variety of illustrations so far given have each illustrated some methodological point and have thus ended up as "bad examples." To provide a "good example," an instance that survives methodological critique as a valid illustration of a successful reform, data from the British Road Safety Act of 1967 are provided in Figure 3.10 (from Ross, Campbell, & Glass, 1970).

The data on a weekly-hours basis are available only for a composite category of fatalities plus serious injuries, and Figure 3.10 therefore uses this composite for all three bodies of data. The "Weekend Nights" comprises Friday and Saturday nights from 10:00 p.m. to 4:00 a.m. Here, as expected, the crackdown is most dramatically effective, producing initially more than a 40% drop, leveling off at perhaps 30%, although this involves dubious extrapola-

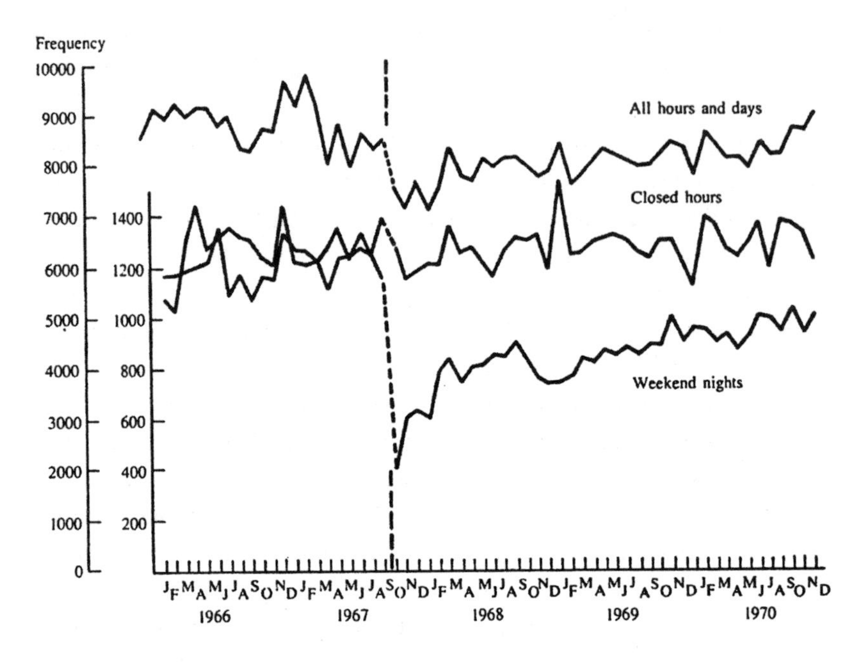

Figure 3.10. British traffic fatalities plus serious injuries, before and after Breathalyser crackdown of October 1967.

tions in the absence of some control comparison to indicate what the trend over the years might have been without the crackdown. In this British case, no comparison state with comparable traffic conditions or drinking laws was available. But controls need not always be separate groups of persons; they may also be separate samples of times or stimulus materials (Campbell & Stanley, 1966, pp. 43–47). A cigarette company may use the sales of its main competitor as a control comparison to evaluate a new advertising campaign. One should search around for the most nearly appropriate control comparison. For the Breathalyzer crackdown, commuting hours when pubs had been long closed seemed ideal. (The commuting hours figures come from 7:00 a.m. to 10:00 a.m. and 4:00 p.m. to 5:00 p.m. Monday through Friday. Pubs are open for lunch from 12:00 to 2:00 or 2:30, and open again at 5:00 p.m.)

These commuting hours data convincingly show no effect, but they are too unstable to help much with estimating the long-term effects. They show a different annual cycle than do the weekend nights or the overall figures, and do not go back far enough to provide an adequate base for estimating this annual cycle with precision.

The use of a highly judgmental category such as "serious injuries" provides an opportunity for pseudo-effects owing to a shift in the classifiers' standards. The overall figures are available separately for fatalities, and these show a highly significant effect as strong as that found for the serious injury category or the composite shown in Figure 3.10.

More details and the methodological problems are considered in our fuller presentation (Ross et al., 1970). One additional rule for the use of this design needs to be emphasized. The interrupted time-series design can provide clear evidence of effect only where the reform is introduced with a vigorous abruptness. In the Breathalyser crackdown, an intense publicity campaign naming the specific starting date preceded the actual crackdown. Although the impact seems primarily a result of publicity and fear rather than an actual increase of arrests, an abrupt initiation date was achieved. Had the enforcement effort changed at the moment the act was passed, with public awareness being built up by subsequent publicity, the resulting data series would have been essentially uninterpretable.

RANDOMIZED CONTROL GROUP EXPERIMENTS

Experiments with randomization tend to be limited to the laboratory and agricultural experiment station. But this certainly need not be so. The randomization unit may be persons, families, precincts, or larger administrative units. For statistical purposes the randomization units should be numerous, and hence ideally small. But for reasons of external validity, including reactive arrangements, the randomization units should be selected on the basis of the units of administrative access. Where policies are administered through individual client contacts, randomization at the person level may be often inconspicuously achieved, with the clients unaware that different ones of them are getting different treatments. But for most social reforms, larger administrative units will be involved, such as classrooms, schools, cities, counties, or states. We need to develop the political postures and ideologies that make randomization at these levels possible.

Pilot project is a useful term already in our political vocabulary. It designates a trial program that, if it works, will be spread to other areas. By modifying actual practice in this regard, without going outside of the popular understanding of the term, a valuable experimental ideology could be developed. How are areas selected for pilot projects? If the public worries about this, it probably assumes a lobbying process in which the greater needs of some areas are only one consideration, political power and expediency being others. Without violating the public tolerance or intent, one could probably devise a system in which the usual lobbying decided on the areas eligible for a formal public lottery that would make final choices between matched pairs. Such decision procedures as the drawing of lots have had a justly esteemed position for many years (e.g., Aubert, 1959). At the present, record keeping for pilot projects tends to be limited to the experimental group only. In the experimental ideology, comparable data would be collected on designated controls.

Another general political stance making possible experimental social amelioration is that of staged innovation. Even though by intent a new reform is to be put into effect in all units, the logistics of the situation usually dictate that simultaneous introduction is not possible. What results is a haphazard sequence of convenience. Under the program of staged innovation, the introduction of the program would be deliberately spread out, and those units selected to be first and last would be randomly assigned (perhaps randomization from matched pairs), so that during the transition period the first recipients could be analyzed as experimental units, the last recipients as controls. A third ideology making possible true experiments has already been discussed: randomization as the democratic means of allocating scarce resources.

This chapter will not give true experimentation equal space with quasi-experimentation only because excellent discussions of, and statistical consultation on, true experimentation are readily available. True experiments should almost always be preferred to quasi-experiments where both are available. Only occasionally are the threats to external validity so much greater for the true experiment that one would prefer a quasi-experiment. The uneven allocation of space here should not be read as indicating otherwise.

MORE ADVICE FOR TRAPPED ADMINISTRATORS

The competition is not really between the fairly interpretable quasi-experiments here reviewed and "true" experiments. Both stand together as rare excellencies in contrast with a morass of confusion and self-deception. Both to emphasize this contrast and again as guidelines for the benefit of those trapped administrators whose political predicament will not allow the risk of failure, some of these alternatives should be mentioned.

Grateful Testimonials

Human courtesy and gratitude being what it is, the most dependable means of ensuring a favorable evaluation is to use voluntary testimonials from those who have had the treatment. If the spontaneously produced testimonials are in short supply, these should be solicited from the recipients with whom the program is still in contact. The rosy glow resulting is analogous to the professor's impression of his teaching success when it is based solely on the comments of those students who come up and talk with him after class. In many programs, as in psychotherapy, the recipient, as well as the agency, has devoted much time and effort to the program, and it is dissonance-reducing for himself, as well as common courtesy to his therapist, to report improvement. These grateful testimonials can come in the language of letters and conversation, or be framed as answers to multiple-item "tests" in which a recurrent theme of "I am sick," "I am well," "I am happy," "I am sad" recurs. Probably the testimonials will be more favorable (a) the more the evaluative meaning of the response measure is clear to the recipient, (b) the more directly the recipient is identified by name with his answer, (c) the more the recipient gives the answer directly to the therapist or agent of reform, (d) the more the agent will continue to be influential in the recipient's life in the future, (e) the more the answers deal with feelings and evaluations rather than with verifiable facts, and (f) the more the recipients participating in the evaluation are a small and self-selected or agent-selected subset of all recipients. Properly designed, the grateful testimonial method can involve pretests as well as posttests, and randomized control groups as well as experimentals, because there are usually no placebo treatments, and the recipients know when they have benefited.

Confounding Selection and Treatment

Another dependable tactic bound to give favorable outcomes is to confound selection and treatment so that in the published comparison those receiving the treatment are also the more able and well-placed. The often-cited evidence of the dollar value of a college education is of this nature. All careful studies show that most of the effect—and of the superior effect of superior colleges—is explainable in terms of superior talents and family connections, rather than in terms of what is learned or even the prestige of the degree. Matching techniques and statistical partialings generally undermatch and do not fully control for the selection differences. Instead, they introduce regression artifacts often confused as treatment effects.

There are two types of situations that must be distinguished. First, there are those treatments that are given to the most promising, treatments like a college education that are regularly given to those who need it least. For these, the later concomitants of the grounds of selection operate in the same direction as the treatment: Those most likely to achieve anyway get into the college most likely to produce later achievement. For these settings, the trapped administrator should use the pooled mean of all those treated, comparing it with the mean of all untreated, although in this setting almost any comparison an administrator might hit on would be biased in his favor.

At the other end of the talent continuum are those remedial treatments given to those who need it most. Here the later concomitants of the grounds of selection are poorer success. In the Job Training Corps example, casual comparisons of the later unemployment rate of those who received the training with those who did not are in general biased against showing an advantage to the training. Here the trapped administrator must be careful to seek out those few special comparisons biasing selection in his favor. For training programs such as Operation Head Start and tutoring programs, a useful solution is to compare the later success of those who completed the training program with those who were invited but never showed plus those who came a few times and dropped out. By regarding only those who complete the program as "trained" and using the others as controls, one is selecting for conscientiousness, stable and supporting family backgrounds, enjoyment of the training activity, ability, determination to get ahead in the world, and so on. All these factors would predict future achievement even if the remedial program is valueless. To apply this tactic effectively in the Job Training Corps, one might have to eliminate from the so-called control group all those who quit the training program because they had found a job. This would seem a reasonable practice and would not blemish the reception of a glowing progress report.

These are but two more samples of well-tried modes of analysis for the trapped administrator who cannot afford an honest evaluation of the social reform he directs. They remind us again that we must help create a political climate that demands more rigorous and less self-deceptive reality testing. We must provide political stances that permit true experiments or good quasi-experiments. Of the several suggestions toward this end, the most important is probably the initial theme: Administrators and parties must advocate the importance of the problem rather than the importance of the answer. They must advocate experimental sequences of reforms, rather than one certain cure-all, advocating Reform A with Alternative B available to try next should an honest evaluation of A prove it worthless or harmful.

MULTIPLE REPLICATION IN ENACTMENT

Too many social scientists expect single experiments to settle issues once and for all. This may be a mistaken generalization from the history of great crucial experiments in physics and chemistry. In actuality the significant experiments in the physical sciences are replicated thousands of times, not only in deliberate replication efforts but also as inevitable incidentals in successive experimentation and in utilizations of those many measurement devices (such as the galvanometer) that in their own operation embody the principles of classic experiments. We social scientists have much greater need for replication experiments than do the physical sciences. This is because we have less ability to achieve "experimental isolation." We also have good reason to expect our treatment effects to interact significantly with a wide variety of social factors, many of which are not yet fully understood.

The implications are clear. We should not only do hard-headed reality testing in the initial pilot testing and choosing of which reform to make general law, but once it has been decided that the reform is to be adopted as standard practice in all administrative units, we should experimentally evaluate it in each of its implementations (Campbell, 1967a).

CONCLUSION

Trapped administrators have so committed themselves in advance to the efficacy of the reform that they cannot afford honest evaluation. For them, favorably biased analyses are recommended, including capitalizing on regression, grateful testimonials, and confounding selection and treatment. Experimental administrators have justified the reform on the basis of the importance of the problem, not the certainty of their answer, and are committed to going on to other potential solutions if the one first tried fails. They are therefore not threatened by a hardheaded analysis of the reform. For such, proper administrative decisions can lay the base for useful experimental or quasi-experimental analyses. Through the ideology of allocating scarce resources by lottery, through the use of staged innovation, and through the pilot project, true experiments with randomly assigned control groups can be achieved. If the reform must be introduced across the board, the interrupted time-series design is available. If there are similar units under independent administration, a control series design adds strength.

OVERVIEW OF CHAPTER 4

In 1966, Campbell and Stanley went a long way toward increasing the validity of out-of-door experiments. First they identified and labeled two types of invalidity that can occur. The first refers to the question of whether the experimental treatment was responsible for the observed outcome. This was labeled *internal validity.* The threats to internal validity were defined as the threats controlled for by random assignment to treatments. The other type of invalidity refers to the question of whether the results are generalizable—that is, that they can be applied to other samples, treatments, and times. This has been called *external validity.* These two types of validity and some of the common threats were discussed in Chapter 3.

Campbell stated that many of his students often confused the distinction between these two types of validity. For example, if a placebo control group is added to a pharmaceutical experiment, does it improve internal or external validity? As Campbell read it, it improves external validity. In the preparation of this book, I grudgingly admitted to him that I still struggled with his answer, although he assured me that he had no doubt that it was correct. This can be a challenge for students and other readers.

Because of the confusion infecting these terms, Campbell, along with Tom Cook, agreed that internal validity needed relabeling. They settled on the term *local molar causal validity. Local* refers to the unique place and time of the experiment. *Molar* refers to the entire treatment, recognizing that in nonlaboratory settings, the researcher cannot isolate particular aspects of the treatment to determine which aspect is responsible for the effect. Despite this effort to clarify the issue, this new terminology was not widely adopted. This article is useful in expanding our understanding of the concept; however, researchers continue to speak in terms of internal and external validity.

Although Campbell emphasized the importance of establishing causal inference, he admitted that social problems do not always lend themselves to methods that provide a great deal of precision in this regard. He suggested that qualitative methods, such as case studies, ethnographies, and hermeneutics, could be used to ascertain the plausibility of threats to validity. Further, rather than abandoning the study of important problems that cannot be studied through experimentation or quasi-experimentation, he advocated that such qualitative methods be used alone. This was seen as a turning point in the legitimacy of qualitative methods.

After the researcher rules out rival hypotheses and determines that an effect occurred, local molar causal validity has been established. Because the original hunch regarding the effectiveness of the treatment had been confirmed, the researcher has some confidence that the treatment may be generalizable. Will it be effective in other settings, on other populations, in nonidentical treatments, and at other times? These are the issues of external validity.

Cook in 1979 classified some of the components of external validity. For example, generalization to nonidentical treatments was classified as construct validity of causes, and generalization to other measurements of the outcomes was classified as construct validity of effects. The concepts that remained had to do with applying the results to other samples, places, and times.

One way to ensure external validity is to use a representative sample of the universe to which you wish to apply the results. Because the researcher wishes to generalize to future times, this is an impossibility.

Another approach is to choose a sample that is feasible and can be defended on a theoretical basis. This approach has been used in the physical sciences, and Campbell cited the experiment by Nicholson and Carlisle as an excellent illustration. A presumption is made that the results of the experiment are universally true until demonstrated otherwise.

Campbell felt that generalizing to other treatments, measures, populations, settings, and times was not enough. Through Tom Cook's notion of construct validity of causes and effects, it was recognized that the validity of our theoretical interpretations of the constructs needed to be considered. All treatments and measures of outcomes are only imperfect proxy variables for latent causes and effects. Campbell attempted to provide a basis for his approach to generalization in his discussion of the principle of proximal similarity. In nature, one is more likely to find similarities in phenomena that are close in space and time than in those more distant. For example, a student's attitude toward school will likely be more similar tomorrow than it would be a year from now (unless the professor returns an exam today with an unexpectedly low grade). Through the year, new classes, professors, and experiences will continually shape the student's attitude. Proximal data tend to be more similar.

We apply this principle of proximal similarity when an ameliorative program is applied to a group that is similar to the pilot group, similar treatments are used, and it is carried out soon after the pilot study was completed. This increases our confidence that the initial results will be repeated. Still, Campbell recommended that treatments should be applied to widely different groups to explore the potential generalizability of the results.

Campbell suggested that the discussion in this chapter was just the beginning of the differentiation between internal and external validity. He was hopeful that students would be more likely to answer correctly that external validity is increased by adding a placebo group to a pharmaceutical experiment after reading more about local molar causal validity, construct validity of causes and effects, and the principle of proximal similarity that forms the basis for external validity.

FOUR

RELABELING INTERNAL AND EXTERNAL VALIDITY

On the one hand, there have been widespread expressions of dissatisfaction with the distinction between internal validity and external validity and suggestions for its revision or elimination. In Cook and Campbell (1979), Tom Cook began a review of this literature, and he now advocates a much more systematic and extensive one (Cook, 1985; Cook & Campbell, 1979). This chapter avoids that task.

On the other hand, both those who have enthusiastically adopted the distinction and those who oppose it have most frequently redefined it to epitomize all the differences between pure laboratory experimentation and field tryouts of ameliorative programs. Are you, dear reader, perhaps one who has done so? Half of my own students fail to answer the following question correctly: When one adds a placebo control group in a pharmaceutical experi-

Campbell, D. T. (1986a). Relabeling internal and external validity for applied social scientists. In W. M. K. Trochim (Ed.), *Advances in quasi-experimental design and analysis: New directions for program evaluation*. San Francisco: Jossey-Bass. Used with permission.

ment, is this done to improve internal or external validity? As I read Campbell (1957) and Campbell and Stanley (1966), the correct answer is external validity.

The meanings compiled in dictionaries are properly based on usage, and the meanings ascribed to specialist terms are based on usage in the relevant specialty. On these grounds, the term *internal validity* now means similarity to the pure treatment (rule-of-one-variable), fully controlled, laboratory experiment. Because that is not what we had in mind, we need to try again, with new terms.

It will help to remind ourselves of the motivation that led to the introduction of the distinction between internal and external validity. In the 1950s, the training in research methods that social psychologists received was dominated by Fisherian analysis of variance statistics, as though random assignment to treatment were the only methodological control that needed to be taught. In dialogue with this lopsided and complacent emphasis, we wished to point out that in out-of-doors social experimentation, there were a lot of threats to validity that randomization did not take care of. We believed the teaching of research design should be expanded to cover these other threats, which we classified as issues of external validity. It was against the overwhelming dominance of Fisher's randomized assignment models and an implicit, complacent assumption that meticulous care in this regard took care of all experimental validity problems that we were reacting. Thus, threats to external validity came to be defined as threats not controlled for by random assignment to treatment. And, backhandedly, threats to internal validity were, initially and implicitly, those that random assignment did control. This overlaps with the laboratory ideal, but it excludes purity of the treatment variable. This was most nearly made clear by the inclusion of generalization to other treatments in external validity from the very beginning.

With my approval, Cook has fuzzed up this simple distinction and added four threats to internal validity that randomization does not control (Cook & Campbell, 1979). It is symptomatic of problems with the concept that we now believe that at least one of these threats, resentful demoralization of respondents receiving less desirable treatments, if not all four, might well be grouped with construct validity of treatments. In the hypothetical case that Cook (Cook & Campbell, 1979) used to illustrate the threat, the posttest difference between experimental and control groups was a product of no change in the experimental group and entirely a result of the resentful demoralization of the control group. In terms of the internal validity concept that I used in scoring the placebo-control-group item that I give to students, one could answer affirmatively the internal validity question, "Did the experimental contrast as a total package in its specific setting cause a real difference?" and relegate to construct validity (external validity in Campbell & Stanley, 1966) the question of how to interpret that valid noted difference.

In June 1984, Cook and I spent 3 days planning a revision or new book. We tentatively agreed that *internal validity* needed relabeling. We went round and round on alternative labels. We are not satisfied with our present one, local molar causal validity. It has "historicist dialectical indexicality" (Campbell, 1982, p. 327)—that is, the choice of terms is a product of the particular argument of this historical moment.

LOCAL MOLAR (PRAGMATIC, ATHEORETICAL) CAUSAL VALIDITY

For the applied scientist, local molar causal validity is a first crucial issue and the starting point for other validity questions. For example, did this complex treatment package make a real difference in this unique application at this particular place and time? By *molar* we connote recognition that the treatment is often a very complex hodgepodge that has been put together by expert clinical judgment, and not on the basis of the already proven efficacy of its theoretically pure components (main effects plus interactions). By *molar* we also connote an interest in evaluating this complex treatment as it stands, rather than first testing its hundred or so components one at a time, or in a hundred-variables-by-several-levels-each randomized ANOVA experiment. The molar approach assumes that clinical practice, participant observation, and epidemiological studies already have accumulated some wisdom, suggesting treatments that are worth further testing as molar packages. If these packages turn out to have striking molar efficacy, we will be interested in further studies, both clinically and theoretically guided, that will help us to determine which of several conjectured major components is most responsible for the effect. These later studies in turn will still be using complex packages, rather than testing theoretically pure variables in isolation or in experimentally controlled higher order interaction. Pure-variable science can, of course, be a source of treatment packages (as in brain metabolite therapy for children diagnosed as potentially schizophrenic because of metabolite abnormality), but in preventive intervention these, too, will inevitably be a part of a complex social system of diagnosis and delivery. They will need to be tested as intervention packages under conditions of eventual application, or under facsimiles of such situations, that have been chosen both for clarity of scientific inference and for similarity to target conditions of application.

By local we indicate the strategy illustrated in pilot testing: Let us see if it really works in some one setting and time. If it does, later on we can explore the boundaries of its efficacy in other locales and with specialized populations. If it does not, we may be appropriately discouraged from further trials, even though it might conceivably work in some other setting.

Although the local molar causal validity of the applied social scientist may be a far cry from the agenda of basic science, most of the problems of validity—and thus the methodology for the establishment of validity—are shared. Thus (if the applied research problem is not thereby abandoned), the two traditions of experimental control—experimental isolation and randomized assignment to treatments—are also ideal ways of establishing local molar causal validity. For applied ameliorative research under the field conditions of application, random assignment to treatments usually is the optimal approach to local molar causal validity, although not for external and construct validities.

Molar and *local* could both be taken as implying no generalization at all, conceptualizing a validly demonstrated cause–effect relationship that we do not as yet know how to generalize. The causal relationship would be known locally and molarly, but there would be no validated theory of it that would guide generalization to other interventions, measures, populations, settings, or times. This is, of course, an exaggeration. The theories and hunches used by those who put the therapeutic package together must be regarded as corroborated, however tentatively, if there is an effect of local, molar validity in the expected direction. Nonetheless, this exaggeration may serve to remind us that very frequently in physical science (and probably in social science as well) causal puzzles (that is, effects that are dependable but not understood) are the driving motor for new and productive theorizing. We must back up from the current overemphasis on theory first.

Basic scientists put a premium on clarity of causal inference and hence limit, trim, and change problems so that they can be solved with scientific precision given the current state of the art. Other causal hypotheses are postponed until the state of the art and theory development make them precisely testable. This strategy is not available to applied scientists. They should stay with the mandated problem, doing the best they can to achieve scientific validity. To stay with the problem, however, it may be necessary to use methods providing less precision of causal inference. Thus, we applied social scientists need not only randomized experiments and strong quasi-experiments but also case studies, ethnography, participant observation, gossip collection from informants, hermeneutics, and so forth. Ideally, these materials will be used to provide the context necessary for valid estimation of the seriousness of the threats to validity and for valid interpretation of the results of formal experimentation, but if need be they may be used alone. Social sciences do not seek a different kind of validity than other sciences do, but to stay with our problems, we must use techniques that improve the validity of our research, yet provide less clarity of causal inference than if we retreated to narrowly specified variables under laboratory control. Although using these techniques of the humanities, staying in real-world, nonlaboratory settings, the critical tools of threats to validity and plausible rival hypotheses are still central (Becker, 1979; Cook & Reichardt, 1979).

RELABELING EXTERNAL VALIDITY

There was even less consensus on a complementary reconceptualization of external validity. The following discussion, however dogmatic it may be, is a tentative and incomplete tryout of the principle of proximal similarity for that role. Although that obviously will not quite do as a heading under which the present threats to construct validity and external validity fall logically, it may help to shake us free of past conceptualizations, as a way station en route to something more satisfactory. Here I move to this concept by stages.

Generalizing From the Local Without Representative Sampling

Akin to the relabeling of internal validity as local molar causal validity is a reformulation of the concept of external validity or generalizability set forth by Campbell and Stanley (1966). This has, of course, already been done by Cook (Cook & Campbell, 1979), who explicitly separated out and reconceptualized the issues of generalizing to other nonidentical treatments, now called *construct validity of causes,* and of generalizing from the outcome measures employed to other measures of effects, now called *construct validity of effects.* Remaining in the Cook and Campbell (1979) residual category of external validity is the validity with which one can generalize to other persons, settings, and times. Such generalizations should also be made on the basis of theory and thus they, too, should be reconceptualized as construct validities. That is, the validity of generalizations to other persons, settings, and future (or past) times would be a function of two factors. The first factor is the validity of the theory involved. The second factor is the accuracy of the theory-relevant knowledge of the persons, settings, and future periods to which one wanted to generalize (for example, to which one wanted to apply an intervention with demonstrated local molar causal validity).

This perspective has already moved us far from the widespread concept that one can solve generalizability problems by representative sampling from a universe specified in advance. Such an approach is obviously impossible for sampling from future occasions. However, the statistical technology and practical possibility is available for persons and for specified setting units, such as schools, schoolrooms, factories, hospitals, retail stores, cities, and counties. Although national samples along these lines are often called for in evaluations mandated by Congress, it turns out that nearly all the high-quality scientific program studies (such as guaranteed annual income, housing allowances, school vouchers, coverage of psychotherapy by health insurance, and so forth) have chosen illustrative samples, exemplifying the target population in informal judgmental ways, employing samples of feasibility if not samples of convenience. For the New Jersey negative income tax experiment (Kershaw & Fair, 1976b; Watts & Rees, 1977a, 1977b), the researcher gave the idea of a nationwide randomly selected sample of low-income households very careful consideration before deciding on a few areas in New Jersey and a portion of Pennsylvania selected for both feasibility and theoretical reasons of an unquantified, general, and informal sort. I strongly endorse this approach. I believe it characterizes physical science as well as the most valid and useful of applied science.

Acquiring a truly representative sample is an unpracticed ideal. But it is so out of keeping with what we know of science that it should be removed even from our philosophy of science. A consideration of the time dimension will help to show that it is utterly unreasonable. In the physical sciences, the presumption that there are no interactions with time (except those of daily, lunar, seasonal, and other cycles) has proved to be a reasonable one. But for the social sciences, a consideration of the characteristics of potentially relevant populations shows that changes over time (for example, a 30-year comparison of college students) produce differences fully as large as those produced by social class and subcultural differences. To sample representatively from our intended universe of generalization would require representative sampling in time, an obvious impossibility.

More typical of science is the case of Nicholson and Carlisle. Taking in May 1800 a local sample of Soho water, inserting into it a very biased sample of copper wire, into which flowed a very local electrical current, they obtained hydrogen at one electrode and oxygen at the other and uninhibitedly generalized to all the water in the world for all eternity. It was a hypothetical generalization, to be sure, rather than a proven fact. There have been by now many studies of the effect of impurities in the water on hydrolysis, but these studies, too, have been based on very biased samples. The idea of a representative sampling of all the waters of the world, or even of all the waters of England, never occurred even as an idea. The very concept of impurities, of distinguishing the contents of water as "pure" stuff and alien materials, is one that would never have emerged had a representative sampling approach to water been employed. In the successful sciences, generalizations have never been inductive in the sense of summarizing what has been observed within the bounds of the generalization, but instead they have always been presumptive, albeit guided by prior laws. The limitations on generalization have emerged from efforts to check on an initial bold generalization in nonrepresentative ways. Scientists assumed that hydrolysis held true universally until it was shown otherwise.

In this light, had we achieved one, there would be no need to apologize for a successful psychology of college sophomores, or even coeds from a particular university, or of Wistar strain white rats. Exciting and powerful laws would then be presumed to hold for all humans or all vertebrates at all times until specific applications of that presumption proved wrong. We already are at this latter stage, but even here a representative sampling of species or school populations is not the answer. Theory-guided, dimensional explorations, as in comparing primates that vary widely in evolutionary development, are in the typical path of science (Campbell, 1969c).

In program evaluation, I at least recommend formal abandonment of the goal of nationally representative sample selection. Once there are interventions of such well-established effectiveness that the decision is made to adopt them nationally, sample census data can be employed for budgetary planning purposes—for example, data on the population distribution, on schools, hospitals, or other distribution facilities to be employed, and on labor and space costs. I do not anticipate that cross-validation of an intervention's effectiveness on a strictly representative national sample would ever be cost-beneficial or needed. If it were employed, it would be for administrative reasons, not for applied scientific validity.

The Principle of Proximal Similarity

The first presentation of external validity (Campbell, 1957) was entirely in terms of generalization to other treatments, measures, populations, settings, and times. As stated earlier, I feel we need something more appropriate than the generalization rhetoric and the solution of it by representative sampling from a universe designated in advance. This rhetoric is greatly reduced in Cook and Campbell (1979) through Cook's notion of construct validity of causes and effects. In this shift, the validity of theoretical interpretation replaces atheoretical generalization to other treatments and measures. A shift has been made from a positivist phenomenalism to a fallibilist realism in which all treatments and measures are regarded as imperfect proxy variables for latent causes and effects.

Under the principle of proximal similarity I would like to provide a metatheoretical basis for justifying a seemingly atheoretical rationale and approach to the generalization of findings. I do this ambivalently, because one of the attractive summaries of our new contrast is to regard local molar causal validity as atheoretical and construct or external validity as theoretical. Perhaps the principle of proximal similarity merely describes the route to theory-based generalization, given the multiattribute contexts with which we must begin and from which we can only be released by degrees of experimental isolation and control that are unaccessible in social settings.

Although I believe that the principle of proximal similarity applies to pure science also, I want to make an argument for it specific to applied social science. In so doing I borrow from some earlier papers (Campbell, 1972, 1973; Raser, Campbell, & Chadwick, 1970). It was Harrod's (1956) effort to justify induction that moved me to this conceptualization. Although I judge that he failed, as all such efforts must fail, his work introduces a profoundly different understanding of the presuppositions underlying scientists' efforts at inductive inference. For the earlier postulate that nature is orderly, Harrod substitutes the presupposition that nature is "sticky," "viscous," proximally autocorrelated in space, time, and probably *n*-dimensional attribute space, with adjacent points more similar (as a rule) than nonadjacent ones.

The most important practical justification of the principle (and of the need for confirming in practice the efficacy of social ameliorations) comes from the fact that our experience in generalizing social science findings shows that higher order interactions abound, precluding unqualified generalization of our principles not only from laboratory to laboratory but especially from laboratory to field application.

It is most convenient to explain this in terms of the model of analysis of variance. Consider multiple dimensions of experimental variation *A, B, C, D,* and so forth, each of which occurs in several degrees of strength, with (in the simplest design) each combination of strengths being employed. (Thus, if there were four dimensions, *A, B, C,* and *D,* each of which had three strengths, there would be eighty-one different treatment packages.) In addition to these treatment or independent variables, there is at least one dependent variable in terms of which the results of the treatments are measured. Let us call this *X*. For our present purposes, two major types of outcome need to be distinguished: main effects and interactions. If a main effect for *A* is found on *X*, then we have what could be called a *ceteris paribus* law, meaning other things being equal. For example, with *B, C, D* being held constant at any level, the same rule relating to *X* is found: perhaps, the more A, the more X. Where interactions are found, the relations are complexly contingent. For example, in an *A–B* interaction, there may be a separate rule relating *A* and *X* for each different level of *B* (for example, if *B* is high, the more *A* the more *X*, but if *B* is low, the more *A* the less *X*). Much more complex (higher order) interactions can also occur, such as an *A–B–C* interaction in which the *A-to-X* rule is different for each combination of *B* and *C*.

Interactions, where they occur in the absence of main effects, represent highly limited and qualified generalizations. It is typical of the history of the physical sciences that many strong main effects have been found—generalizations conceivably true independent of time and place and the status of other variables. Although eventually, in fine detail, the laws were found to be more complex, there was nonetheless a rich experience of discovering approximate laws of nature that could be stated without specifying the conditions on the infinitude of other potentially relevant variables.

There is no compelling evidence so far that the social sciences are similarly situated. If we take the one social science that uses the analysis of variance approach, experimental social psychology, the general finding is of abundant higher order interactions and rare main effects. Even where we get main effects, it is certainly often a result of the failure to include the additional dimensions (*E, F, G,* and so forth) that would have produced interactions. Frequently we are unable to replicate findings from one university laboratory to another, indicating an interaction with some unspecified difference in the laboratory settings or in the participants.

If such multiple factorial experiments can be regarded as experiments in generalization, they give us grounds for great caution, particularly when we generalize the expectation that, had we included dimensions *E, F, G,* and *H* or *Y* and *Z* in our experiment, the *A–X* relationship might well have shown interactions with some or all of them, too. The high rate of interactions on the variables that we have explored must make us expect something similar for the many variables that we have not explored.

Any given experiment can be regarded as holding constant at one particular level every one of the innumerable variables on which no experimental variation is introduced. This is like choosing a single level for variable *B*, perhaps, from a potential experiment in which two or more levels of variable *B* were systematically employed. We can guess with confidence that the farther apart the two values of *B* (or *E* or *Z*), the more likely it is that *B* will interact with the *A–X* relationship. (An empirical exploration of this might well be worth making. Data from complex experiments using three or more levels of a given treatment could be reanalyzed as two-level experiments, some as wide-range, using the two extreme levels and disregarding the intermediate, others as narrow-range, using adjacent levels from the original experiment.)

In anticipation of the outcome of such studies and in common with the intuition of most scientists, let us assume as a general rule that the larger the range of values on the background variable, the more likely these variables are to have strong interactions with the *A–X* relationship under study. To put it more simply, we generalize with most confidence to applications most similar to the setting of the original research. When generalizing from our laboratory to a real-world social–ameliorative program, the values on all dimensions differ widely. We can expect that new interaction effects, as yet unexplored, are extremely likely.

Intuitively, we already use this principle of proximal similarity in many ways, and we can self-consciously use it in more. When it comes to disseminating a new ameliorative program of local molar causal validity, we will apply it with most confidence where treatment, setting, population, desired outcome, and year are closest in some overall way to the original program treatment. In contrast, for research on the limits of generalization, exploratory contrasts should be sought out for cross-validation that differ as much as possible from the first intervention in population, setting, and so forth. Purposive sampling for maximum exploration of generalizability on conceptualized dimensions will be substituted for a population-representative sampling.

In the new contrast, external and construct validities involve theory. Local molar causal validity does not. Although this contrast is weakened in the principle of proximal similarity, I still want to retain it. The principle of proximal similarity is normally implemented (as it should be) on the basis of expert intuition. The use of the term *construct* in the expression *construct validity of causes and effects* (Cook & Campbell, 1979) may too strongly connote formal theory. Nevertheless, most philosophers, or at least most logicians, may well agree with Nelson Goodman (1972) that any concept of overall similarity is meaningless or incoherent because there are potentially an infinite number of attribute dimensions on which such similarity could be computed. Our intuitive expectations about what dimensions are relevant are theory-like, even if they are not formally theoretical. Moreover, clinical experience, prior experimental results, and formal theory are very appropriate guides for efforts to make the exploration of the bounds of generalizability more systematic.

NONCONCLUSION

The material presented in this chapter is self-consciously inconclusive. It is a dialectical reaction, or overreaction. Let us hope that the overall iteration is headed for convergence.

PART III

THEORY OF SCIENCE FOR SOCIAL EXPERIMENTATION

In the last two decades of his life, Campbell devoted a large measure of his time to epistemological and theory of science issues. There had been three marked shifts in epistemological discussions over the past 40 years. Campbell felt that an understanding of these issues, especially those relating to sociologists of science, were relevant and useful both to students and teachers of applied social science.

First, logical positivism, which had been the dominant philosophy from the 1920s to the early 1950s, has been abandoned. Logical positivism was based on the premise that there is no reality beyond that which can be empirically measured or observed. It denies underlying causes by unobserved variables. It implies a correspondence theory of reality that would suggest a one-to-one relationship between our beliefs about reality and reality itself. Definitional operationalism, which was a characteristic feature of this philosophy, suggested that a single quantitative indicator of a construct defined that construct. In Part III of this book, Campbell wished to point out the aspects of social science methodology that have been affected by this shift away from logical positivism.

One of the competing philosophies (actually, a large set of philosophies) to replace logical positivism was postpositivism, in which tradition Campbell placed himself. In his early work with Julian Stanley, he supported a coherence-like theory in which reality could not be directly observed, but scientists could make statements about reality that made logical sense. This tradition advocated triangulation, which uses diverse imperfect measures to probe reality. Campbell believed that true experiments offered the best method for increasing our understanding of reality, although he adamantly held that experiments could not offer proof, but merely add to the coherence-like criteria. He employed what he referred to as evolutionary epistemology, which applied the concepts of biological evolution to the study of how we come to know and to understand the world around us. This concept held that "blind variation and selective retention" described the model through which scientists gain more valid knowledge. He believed that the disputatious community of science, through competition and mutual criticism, could arrive at better approximations of reality. He admitted that this was a hypothetical reality, because there is no proof that such a reality exists. Still, he held that this hypothetical reality plays a role, albeit a very small one, in increasing validity by constraining beliefs. He maintained that the scientific community could and should develop norms that would help them improve the fit of scientific belief with hypothetical reality. Based on these assumptions, Campbell adhered to a somewhat minority view that knowledge systems could be compared to determine which systems generate more valid knowledge.

The second shift involved some more recent social science epistemologists in the humanistic social science theory of interpretation that have developed what Campbell referred to as a *nihilistic hermeneutics,* or to use his more common term, *ontological nihilism.* An example is the Strong Programme, which began in Edinburgh in the late 1970s and whose adherents are known as social constructivists. These sociologists of science believed that the same kind of psychological and sociological analyses that were applied to unsuccessful theories in science should be applied to those theories believed to be true today. That is, one could not use the "correctness" of the theory to explain why it was adopted by the scientific community. Those that Campbell referred to as ontological nihilists held that reality or nature does not play a role in scientific knowledge and emphasized the social construction of scientific beliefs. This is sometimes accompanied by a general skepticism about the possibility of any validity in science, physical or social. These epistemologists applaud novel interpretations but do not believe that the relative validity of these interpretations should be argued. Although Campbell recognized that science is flawed, that it is culture-bound and time-bound, still he believed that it is a progressive activity that provides increasingly better ideas about reality. In the following chapters, he wished to present arguments to address the criticisms of the newer epistemologists.

Finally, the third shift involves the debate surrounding the issue of social science's quantitative–experimental methods versus the qualitative methods of the humanities. Over the past two decades, interest in the qualitative methods has increased, and Campbell applauded such efforts. Some advocates of these humanistic methods believe they are incompatible with the experimental approach. In Part III, Campbell argued that the approaches were complementary and that employing both would enhance the search for better understanding of social processes and settings.

These chapters are intended to sensitize social scientists to the social aspects of their work by increasing their awareness of the social impacts on the operations, products, and consequences of science. They also provide support for a validity-seeking theory of science and argue against those who see no value in comparing knowledge systems to determine which would enhance validity. In Chapter 5, Campbell reviewed the developments in the philosophy of science, sociology of science, and sociology of knowledge to address the question of the degree to which the social sciences can achieve scientific status. Chapter 6 presents Campbell's response to an article by W. N. Dunn that suggested that a new paradigm for epistemology replace the one outlined in The Experimenting Society. Although Campbell agreed with Dunn on many points, still he defended certain concepts, such as causality, experimentation, ontological reality, and the scientific method in general. In Chapter 7, he suggested that philosophers of science develop a sociological theory of scientific validity so that the norms emanating from this theory could provide advice on science policy. Chapter 8 compared the physical and social sciences in terms of the social, psychological, and ecological requirements for science, and posed reasons for their differential success. In Chapter 9, Campbell offered recommendations to scientific institutions regarding the best practices for enhancing scientific validity. Finally, in Chapter 10, he urged sociologists of scientific knowledge to further develop their theories to demonstrate the plausibility that the physical world plays a part in selecting scientific beliefs and to examine the social system that maximizes that role.

OVERVIEW OF CHAPTER 5

Can we be scientific in applied social science? Campbell was not able to give a definitive yes or no answer. He believed that we could become more scientific if the political, financial, and administrative climates changed somewhat. Still, he did not believe that applied social science had earned the authority to override conventional wisdom if the results disagreed, and he worried about the corruption of measures used to make policy and budgetary decisions. He came to these conclusions after considering developments in the philosophy of science, sociology of science, and sociology of knowledge.

Logical positivism, the dominant philosophy of 40 years ago, has been abandoned. This philosophy is based on knowing what can be observed, denying underlying causes, phenomena, or ultimate origins. With the publication of his book, *Experimental and Quasi-Experimental Designs for Research* with Julian Stanley, Campbell considered himself a postpositivist. He believed that reality cannot be captured or fully understood, but only probed. Methodology is important in this philosophy because each method yields only an imperfect picture of the processes being studied. Campbell offered five postpositivist points.

First, in social scientific inquiry, there are many junctures at which the social scientist must make subjective judgments. Although logical positivism tried to remove all discretion, postpositivists see it as unavoidable. They recognize theories cannot be proven through experimentation, but only probed, and that uncontrolled threats to validity are only problematic if they are plausible threats. Second, the postpositivists recognize that science is conducted within a paradigm in which we depend on a number of presuppositions. Campbell applied what he termed *evolutionary epistemology* as a basis for the scientist's continued search for validity. Evolutionary epistemology applies the concepts of biological evolution to the study of knowledge—in this instance, the concepts of variation and retention, which must be balanced to acquire knowledge. Campbell advocated what he called the 99-to-1 trust–doubt ratio—that is, trusting 99% of the beliefs in the paradigm and doubting 1%. By challenging and examining the untrusted beliefs, scientists gradually revise their theories, sometimes resulting in great intellectual revolutions. Both the quantitative methods of the physical sciences and qualitative methods of the humanities are important to social science. The qualitative approaches are used to qualify and intuitively confirm quantitative data and to rule out plausible rival hypotheses. Third, experiments and theoretical arguments are historically embedded. This means that an understanding of the past is essential to understanding new terms and experiments. Fourth, although our knowledge of reality is imperfect and relative, this philosophy assumes a real world with which organisms interact. *Ontological nihilist* was Campbell's characterization of the social constructivist who believes that nature plays no role in scientific knowledge. Campbell disagreed with the ontological nihilists and other sociologists of science who refused to study the social processes that could produce improved scientific beliefs. Finally, science is performed by human beings with all their flaws and idiosyncracies. Scientific validity depends on mutual criticism. A social system of science, which has scientific validity as its goal, should foster competitive replication, reward competitive innovation, and discourage dishonesty. Although on one hand providing incentives for scientists, this sociology of science must ensure that the collective validity of science is not undermined.

Relating this postpositivist approach to social science, we might ask with understandable cynicism how it applies, for example, to the evaluation of a popular social program believed to be effective in helping families overcome poverty.

Trying to determine whether the social program was responsible for the improved economic situations of the families can be very difficult. Because this problem cannot be studied in a laboratory, rival hypotheses must be identified and ruled out. Campbell pointed out that the rival hypotheses and designs outlined in *Experimental and Quasi-Experimental Designs for Research* were not an attempt to borrow methods from the physical sciences, but rather were a result of the mutual criticisms of researchers in the field and their attempts to rule out alternative hypotheses. Thus, the researcher evaluating the program to help impoverished families has the benefit of building on the experiences of other evaluators in choosing an appropriate design and increasing the validity of the results.

There is a further problem when the results of the evaluation will be used to make a decision regarding whether funding for the poverty program will be continued, increased, or decreased. When policy will be affected, the discretionary judgments of the researcher may be biased toward choices that will produce a desired effect.

Campbell emphasized the importance of qualitative methods to enhance and expand on the quantitative measures. Talking to program administrators, workers, and clients, for example, would provide the foundation for a more quantitative study and increase the understanding of the eventual results. Campbell stated that definitional operationism was "positivism's worst gift to the social sciences." This is because, when applied to evaluation research, the single quantitative measure embraced by this tradition is fallible and can be manipulated and corrupted. For example, if the goal of the program is to increase the number of individuals in the serviced families who secure employment, the target number might be reached by expanding the definition of what is meant by "employment."

The following points out some of the errors that social science methodologists offered in trying to guide early evaluation research. One was the idea that a single evaluation was all that was required to reach a decision on the effectiveness of a program. This does not conform with what we know of how science is performed—that is, a scientific solution is pursued by many scientists working individually. Also, it was believed that evaluations should be external to the program. This does not parallel the physical sciences, where the research is designed and carried out by those who are passionately involved in the problem. Detachment of the researcher does not improve the integrity of science; rather, it is improved by competitive cross-validation carried out by other scientists.

Other mistakes in early evaluation methodology involved carrying out the evaluation too early in the program to decide the effectiveness of the program (a classic example of this is the Head Start evaluation). Also, too much emphasis had been placed on external validity by requiring representative samples for nationwide evaluations. Clearly, this was not a great concern to the physical sciences. If the results were not internally valid, there was no point in being able to generalize to other populations. Finally, early methodologists erred in believing that a single evaluator could observe the process without affecting it and would have no motivation to influence research results. This is no longer believed to be true. It is now recognized that the community of scientists is a vital factor in the search for scientific truths, with their contributions to, their assessment of the significance of, and their critical review of experimental results.

FIVE

LEGACIES OF LOGICAL POSITIVISM AND BEYOND

Can we be scientific in applied social science? My ability to take a middle-of-the-road, sensible position in a militant argumentative way makes you know that my answer will be both yes and no. Certainly it is much harder to be scientific where financially enormous policy decisions hang on our fragile social science tools. Let me give you one preliminary yes and two preliminary no's.

A feeble yes: We can be somewhat more scientific than we are now or have been (in educational program evaluation, for example). Changes feasible within the current financial, administrative, and political climate could make us able to be more scientific. An equally feeble no: If we present our resulting improved truth claims as though they were definitive achievements comparable to those in the physical sciences, if we maintain that they deserve to override ordinary wisdom when they disagree, we can be socially destructive.

Campbell, D. T. (1984a). Can we be scientific in applied social science? In R. F. Conner, D. G. Altman, & C. Jackson (Eds), *Evaluation studies: Review annual* (Vol. 9). Beverly Hills, CA: Sage.

We can be engaged in a political misuse of the authority of science that has not been fully earned in our own field. Another no: Using quantitative social science measures for administrative control and budgetary decision making (as in the accountability movement) can be destructive of the institutions and processes over which control is intended, and destructive as well of whatever prior validity the social science measures employed once had.

I want to come to these conclusions, or get close to them, by briefly reviewing recent developments in the philosophy of science, sociology of science, and sociology of knowledge, including the argument within our own program evaluation community as to whether we should employ the methods of physics or the methods of the humanities. In light of a fragmentary, modern, postpositivist theory of science, I will then discuss the special problems of policy-relevant social science research. One such problem resulted from the politicization of our own mistaken view as to what an applied social science should look like. This was the vision that we offered in the heyday of the Great Society Program of the 1964–1968 period, under the regime of one of our two presidents named Lyndon Johnson (that is, "Lyndon Johnson the Good"). I am thinking of the Office of Economic Opportunity, Program Planning and Budgeting Systems, and program evaluation.

POSTPOSITIVIST THEORY OF SCIENCE

Twenty years ago logical positivism dominated the philosophy of science and, through concepts such as *operational definition*, dominated our thoughts about research methods. Today the tide has completely turned among the theorists of science in philosophy, sociology, and elsewhere. Logical positivism is almost universally rejected. This rejection, in which I have participated, has left our theory of science in disarray. Under some interpretations it has undermined our determination to be scientific and our faith that validity and truth are rational and reasonable goals. What we should have learned instead was that logical positivism was a gross misreading of the method of the already successful sciences. Logical positivism was wrong in rejecting causal processes imputed to unobserved variables. Logical positivism failed to recognize that even at its best, experimental research is equivocal and ambiguous in its relation both to the real physical process involved and to scientific theory. This equivocality calls for use of multiple methods, none of them definitional, triangulating on causal processes that are imperfectly exemplified in our experimental treatments and measurement processes. Properly interpreted, the dethronement of logical positivism should have led to an increase in methodological concern rather than its abandonment. Positivism's worst gift to the social sciences was definitional operationism, and this still persists in applied social science, as in the accountability movement in which goal statements and achievement claims are rigidly defined in terms of singular, quantitative indicators. (In Chapter 1, we discussed how the use of such indicators as evidence of performance corrupts the indicator and diminishes its validity.)

Campbell and Stanley's *Experimental and Quasi-Experimental Designs for Research* (1966, written in 1961 and 1962, first published in 1963), was lucky to be already postpositivist. (At least in a whiggish rewriting of history, I can claim that. In Cook & Campbell [1979, pp. 10–14] the assessment is more mixed.) First of all, we cited N. R. Hanson (1958), who was the first in the Hanson–Kuhn–Feyerabend tradition to emphasize the theory-ladenness of the factual observations of science. It cited Popper (1959) with approval (although it did not cite my favorite slogan of his: "We don't know, we can only guess"). We emphasized the equivocality of both the treatment implementations and the observations. We gave a section head and two paragraphs to evolutional epistemology. Most important, postpositivist was the concept of "plausible rival hypotheses," putting so much scientific weight on that squishy concept of plausibility.

I would like to point out five postpositivist points with which I agree and with which I think you also should agree. I am borrowing from Hanson (1958), Polanyi (1958), Popper (1959), Toulmin (1961), Kuhn (1970), Feyerabend (1975 and before) and other wild characters including Quine (1951, 1969b).

1. Judgmental, Discretionary Components Are Unavoidable in Science

These components appear in the choice of experimental design, the choice of a specific apparatus, the wording of the particular questions in our questionnaires, in the interpretation of results, and in the choice between competing theories. These subjective discretionary links cannot be avoided. Logical positivism wanted to remove all discretion. This effort to achieve foundationalist explicitness took two forms: completely explicit observational foundations (meter readings, sense data, and so on) and logical deductive manipulation of these sense data. Logical positivism failed at both levels.

Campbell and Stanley (1966, p. 35) joined in this rejection of logical positivism when we said that "true" experiments at their very best only *probe* theories; they do not prove them. But the rejection was most important in our emphasis on the role of *plausibility*. We took the position that there could be lots of threats to validity that were logically uncontrolled but that one should not worry about unless they were plausible. The general spirit was that any interpretation of a body of data or research procedure should be regarded as innocent until judged guilty for plausible reasons, as determined through the scientific method of mutual criticism.

I have often wondered why there were no hostile logical–positivist reviews of Campbell and Stanley, accusing us in this paper of undermining scientific standards. We failed to get one as far as I know. It is with mixed pride that I note we are now regularly being used as an exemplar of logical positivism, and of the mistaken effort to import into the social sciences the inappropriate methods of the physical sciences. (Although I am grateful for every citation, I think this is a misreading, as will be argued.)

2. The Paradigm Theme

We are inevitably encapsulated in some paradigm of presuppositions, inexplicit or explicit. Historically, we can look back and see how provincially we were embedded. We cannot do without presuppositions. We cannot pull each presupposition out individually and prove them one at a time. In every expansion of scientific knowledge we have to expand the number of things we assume are true and that have to go unproven. In the evolutional–epistemology version of this, with the recipe of variation, selection, and retention, there is emphasis on the presuppositions about the nature of the world that are built into our retinas, the nerve wiring of our brains, our language, and our own research tradition. From evolutionary epistemology comes the crucial question of balance between variation and retention. These are incompatible, and knowledge becomes impossible if either totally dominates.

In accepting paradigm-embeddedness, again we are rejecting the foundationalism that was so central to logical positivism. There are no untouchable axioms: All are criticizable and revisable. Nor are there any foundational observations or facts. There are indeed at any historic period of time in any successful science a vast array of trusted facts, but none is immune from revision. For the atomistic (sense data, observations, or axioms) foundationalism of the positivists, we must substitute a holistic, squishy, quasifoundationalism, a composite foundationalism that I call the 99-to-1 trust–doubt ratio. This is like the holistic network imagery of Quine (1951), but I will give it to you in my version.

For the cumulative evolutionary process of knowing, our only available tactic is to trust most of our current beliefs. This body of beliefs provides the support we need to revise a few of them. The ratio of the trusted to the doubted has got to be in the order of 99% trust to 1% doubt. In biological evolution, 99% of the genes are trusted while mutation and recombination vary 1% of them. However wrong-headed the initial beginnings, nature is stuck with this great mass of presuppositions on how to design an animal. Similarly, in a science such as physics, the great revolutions have been achieved by trusting 99% of the cumulated facts and using that basis to revise 1% of its beliefs and their theoretical integration. This produces a kind of gradualism at the level of facts.

Don Moyer (1979) has studied the belief changes following the 1919 eclipse observations, where English physicists and astronomers moved from 5% adoption of Einsteinian general relativity to 99% adoption in a 5-year period. He documented the ways this revolution was based on profound trust of previous physics, which provided the factual leverage for overthrowing the dominant Newtonian theory. It was palace revolution in conceptual organization and theory, in which most of the facts (all being theory-laden "facts") were retained.

Before going on to the other three points, I would like to use these first two points (paradigm dependence and discretionary human judgments) to discuss the qualitative versus quantitative agenda that is so important right now in educational research and program evaluation. Should we be using the methods of the humanities or the methods of the physical sciences? I would like to argue that if we had not misread the record of the physical sciences, we would recognize that these methods are very similar.

Let us start out with that old tradition, at one time called *philology* and now called *hermeneutics*, which asks such questions as, "What did Homer mean by this particular phrasing?" or, "What did Saint Paul mean in this particular verse?" In philology and hermeneutics, one had generations of scholars quarreling about these issues, but remaining within the same social communication net, a quarreling collective committed to getting the truth. Now part of this hermeneutic tradition is this presuppositional and contextual dependence that I have called the 1-to-99 doubt–trust ratio. The translators generate and criticize plausible rival hypotheses as to alternative interpretations, including the hypothesis that some copyist had made a clerical error that was subsequently transmitted by other loyal copiers, and so on. This self-critical community of interpreters, by looking at a wider range of manuscripts from this same time, and thus extending the grounds of judgment, often eventually arrived at consensual decisions as to the best interpretations of a particular manuscript.

Or look at the method of the historians as taught and exemplified by Collingwood (Levine, 1978), who was a historical relativist with a historical paradigm theme. His method was explicitly the method of a detective in a detective story. The method is epitomized by trying to rule out plausible rival hypotheses.

When we get down to our own practical work, a plausible-rival-hypothesis approach is absolutely essential, and must for the most part be implemented by commonsense, humanistic, qualitative approaches. In program evaluation the details of program implementation history, the site-specific wisdom, and the gossip about where the bodies are buried are all essential to interpreting the quantitative data (Campbell, 1975a, 1978, 1979a: Cook & Reichardt, 1979).

3. Historicism

At any given time, even in the best of science, we are in a historical context and our experiments and our theoretical arguments are historically embedded. They have a historical provincialism; they are reactions to what has gone before; they are dated and uninterpretable outside of that context. The contrasts with the past are, in some kind of a problem-solving way, almost necessarily exaggerated. So we have a dialectic of contrast, in which exaggerated, oversimplified corrections for what has gone before are an essential part of the process, and the past that has gone before is essential for understanding the new terms and new experiments that are introduced. In an effort to speak in the extreme forms of postpositivist jargon, I have called this the "dialectic historical indexicality of scientific terms" (Campbell, 1982). Gergen (1982) has presented the historicist argument for social psychology.

4. Relativism

This treasure of postpositivism encompasses epistemological, historical, cultural, and paradigm relativism. In the evolutionary epistemology tradition (Campbell, 1974a), my slogan is "blind variation and selective retention." This is an emphasis on exploring in the dark, with the fumbling of a blind person being a better model for epistemology than clairvoyant vision. All of this commits me to a profound epistemological relativism.

Now, I am a thoroughgoing epistemological relativist. Still, I reject an ontological relativity, or, because Quine (1969b) has used that term in a different sense, an ontological nihilism. Evolutionary epistemology has in it an unproven assumption of a real world external to the organism, with which the organism interacts. I spent a lot of time reading and meeting with (Campbell, 1981) exciting sociologists of science such as Barry Barnes (1974), David Bloor (1976), and Michael Mulkay (1979), Karen Knorr-Cetina (1981), Bruno Latour and Steven Woolgar (1979), and Harry Collins (1981b). Also relevant is the book Robert Merton and Thomas Kuhn have resurrected, Ludwig Fleck's 1935 *Genesis and Development of Scientific Fact* (1979). Harry Collins called this the "relativist" program in the sociology of science. Latour and Woolgar and Knorr-Cetina called it the "social constructivist" program. David Bloor and Barry Barnes (Barnes & Bloor, 1982; Bloor, 1976) called it the "strong programme," meaning that in doing sociological, historical studies of science (asking the question, "What were the causes for their changing their scientific beliefs?"), it is illegitimate to use our current confidence in the truth of the belief as an explanation for why, back then, they came to believe it.

This agnosticism I find methodologically correct. After all, those past scientists were not clairvoyant, and many of the changes we now regard as in the mainstream of scientific development we do not now regard as "true." But these new sociologists of science carry this agnosticism too far. They refuse to speculate about what kinds of social processes, what kinds of systems of interaction among scientists and between scientists and society, could produce improved beliefs. They refuse to undertake what I call an epistemologically relevant internalist sociology of science (Campbell, 1979c, 1981).

5. Sociologism and Psychologism

Science is a social process, and scientists are thoroughly human beings: greedily ambitious, competitive, unscrupulous, self-interested, clique-partisan, biased by tradition and cultural memberships, given to mutual backscratching, and the like. James Watson's *The Double Helix* (1968; but see Olby, 1974, for Crick's perspective) is one of the most used texts in the sociology of science relevant to this.

Out of this, I want us to keep the goal of *truth*, and to attempt to understand and foster a social system of science in which it becomes sociologically plausible that the processes would lead to beliefs of increasing validity. The scientific method itself is a social system product. Science is itself a social system, it is "tribal" in that sense (Campbell, 1979c); but strangely, its norms preach against that very tribalism: against deference to authority, against deference to majority rule. A key part of this sociology of successful science is a mutual criticism that keeps those who are criticizing each other still remaining in the same group, rather than splitting off into their own insulated cults. Competitive replication, threat of replication, and a reward system that encourages competitive innovation but punishes dishonesty in the resulting competition (Merton, 1973) are all parts of it (Campbell, 1986c).

From this sociological point of view, combined with an evolutionary–epistemology point of view, it follows that large numbers of independent decision makers are essential for objectivity in science. It follows too that we must maintain scientists' collective interests in the trust given the system of science by the larger public (Merton & Gieryn, 1982). We must maintain the individual scientist's interest in reputation, recognition, and fame, without allowing these interests to undermine the self-interest in science's collective validity. We scientists cannot avoid being dependent on the trust of fellow scientists. We must avoid creating a motivational system that generates truth claims or belief assertions that we distrust. We need a scientific method (as a social invention and social process) that will counteract the ill effects that a cynical and nihilistic interpretation of point 4 (relativism) and point 5 (sociologism and psychologism) can produce.

This epistemologically relevant internalist sociology of science will not deny the scientist's paradigmatic provincialism, self-seeking competitiveness, and human fallibility, but will rather propose a social system designed to curb side effects that produce invalid beliefs. Inevitably our model of science will show science as a fragile and vulnerable social institution, one that is capable of flourishing only now and then, only here and there, on the face of the earth. A validity-producing social system of science is nothing we should take for granted.

APPLICATION TO APPLIED SOCIAL SCIENCE

If we move such a postpositivist theory of science into the problem of the validity of applied social science, we find that we need all of the social system features of pure science (e.g., physics, laboratory psychology, and biology). From this perspective, when we move into the arena of policy making, there are some regular features of applied social science for policy purposes that come to our attention.

First is clearly the *greater equivocality of causal inference for research done in policy settings.* There are many, many more plausible rival hypotheses. There is much less control. Looking back at the "artificiality" of physical science laboratories (their soundproof walls, atmospheric controls, insulation against electromagnetic and magnetic fields, achievement of vacuums, and all of the other accoutrements of "experimental isolation"), we can see that all of this laboratory apparatus is designed to control or to rule out *plausible rival hypotheses,* or at least to render them "implausible," thus achieving an artificial situation in which causal inference can be done more competently.

When biologists left the insulated laboratory, in which apparatus and walls are the essence of the scientific method, to move out into the agricultural experimental station in which the winds blew and the rains rained, they invented another type of artificiality to render implausible large classes of plausible rival hypotheses. This was the randomized experiment. We should note that slightly before that, educational researchers such as E. L. Thorndike and his students, moving from the insulated psychology laboratory out into the classrooms, independently invented randomized assignment to experimental treatments and Latin square designs, again as artificialities that operated somewhat like experimental isolation in generating controls, in reducing the plausibility of rival hypotheses such as selection, selection–treatment interaction, practice effects, and the like. Although we educational psychologists did not do it with Fisher's mathematical elegance, we were *first* with these great tools of artificiality. McCall's (1923) *How to Experiment in Education* summarized this early achievement.

Today, as so many of us react to the frustrations of social science research with the hope that humanistic methods will turn out to be more appropriate than physical science ones—an exploration that I too favor (Campbell, 1975a, 1978, 1986c), our troubles are often blamed on a prior, mistaken, subservient borrowing of physical science methods. Indeed, Campbell and Stanley (1966) are often accused of this fallacy. Close analysis will, I believe, show that this is unfair. Thorndike and McCall were *not* borrowing random assignment and the "rotation experiment" (Latin square) from the physical sciences, nor from R. A. Fisher and his agricultural experiment stations. Instead, they were reacting to the mutual criticisms of their own educational–psychology research community and inventing research designs that would help rule out the recurrent very plausible rival hypotheses generated by their fellow critics.

So too, Campbell and Stanley's list of threats to validity is an accumulation of our field's criticisms of each other's research. The list of quasi-experimental designs is a cumulative listing of our discipline's inventions of ways of ruling out some of the very plausible rival hypotheses. We can thank Campbell and Stanley for being conscientious collectors of the achievements of this tradition of collective self-criticism. (That is what they were: collators, bookkeepers, reviewers of the literature.) Their collection of designs is not at all drawn from physical science. Of course, from the quasi-experimental perspective, just as from that of physical science methodology, it is obvious that moving out into the real world increases the number of plausible rival hypotheses. Experiments move to quasi-experiments, and on into queasy experiments, all too easily.

A second difference between applied social science and laboratory research is the greater likelihood of *extraneous, nondescriptive interests and biases.* These biases enter through the inevitable discretionary judgmental components that exist in all science at the levels of data collection, instrument design and selection, data interpretation, and choice of theory. As we move into the policy arena there is much less social-system-of-science control over such discretionary judgment favoring descriptive validity. There are much stronger nondescriptive motives to consciously or unconsciously use that discretionary judgment to, so to speak, break the glass of the galvanometer and get in there and push the needle one way or the other so that it provides the meter reading wanted for nondescriptive reasons (Campbell, 1979a, pp. 84–86; 1982).

The next few points about moving the theory of science into the applied social science arena stem in considerable part from the seriously mistaken model of applied social science that we social science methodologists offered to ourselves and to government in the 1960s, in the period of the Great Society, in the era of the Office of Economic Opportunity, and Program Planning Budgeting and Systems.

My third point is the *mistaken belief that quantitative measures replace qualitative knowing.* Instead, qualitative knowing is absolutely essential as a prerequisite foundation for quantification in any science. Without competence at the qualitative level, one's computer printout is misleading or meaningless. We failed in our thinking about program evaluation methods to emphasize the need for a qualitative context that could be depended on. One example is frequent separation of data collection, data analysis, and program implementation that was once characteristic of Washington's funding of programs. One firm would collect the pretest, another firm would collect the posttest, and a third firm would analyze the data. This easily led to a gullible credulity about the numbers on the computer tape, with the analyst in total innocence about what was actually going on in the program implementation and testing situations.

To rule out plausible rival hypotheses we need situation-specific wisdom. The lack of this knowledge (whether it be called *ethnography,* or *program history,* or *gossip)* makes us incompetent estimators of program impacts, turning out conclusions that are not only wrong, but are often wrong in socially destructive ways.

Fourth, the evaluation model we offered mistakenly bought into the logical positivist's *definitional operationism*, specifying as program goals fallible measures open to bureaucratic manipulation (Campbell, 1969a, pp. 414–417, 1979a, pp. 84–86).

Fifth, *a one decision/one research ideal* was a central feature of our original program evaluation model. (This is diametrically opposed to the social system of the successful physical and biological sciences.) Each program evaluation was to be done to support a specific administrative decision. One researcher–evaluator was to have a monopoly on the resulting truth claims. This one study was to be the basis for the decision. With this often went a disregard of prior wisdom and prior science in making the decisions about the future of the program (Lindblom & Cohen, 1979). The program evaluation was conceptually tied to refunding, to be the sole or an important base for expanding or contracting the program.

Such a policy violates common sense as well as the sociology of knowledge. Had we sat down and thought, "What will it do to all of those discretionary points in data collection if next year's funding is going to ride on them? Where are the discretionary points and how can they be distorted?" we would have recognized the problem. This program evaluation model belied our common experience, the sociology of bureaucracy (Blau, 1955, 1956; Ginsberg, 1984), and our knowledge as psychologists regarding the multiple motives the individuals implementing programs have, including the motive of being able to feed one's children next year. ("Where will another job come from if this program is discontinued?" or, "If we report to our client our unpleasant results, where will next year's contract come from?" and so on.) These considerations add into the recurring conflict we all have observed between the evaluation staff and the program delivery staff. Program evaluation became destructive of program delivery morale.

A sixth mistake in the model that we in the 1964–1968 period recommended to government was the emphasis on external evaluation of programs rather than evaluation by the delivery team itself. This again is the complete opposite of the social customs of the physical sciences, in which passionate believers in new theories design the research and carry it out. The objectivity of physical science does not come from turning over the running of experiments to people who could not care less about the outcome, nor from having a separate staff to read the meters. It comes from a social process that can be called competitive cross-validation (Campbell, 1986c) and from the fact that there are many independent decision makers capable of rerunning an experiment, at least in a theoretically essential form. The resulting dependability of reports (such as it is, and I judge it usually to be high in the physical sciences) comes from a social process rather than from dependence on the honesty and competence of any single experimenter. Somehow in the social system of science a systematic norm of distrust, combined with ambitiousness, leads people to monitor each other for improved validity. Organized distrust produces trustworthy reports. In contrast, in program evaluation, the monopoly of a single evaluation for each program, with but one decision maker to use it, and the dogma of external evaluation, all combined to make impossible this crucial aspect of the social system of the successful sciences.

The seventh mistake involved *immediate evaluation,* evaluation long before programs were debugged, long before those who were implementing a program believed there was anything worth imitating.

Eighth, a totally unnecessary feature was recommending a single national once-and-for-all evaluation that would settle the issue forever.

Ninth, there was *gross overvaluing of, and financial investment in, external validity,* in the sense of representative samples at the nationwide level. In contrast, the physical sciences are so provincial that they have established major discoveries like the hydrolysis of water by a single water sample taken from the Soho neighborhood of London in 1903, never cross-validating the discovery on a "representative sample" of all of the water of the world.

The so-called Northwestern School has been criticized for overemphasis of internal validity at the expense of external validity. This accusation must be, in a historical sense at least, wrong. Who, after all, introduced the great emphasis on, itemized all of the threats to, and assembled the controls for external validity (Campbell, 1957; Campbell & Stanley, 1966, Cook & Campbell, 1979)? Of course, we are interested in external validity, but we see no point in having a representative national sample of a repeated regression artifact, or of some other internally invalid research design.

Tenth is *the neglect of the fact that scientific truths are a collective product of a community of scientists at any given time.* Such a community is self-critical, gets into the guts and looks under the cover and tries to decide what was going on in specific experiments. There was a neglect of this insulating layer of human judgments that are well informed and mutually disciplined. We somehow assumed in our OEO-PPB&S model that a single computer output could speak directly to the administrator. Now, however, as postpositivist fallibilist critical realist, we want our realism to include the real and fallible processes of data collection and conclusion drawing. We can see vision as the product of imperfect lenses, imperfect nervous systems, and oversimplified presumption systems, which lead to generally valid perceptions but also to optical illusions (Campbell, 1987b).

This physicalization, this materialization of the process of knowing, is a very important part of the historical development of epistemology. Extended to science we should have seen from the very beginning that social data collection and social experimentation were social system intrusions into the ongoing processes, and that putting policy-decision pressure on them would distort every crushable, squishy, little discretionary link. We were guilty of a doctrine of "immaculate perception," guilty of assuming a noninteractive acausal observational process in which all of our questionnaires and arrangements could describe without disturbing what they were describing, and in which the people being described as well as the describers would be unmotivated to bias the meter readings.

OVERVIEW OF CHAPTER 6

In his 1982 article, "Reforms as Arguments," W. N. Dunn attempted to describe a new paradigm for policy research. This new paradigm addresses not only explanatory hypotheses, but considers competing ethical hypotheses as well. Dunn states, " . . . reforms are appropriately viewed as a process of reasoned argument and debate where competing standards for assessing the adequacy of knowledge claims include, but are not limited to, rules for making valid causal inferences" (p. 302). Because the concepts of a new paradigm are indexical—that is, they can only be understood by examining them in the contexts in which they were formed, Dunn contrasted his new perspective to Campbell's "experimenting society" paradigm, which he argued should be replaced. Campbell's article, "Experiments as Arguments," was in response to Dunn's article. Campbell defended many of the concepts Dunn credited to him, although he conceded that his ideas were also evolving and, in many ways, were close to Dunn's point of view.

In referring to "reforms as placebos," Campbell distinguished between the idea of "truth" and the political use of that term. Both he and Dunn agree that most reforms are attempts to demonstrate that the government recognizes the seriousness of the problem. The programs are often seriously underfunded, and scientific evaluations are simply an effort to show that the program is being implemented and that the money is being spent responsibly.

Campbell strongly disagreed with Lindblom and Cohen, quoted in Dunn's article, who deny an ultimate reality and merely conceptualize "truth entirely as a social construction." He agreed with these philosophers, however, that a single quantitative study cannot claim to capture the truth and should not be used as the sole basis for policy decisions. Although an evaluation that demonstrates positive results does not guarantee the validity of those results, it contributes to the body of coherence-like criteria that is used to test beliefs. Although correspondence theory implies a one-to-one relationship between our beliefs about reality and reality itself, the coherence notion merely constructs statements about reality in a way that makes logical sense. Because experiments are not about facts in themselves, but rather a set of statements that make logical sense and tend to be persuasive, experiments are arguments.

Agreeing with Dunn, Campbell admitted that the social sciences are and should be "practice-driven." There are no theories of social science that are so well validated that they can define the appropriate practice to alleviate social problems. Instead, information about the results of the praxis help to validate the evolving theories. *The American Heritage Dictionary of the English Language* (3rd edition, 1994, p. 1423) defines *praxis* as "the practical application or exercise of a branch of learning." Unfortunately, the interpretations of outcomes of praxis are not clear-cut. When results are undesirable, they may be credited to incomplete or nonuniform implementation. When the program appears successful, specific aspects are examined for their contributions. There are many plausible rival hypotheses that cannot be ruled out by relying only on praxis. Experiments, on the other hand, attempt to control for those factors that may provide alternative explanations. Still, experiments only probe theories and cannot prove them.

A quasi-evolutionary perspective provides a more accurate description of the history of experimental physics than viewing experiments as direct observations of nature. The measures of independent and dependent variables used in experiments are not fully defined or meaningful apart from their use in previous experiments and theories. In this way, experiments are arguments.

Early scientists believed that disputation and criticism among the scientific community would uncover and clarify theories. The scientific method was meant to go beyond present knowledge and settle issues left unsolved through disputation. Although experiments were unable to settle the arguments, they added a great deal to the body of knowledge. This is in keeping with Dunn's thesis that initiating reforms and measuring their impact is an argument that increases our understanding of social processes. Healthy criticism is still an important ingredient when using any intentional intervention to bring about social change, whether using reforms or experiments.

The disagreement between Dunn and Campbell concerned the role of social experimentation as a method for reaching logical conclusions about the nature of social processes. Campbell saw the necessity of repeating experiments and improving them so that each effort contributes to the validation process by corresponding, or differing, with previous efforts. He adamantly held that no single evaluation should affect policy decisions by claiming to have achieved a definitive answer.

Although Campbell believed that a dialectic of experimental arguments is possible, he recognized that pressures exerted by political forces might prevent the open and honest criticism that is so vital to scientific validity. He agreed with Dunn that this should be opposed; however, if Dunn was claiming to offer a more valid epistemology that precludes the need for better experimentation, Campbell disagreed with that perspective.

The final point discussed by Campbell in his response to Dunn involved the "fact–value" distinction. This refers to the tendency of humans to be influenced by their values and biases. The public ascribes more validity to research results presented as if they were produced in a value-free environment. For example, polling results for a president's performance may show 48% feel the president is doing a good job, and 46% feel he is doing a poor job. A supporter might interpret this to mean that the public feels the president is doing a good job, whereas instead it is a matter of the glass being half full or half empty. Scientists cannot help but be influenced in some ways by their biases. Because of the competition among scientists, it may be in one's best interest to keep one's beliefs private. Unfortunately, in the social sciences, there are not abundant safeguards in place to prevent intentional withholding of values in order to manipulate belief.

Instead of insisting that all research is merely persuasive advocacy, Campbell offered more hope. He believed that the scientific community should come to accept and encourage reports representing dissenting viewpoints of other members of the research team or research community. In other words, being aware of intentional or unintentional co-optation should inspire scientists to continue in their goal of attaining objectivity, and not simply to give up on the goal of truth.

SIX

ON THE RHETORICAL USE OF REPORTS OF EXPERIMENTS

In his important article, "Reforms as Arguments," Dunn is attempting to help us switch to a new perspective on policy research. He is uncovering the implications that postpositivist theory of science has for applied social science. He has assembled an impressive array of relevant conceptual tools that are, for the most part, new to this application. These are missions I applaud and would like to participate in.

Scientific terms are typically indexical (Barnes & Law, 1976; Knorr-Cetina, 1981, pp. 33–48; Putnam, 1975, pp. 229–235) with meanings that are determined by and communicated effectively only in contexts of use. The key concepts in a conceptual revolution are dialectically and historically indexical. That is, they can only be understood in the context of the concepts they are intended to replace. To get across his new perspective, Dunn needs an example of the old paradigm and has done me the honor of choosing me for that role.

Campbell, D. T. (1982). Experiments as arguments. *Knowledge: Creation, diffusion, utilization,* 3(3), 327–337.

Sharing a belief that we must try out and discard many new conceptualizations, and convinced of the clarifying role of vigorous dialogue, he has invited me to provide a comment. To best fulfill this role, I will somewhat suppress my own desire to be ultramodern and will defend such time-honored theses as the concept of cause, experimentation, a realist ontology, and "the scientific method" more generally. However, having been vigorously anti-logical–positivist for at least 25 years, and having a theory of knowledge that has always emphasized the indirectness, presumptiveness, and fallibility of all modes of knowing, I cannot be (as he recognizes in most passages) the exemplar for all that he wants to use as a contrasting background in current applied social science. Nor can I claim to be unmoved by the revolutions in theory of science that he represents. For example, my willingness to use dialectic concepts in describing the scientific method may be regarded as making a major concession to Dunn's point of view.

REFORMS AS PLACEBOS

Dunn's title, "Reforms as Arguments," is a variant on my article "Reforms as Experiments." My overall title for this commentary, "Experiments as Arguments," is a variant on his, intended to connote sympathetic agreement with his general epistemology. "Reforms as Placebos" is a second variant, intended as a warning against confusing the goal of truth with political linguistic usage. Dunn and I agree that most of what most governments offer in the name of "reforms" and "new programs" are symbolic gestures designed to indicate governmental awareness of problems and sympathetic intentions, rather than serious efforts to achieve social change. Abetted by us social scientists, there has been an escalation of rhetoric in Congress and the administration so that funds needed for continuing welfare services to the needy have to be presented as new panaceas that are certain to cure the problems they alleviate. The underlying congressional intent is, for the most part, achieved when appropriations get spent locally. "Revenue Sharing" or "Problem-Specific Revenue Sharing" has long been a more appropriate description than "new program" or "reform." Because of competing needs and limited budgets, most of the few genuine novelties get drastically underfunded and have no chance of producing demonstrable effects. The legislative requirement that such so-called programs be scientifically evaluated becomes just more empty rhetoric, a token to indicate that the money is being spent responsibly, on par with the requirement of audited financial statements.

In stating this I have fallen into the popular usage of combining "rhetoric" with deceptive persuasive efforts and vacuous promises. This should not be taken as a rejection of the current revival of the ancient disciplines of rhetoric, disputation, formal argumentation, and jurisprudence as models in theory of science. The negative connotations of rhetoric are more than vulgar misunderstandings. They warn against uncritical acceptance of current practices of "symbolically mediating evidence." Conceptualizing "truth entirely as a social construction" reduces one's grounds for criticism of specific socially constructed beliefs as false consciousness, exploitative mystification, unwarranted reification, and the like (Keat & Urry, 1975). Substituting the goal of persuasion for the goal of truth, or defining truth as consensus, undermines our effective motivation for criticizing and changing the existing social order. The ontological nihilism of the two quotations from Lindblom and Cohen (1979) seems to me particularly unfortunate. Projecting onto them my otologically realistic perspective, I had sympathetically understood them as criticizing the naive–realist arrogance of those quantitative social scientists who regard their latest study as standing alone as the unambiguous and sole basis for the relevant policy decision. Granted that we can never implement "correspondence" as a criterion for testing the validity of a specific belief. Granted that we must use coherence-like criteria testing beliefs only by comparison with the implications of other beliefs. "Correspondence" still best expresses our goal of valid reference, the distinction between belief and valid belief. Once Dunn gets into his section on "Testing Knowledge Claims," he is using as much of the correspondence theory of truth as I have intended.

THE AMBIGUITY OF THE OUTCOME OF PRAXIS

Even though I voluntarily "squat in the shade of the natural sciences" (in the sense of regarding the social and physical sciences as sharing yet-to-be-fully-explicated principles of scientific inference best exemplified in the physical sciences), I applaud Dunn's opening paragraph. The social sciences we are concerned with are and should be "practice-driven." There is no purified, validated social science from which we can derive sound principles for social engineering. In those cases in which social change efforts have been inspired by a "pure" social theory, it is more true that the theory acquires validation from the outcomes of such theory-inspired practice than that the praxis acquires advance validity from the theory.

Intentionally induced change is close to the core meanings of both praxis and causality (Wright, 1971, and others reviewed by Cook & Campbell, 1979). Praxis is rightly conceived of as the optimal validation process, the optimal truth test for theories of society. Nonetheless, the outcomes of praxis or practice are profoundly equivocal and ambiguous in normal circumstances. The mode of implementation is never quite what theory called for, and these imperfections will be invoked if outcomes seem undesirable or absent. If practice seems successful, there will be interest in which aspect of the conglomerates was most responsible. "Seeming" success or "seeming" failure will be rendered equivocal because of other plausible change agents asserted to be present. The profound equivocality of praxis, be it revolutionary change or minor reforms, can be represented and clarified in argumentation, rhetoric, dialectical reasoning, and adversary process, particularly if well-funded parties in genuinely adversarial roles are permitted to participate (as in Habermas's "ideal speech community"). But these procedures do not in and of themselves remove the intrinsic ambiguity.

Logically this same equivocality exists for intentional causal interventions in the physical aspects of the world, even if not in such extreme degree. "Experiments" are deliberate intentional interventions in situations selected and contrived so as to reduce this equivocality. At their best, experiments do reduce equivocality, at least in regard to those plausible rival hypotheses articulated in the previous dialogue of competing theoretical interpretations and criticism. Even so, experiments at their best merely "probe" causal theories, they do not "prove" them (Campbell & Stanley, 1966, p. 35).

But even so again, deliberate intentional intervention greatly reduces equivocality of causal inference in contrast with efforts to infer causality from passive observation. Beginning explications as to why this is so are to be found in Bhaskar (1975) and Cook and Campbell (1979, chaps. 1 and 7).

In the physical sciences, experiments advance the argument but are never definitive. Dunn and I agree that a dialectical perspective does more justice to the history of experimental physics than does an image of the experiment as a window through which nature is seen directly. At each stage the "experimental variables" and the "outcome measures" are never "defined" for out-of-context or all-context meaningfulness. Instead, they are historically and dialectically indexical, acquiring their meaningfulness in the context of previous experiments and theories. In this important sense, *experiments are arguments* in a historical dialectic for the physical sciences and perhaps potentially for the applied social sciences.

This authentic continuity with the classic and medieval disciplines of dialectics, disputation, argumentation, and rhetoric should not lead us to undervalue the ideology of experimentation that founded the scientific revolution nor undervalue the importance of its self-conscious rejection of the disputation tradition. That older tradition was a part of a world view that saw all of the ingredients for rational understanding as already present, needing only the spelling out of implications and clarifications that verbal disputation could produce. Such scholastic argumentation had reached a stalemate (of boredom if nothing else) and it was against this sterility that Galileo, Bacon, and the others rebelled. The new experimental science was to be exploratory beyond the limits of present knowledge and was to introduce physical acts and observations so as to settle the endless verbal argumentation. The fact that the experiments and empiricism did not settle the arguments, and the fact that they did not replace, but merely augmented, the verbal disputations, should not blind us to the important advances made. Marx's call for a dialectic of praxis, not mere words and passive observations, and Dunn's call for employing "reforms as arguments" can be read as in keeping with the ideology of experimental science in this important regard.

Both Dunn and I are in favor of keeping open indefinitely the dialectic of practice through intentional efforts to change our social order accompanied by open criticism of these efforts. Our small disagreement centers around the role of the ideology of "social experimentation" in promoting this dialectic. For me, "experimentation" connotes a continuing iterative process, a self-conscious tentativeness precluding pretenses to having achieved finalization. But I join him and Lindblom and Cohen (1979) in decrying an image of social experimentation in which a single big national program evaluation or demonstration experiment settles a policy decision once and for all (Campbell, 1979a). We are also at one in regretting the miscarriages that occur when social movements supposedly committed to the dialectics of praxis try to stop this dialectic upon gaining political control.

Will a dialectic of experimental arguments ever be feasible in applied social science? I am in favor of our intentionally trying and of informing our efforts by an understanding of the social processes that have made science possible in the physical sciences. But although I favor trying, I share doubts that political conditions will ever make this possible. Even though in physics, too, the experimental data are also "mediated by assumptions, frames of reference, and ideologies," the stakes and motivational structures of political domination place the powerful participants farther from the "standards of appraisal and the incentive structure of the idealized scientific community." If, in his section, *The Experimenting Society,* Dunn is opposing political mystification, I am happy to join him. This mystification comes from claiming scientific status for partisan rhetoric that is not subjected to competitive scientific criticism. If, on the other hand, he claims to be offering a more valid epistemology that uses means other than invoking more authentic experimental and scientific perspectives, I am in disagreement.

THE "FACT–VALUE" DISTINCTION

To fill out my role as an old-fashioned epistemologist, I would like to make a case for maintaining, rather than abandoning, the fact–value distinction. Dunn's position on this is mixed, so I may be answering other young moderns and not him directly. Although for at least the past 25 years I have been antipositivist, in particular where positivists have wanted to do away with the distinction between evidence and the reality to which the evidence refers, on the issue of the fact–value distinction, I will use the analysis of the logical positivists and their predecessors, but without the value nihilism that often accompanied its expression (Campbell, 1979b). The tools of descriptive science and formal logic can help us implement values that we already accept or have chosen, but they are not constitutive of those values. Ultimate values are accepted but not justified. Most disagreements, however, are not about ultimate values but instead about mediating values. For these, the facts of the matter, the nature of the world, and the nature of human nature, are all highly relevant. The theory of biological evolution predicts that we as a species will value human survival, and that is certainly a part of an ultimate value package that enough people will accept to make it worth the effort to develop mediational values in its service. Species characteristics are, however, so little fixed that probably in our ultimate value package we need more than genetic survival per se and need something more like "human survival under humane conditions" (Campbell, 1979b). For me, at least, this would include the goal of survival with something like our present population level and urban residential patterns. In this perspective, one avoids value nihilism by assuming ultimate values or by discovering what values one has inherited and accepting them for purposes of mediational value clarification.

We are such pervasive valuers that almost none of the facts of the world can be apprised without valuational connotations, but this does not negate the fact–value distinction. It is characteristic of successful science, even in its dialectically and historically transient form, that its descriptions of the world are usable to implement widely divergent values. Thus, bacteriology can be used to control plagues or to design instruments of germ warfare. Thus, as immature a project as Adorno, Frenkel-Brunswick, Levinson, and Sanford's (1950) *The Authoritarian Personality* (which both Dunn and I treat with respect) is usable not only to design modes of reducing ethnic prejudice but also to provide an excellent guide for a fascist rabble-rouser who would like to foment ethnic prejudice. The scientists who did the basic work may have been motivated by an extreme hostility to the disease, but this strong motivation will not restrict the usefulness of their approximative scientific achievements solely to those who share their motivation.

Our predicament as social animals who must achieve our sociality without inhibition of genetic competition among the cooperators (Campbell, 1972, 1975b), and as social animals with a fundamental disposition to individual and clique selfishness, puts us in a special predicament insofar as public belief assertion is concerned. Were science designed only to guide our own behavior, then the value neutrality of our scientific conclusions would be complete. The model of the world best-suited to implementing our own values would also have the validity optimal for guiding others with different values. However, the fact that we are in varying degrees in competition with those others provides a motive to keep our knowledge of our beliefs private. This motive to achieve secrecy is inevitably characteristic of applied science, social or physical, industrial or national. The norms of science and democracy are both against such secrecy. In paying lip service to those norms, we generate public belief assertions. Individual-competitive and clique-competitive motives introduce in us a tendency to modulate these belief assertions so as to manipulate the beliefs of others in a direction favoring our interests. Dunn and I agree that in the social sciences, the norms that would inhibit this are particularly weak, and the feedback from experimental, quasi-experimental, passive descriptive, quantitative, or qualitative probes of reality are so ambiguous and inconsistent that they provide no discipline.

The role of the researcher as a consultant rather than an actor guiding his own action maximizes the belief manipulation interest in research reports. An established power structure with the ability to employ applied social scientists, the machinery of social science, and control over the means of dissemination produces an unfair status quo bias in the mass production of belief assertions from the applied social sciences. The naively idealistic scientifically trained social scientist who enters this arena unaware of the belief manipulation component to the belief assertions produced by the research establishment that he is entering may indeed become an unwitting coconspirator in this mystification. This state of affairs is one that both Dunn and I deplore, but I find myself best able to express my disapproval through retaining the old-fashioned construct of truth, warnings against individually and clique selfish distortions, and a vigorously exhorted fact–value distinction, whereas he is tempted at points to abandon these old-fashioned tools of moral righteousness and to accept or even welcome the combining of persuasive bias with descriptive accuracy. If not he, then there are others of his generation who, recognizing that published social science often constitutes advocacy of a partisan nature, accept this model for their own research. Rather than accepting for themselves the model of advocacy science focused on persuasion of audiences, out-of-power minorities would do better to maintain the traditional distinction, and expose as false the value-biased distortions of establishment-belief assertions made in the name of science. These minorities should devote their own research efforts to correcting those biases in the name of truth.

The occurrence of whistle blowing by members of bureaucratic structures gives testimony to the spontaneous occurrence of exposés in the name of truth by cogs in the establishment machine who have no ideological commitment of an antiregime nature. We have yet to fully explore the sources in human nature of the pain of continually lying, but the phenomenon no doubt exists, as Polanyi (1966a) has noted in his analysis of the role of the privileged journalistic elite in the Hungarian revolution of 1956. Such pain was very plausibly involved in the leadership in the Dubcek regime in Czechoslovakia (Campbell, 1971b). In our more pluralistic society, we would do better to expand the opportunities for whistle-blowing than to announce that all research is but persuasive advocacy. Multiple minority reports are appropriate for every program evaluation and policy study, and should be open to every neurotic and disgruntled research assistant under extended freedom of information requirements.

Dunn agrees with the importance of the problem and wants to go beyond attention to deliberate deception in order to increase the social science research community's awareness of its unconscious co-optation into partisan paradigms. This is also the critical message of cultural relativism in the Boas–Herskovits tradition and the social class relativism of Marx and Mannheim (Campbell, 1972, 1977a). This effort to make us aware of biased-paradigm co-optation is again one best done by retaining a traditional fact–value distinction; it is a matter of becoming self-critically aware of our profoundly relativistic epistemologic predicament and using this awareness in the service of a more competent effort to achieve objectivity, rather than employing it to justify giving up the goal of truth.

SUMMARY

Faced with Dunn's challenging and persuasive effort to introduce a new and broader epistemological paradigm for applied social science, I have nonetheless been able to convince myself that I should hang on to a number of old-fashioned points of view. These include acceptance for social research of a model of science shared with the physical sciences, an ideology of social experimentation, the concept of cause, a correspondence meaning for the goal of truth, a critical realist ontology, and the fact–value distinction. However, because I have long held a radical fallibilism and an emphasis on the unproven presumptions underlying science, and in this chapter have been willing to state this in the language of dialectics, I may in fact have conceded most of the points Dunn is making.

OVERVIEW OF CHAPTER 7

In this speech to philosophers of science, Campbell exhorted the scholars in this field to become more involved in providing science policy advice to the government and other scientific institutions. In order to do this, a "naturalized" epistemology must be outlined. Campbell suggests that science, as practiced, has been successful in improving the validity of its beliefs and proposes a theory of why this is so. Although biological evolution offers a plausible justification for why we can believe the processes that we observe, unobserved processes cannot be justified in this way. Justification for such processes must be extended to include a sociology of scientific validity. Then norms emanating from this theory of scientific validity should be made clear so that advice can be given. Instead of a descriptive philosophy that merely theorizes about how we come to valid scientific beliefs, Campbell advocated a prescriptive philosophy, which not only poses a theory but postulates the methods or "best practices" by which scientific validity would be enhanced. He believed that it was from the philosophers of science that an "epistemologically relevant" sociology of science would emerge.

If we put a tentative trust in the theory of evolution, we can see a real-world example of the scientific standard that we are justified in believing what we see. This provides a reliability-type justification for the idea that "knowledge equals justified true belief." Although the evolutionary argument may justify beliefs in observed objects, it does not automatically justify the unobserved scientific processes we believe to be true. Yet the progress of science is dependent on continual observations and checking of predictions, and it is the similarity standards of common sense that lead to acceptance or rejection of a hypothesis.

Campbell admitted that the advice that philosophers of science give policy makers is not justified by visual perception or is not justified empirically. Instead, the foundation and standards of their theories of science are assumed, transcending conventional science's insistence on empiricism. Still, he believed that a number of the postpositive theories of science could provide the basis for advising science policy. The theory should provide direction on how to improve scientific validity, assuming the truth of our tentative theory that present scientific practice does work to enhance validity, and that our beliefs about the nature of the world are correct.

Logical positivists accepted that it was appropriate to make generalizations from observations and believed that the visually perceived objects do exist. By this understanding of how scientists come to know, they also outlined a method for optimal scientific practice. Developing fields like psychology and sociology grasped logical positivism as a way of practicing science. This has proven valuable in many ways; however, Campbell believed that one facet of logical positivism was clearly a disaster for the social sciences. This involved operational definitions for theoretical terms. The dogma of definitional operationalism encouraged the use of a single measure of a theoretical construct for anyone focusing on the same problem. It ignored the theoretical complexity of measures and the fact that irrelevant components of the measure may have been responsible for perceived changes. It did not recognize that all measures have less than perfect validity. This definitional operationalism leads to corruption of measures. In 1959, Campbell's rejection of logical positivism was, in part, his recognition that measures contained sources of variance unrelated to the construct being measured. He also realized that scientists were mistakenly determining the validity of measures by comparing them against other equally invalid measures.

Campbell went on to list four recommendations on science policy in the area of social program evaluation that emanate from one of the competing sociological theories of validity-enhancing scientific belief change. If his advice were followed, it would engender a disputatious and mutually monitoring applied scientific community. Three recommendations were then given to the funders of pure science. Once again, the advice encourages competition and mutual criticism, and would be more supportive of studies that have the potential to result in belief change among fellow scientists.

Campbell believed that a naturalistic epistemology for science must be sociological. He argued that although natural selection would not have left us with eyes that regularly deceive us, this cannot be applied to scientific beliefs, because there are other selection processes operating in science. One cannot say that natural selection would not have left us with untrustworthy scientists. It is the social systems in science that ensure honesty and objectivity.

Finally, Campbell explained that the type of sociological theory, which led to the two sets of science policy advice given earlier, assumes that the ideology of the scientific revolution is an optimal social system for science. Its antitraditional component, however, presented a problem in his sociology of science having to do with two aspects of evolutionary epistemology—that is, retention and variation. The required variation always takes place at the expense of the required retention. Presently, the norms of science are "antitribal" (variation), yet science requires a certain degree of solidarity to preserve special traditions, and other such things (retention). There must be a balance between these two dimensions. However, Campbell was optimistic that epistemological sociologies for science could be achieved that could then provide credible, or at least plausible, advice on science policy.

SEVEN

SOCIOLOGICAL EPISTEMOLOGY

Philip Kitcher and his program committee hoped to stimulate a lively debate this evening on the role that philosophers of science might play as advisors to government and the professional institutions of scientists. It is my guess that they had two changes in mind. They wanted more philosophers of science advising in Washington's science policy decisions. They also want philosophers of science to develop those aspects of the field that are relevant to such decisions. They believe (I would guess) that not all of the scholarship of this association has such relevance, but that much of it does; that the relevant parts can still retain a genuinely epistemological focus; and that science policy advice is better done by philosophers of science rather than by any other profession (e.g., the sociology of science). Perhaps I have merely projected these beliefs on to the program committee. They are, in any event, ones I hold and hope that they share.

I must confess to drawing the inspiration of this talk from a closet role as an amateur, marginal, philosopher of science, who would love to be accepted

Campbell, D. T. (1986b). Science policy from a naturalistic sociological epistemology. *Philosophy of Science Association, 1984*, 2.

as an in-group participant—if this could be achieved without giving up cherished partisan positions on some of the issues involved.

Here are several themes I would like to get to, although not necessarily in this order: I want you to advise government (and also scientific journals, degree programs, etc.) on optimal science policy for achieving *scientific validity.* Your advice is most needed where science goes beyond the classic settings of its most impressive past practice. This may include big-instrument physics and astronomy characterized by one-instrument, one-lab, monopoly. Most important, however, are the needs for your advice when there are deliberate efforts to extend science to areas in which it is not yet effectively practiced, such as the field of social program evaluation, and also in pure sociology, political science, economics, and nonlaboratory social psychology.

Such advice will have to come from an epistemology "naturalized" at least to the extent of assuming that, in exemplary instances, science has been successful in improving the validity of its beliefs. It will have to involve developing a theory to explain how this could be so, employing a "hypothetically normative" stance in using this theory to generate advice. A naturalized epistemology, in which I participate, has achieved some plausibility in justifying the reliability of beliefs based on the visual perception of ordinary objects, by passing the justificatory buck to biological evolution (Campbell, 1959, 1974a. The 1959 paper is the most explicitly hypothetically normative, as signaled by its title.) This cannot be directly extended to justifying scientific beliefs in unobservable processes and entities. That extension will have to involve a plausible sociology of scientific validity (Campbell, 1984e; Giere, 1984, 1985a, 1985b). It is from you philosophers of science—rather than sociologists of science—that such an epistemologically relevant sociology of science is most apt to emerge. Taking this sociology as hypothetically normative will provide the most important advice to science policy makers.

1. HYPOTHETICALLY NORMATIVE NATURALISTIC THEORY OF SCIENCE

Quine ends his great "Epistemology Naturalized" (1969b) too soon. He ends it saying that all we have left to do is *psychology.* He fails to state explicitly in what sense a very specialized type of psychology can be *epistemological*—that is, that it can be relevant to the traditional issues of how knowledge or justified belief might be possible. This lack he soon supplies, as in *The Roots of Reference* (1974). The general strategy employed by him and others doing naturalistic epistemology I wish to make explicit under concepts like *hypothetically normative* and *hypothetically foundational.*

Quine's "The Nature of Natural Knowledge" (1975) provides the epistemological relevance in explicit form. And it turns out that evolutionary biology, rather than psychology per se, carries the justificatory import.

> Our question, "Why is science so successful . . ." [is to be] taken . . . as a scientific question, open to investigation by natural science itself. . . . Individuals whose similarity groupings conduce largely to true expectations have a good chance of finding food and avoiding predators, and so a good chance of living to reproduce their kind. . . . I am not appealing to Darwinian biology to justify induction. This would be circular, since biological knowledge depends upon induction. Rather, I am granting the efficacy of induction, and then observing that Darwinian biology, if true, helps explain why induction is as efficacious as it is. (Quine, 1975, p. 70)

Thus for beliefs supported by the visual perception of ordinary objects, in the modern version of "knowledge equals justified true belief," tentative trust in the theory of evolution provides a reliability-type justification.

This biological evolutionary argument does not extend automatically to justifying scientific beliefs, for these go far beyond the beliefs in objects given in ordinary perception. Although for trust in simple intuitions of rational inference the justificatory buck may also be passed to biological evolution (Burks, 1977; Ellis, 1979; Sober, 1981), biological natural selection could not have provided the trusted reifications of unobserved processes and entities added to our ontology by modern science, even though these new reifications may turn out to be in part dependent on trust in the givens of vision. Again from Quine, "For all of their fallibility, our innate similarity standards are indispensable to science as an entering wedge. They continue to be indispensable, moreover, as science advances. For the advance of science depends on continued observation, continued checking of predictions. And there at the observational level, the unsophisticated similarity standards of common sense remain in force" (1975, p. 71; see also Paller & Campbell, 1989).

Not much of the advice I want philosophers of science to give to science policy makers, science funders, editors of scientific journals, and designers of PhD programs, will come from the epistemology of visual perception per se, but rather from a theory of science that is hypothetically normative, hypothetically foundational. With Quine, it will have to hypothesize the efficacy of science in major instances at least, and will have to hypothetically use aspects of that science reflexively to explain science's success. Unfortunately, for crucial aspects of the problem, the theory employed will not for the time being have the persuasiveness of the contemporary neo-Darwinian orthodoxy. Nonetheless, within the philosophy of science today there are a variety of postpositivist alternatives, which—if interpreted as theories of scientific efficacy—could provide the hypothetically normative basis for advising science policy.

Even though much the same content is employed, the shift of stance from philosophy of science to theory of effective scientific practice should be made self-conscious and explicit. This can be hypothetically and mediationally normative: If one wants to improve the validity of beliefs; if we are right about our tentative theory of validity-enhancing scientific practice; and if our tentative beliefs about the general nature of the world-to-be-known are correct; then here are the procedures one should follow.

2. THE POSITIVISM OF OPERATIONAL DEFINITIONS FOR THEORETICAL TERMS VERSUS FALLIBILIST REALISM

In their pervasive admiration for the achievements of science at its best, logical positivists had already made the crucial assumption of the efficacy of induction as practiced by science. In the early shift from an uninterpreted phenomenalism to sentences about the ordinary perception of real objects, they implicitly conceded that ordinary perceptual reifications were generally competent. These two "beggings of the problem of induction" were enough to have turned their reconstruction into a descriptive theory of optimal scientific practice. It is my impression that implicitly, if not explicitly, major modern philosophers of science have made this shift, or else never accepted the logical positivists' claims to philosophical purity and advisory irrelevance for science practice.

It was inevitable that underdeveloped fields such as psychology and sociology should take logical positivism to be a source of advice on the practice of science. Some of this importation I regard as still valuable, for example the Bergmann–Spence exhortation against R–R laws—that is the need of separate operations to measure the learned response patterns and the environment learned. I also believe that practicing scientists need some version of the fact–theory distinction even though facts are laden with other theories. But the major import from logical positivism I regard as an unmitigated disaster, and that was the advice to employ designated *operational definitions for theoretical terms.*

Operational definitions were, in fact, already being recommended by physicist Percy Bridgman (1927) and his psychologist friend Edwin Boring in the 1920s. (Boring, in 1923, argued that the intelligence test itself was the definition of intelligence. Thank goodness Louis Terman did not take him seriously but went ahead revising the 1916 Stanford–Binet, to remove its "known" imperfections, known in spite of there being no better measure to check it against, no criterion measure, no truer definition.) Once logical positivism arrived in the United States in the 1930s, its emphasis on meter readings became fused with definitional operationalism, and for us sociologists and psychologists, at least, it was the leading practical expression of logical positivism.

This dogma led to the designation of a specific measure as the favored or solely necessary dependent variable and to the recommendation that all researchers in a problem area use the same one measure. It led to suppression of the obvious theoretical complexity of measures and of the frequently plausible argument that the measured change was a result of irrelevant components of the complex. It discouraged support for exploring multiple approaches to measuring "the same thing" at the level of funding, dissertation approval, and publication. It repressed awareness (of trivial importance for the physical sciences, but flagrantly relevant for the social sciences) that all available measures have less than perfect validity, being systematically biased.

This definitional operationalism persists long after the substantial revision or rejection of positivism within the philosophy of science. It persists most perniciously in social policy science, in the accountability movement, or in managerial control efforts employing single explicit quantitative criteria. All of this leads to the corruption of these designated criterial measures, resulting in social processes that correspond, as it were, to breaking the glass of the galvanometer and moving the needle to the reading that will generate the administrative decision that one seeks. The sociology of bureaucracy is replete with such illustrations (reviewed in Campbell, 1979a, 1984a; Ginsberg, 1984), and we ourselves connive in such "corruption" in our ratings of students and employees.

Beginning in 1959 (Campbell & Fiske, 1959; and also Campbell, 1960), two convictions led me to abandon logical positivism for a fallibilist or critical realism. The first conviction was that the best of measures were equivocal because of confounded sources of variance; the other conviction was that the best practice in social and psychological measurement required checking the validity of intentionally designed measures against other measures at least equally invalid. Although the scientific practice resulting has been much more discouraging or humiliating than definitional operationism (Campbell & O'Connell, 1982; Fiske, 1982), I take that outcome as reflecting the genuine state of the psychological science of individual differences rather than as a result of the inappropriateness of the fallibilist–realist theory of optimal scientific practice on which the criteria of test validity were based (Wimsatt, 1981).

Thus interpreted, for methods of measurement, the realist–positivist debate has hypothetically normative implications for scientific practice that make a difference. Not so, however, the current realist–antirealist debate about the status of the theoretical entities, forces, and processes of physics. Here the most articulate antirealists, such as van Fraassen (1980) and Laudan (1981a, 1984), approve of the scientific practice of positing unobservable particles and forces, of quarreling about which of such posits best fits the evidence, and of trying to design experiments in which competing posits predict different observations. Were their theories of science to be made hypothetically normative, generating recommendations to scientists for optimal justificatory *practice*, they would not disagree with fallibilist realists. (See Paller & Campbell, 1989, for more details.) At the level of advice to scientists, Karl Pearson at least clearly recommended an antirealist, meter-reading phenomenalism, advising that no effort be wasted quarreling about unobservables. But did Ernst Mach object in principle to positing and disputing about unobservables? Laudan (1981b, pp. 202–225) convinces me he did not.

This discussion of the currently exciting realist–antirealist debate should serve to remind us that it is only a portion of your concerns that are relevant to advising science policy. Although many of you are doing theory of science in a relevant way, and I want more of you to become engaged in it, and although I believe that your scholarly traditions are most appropriate to the task, I do not want to lose the distinction that makes advice on science policy a matter for a *hypothetically normative* naturalistic epistemology.

3. SOCIOLOGICAL THEORY OF VALIDITY-ENHANCING SCIENTIFIC BELIEF CHANGE

Let me introduce a change of pace by listing some of the advice that I would like to see given our governmental science policy makers. I believe this advice comes from one of the potential competing sociological theories of optimal scientific belief change. The following four recommendations apply to the social program evaluation area. All go counter to current practice.

1. For a new program or policy, give up the demand for a nationwide, once and for all, uniform evaluation, delegated to a single-evaluation contractor. Substitute instead support for a heterogeneity of programs, each evaluating themselves until they feel they have a package worth others borrowing, and support those who borrow to cross-validate the efficacy—that is, adopt a "contagious cross-validational model of program dissemination and validation."

2. Give up the notion of a single new evaluation designed to support a single administrative decision regarding expanding or curtailing a program. Substitute for this the development of a disputatious mutually monitoring, applied scientific community that will advise governmental decisions on specific programs from its general wisdom about research in the problem area.

3. Where the new program and its pilot study cannot be decentralized into large numbers of local innovators and independent adopters, split the pilot

study into two or more independent experiments, independently making the many ad hoc decisions about implementation and measurement. The staffs of these two implementations would then be allowed full details of the other and would be funded for mutual criticism. Also, legitimize whistle-blowing on the part of all members of each research team by creating the duty of, funding for, and data access for dissenting research reports.

4. Avoid using the program evaluation methods and measures to evaluate the current administrators and staffs of the program-implementing agencies. Instead, evaluate only program alternatives available to these staffs.

For science funders of pure social science, or for science in general, I would like to see the following advice given, at least some of it contrary to current practice.

5. For currently important problems, never let a single laboratory have the sole funding. Deliberately fund competitors.

6. Be wary of funding research being done by scholars in fields in which no other scholar will challenge their conclusions. Ask each grant applicant, "If you are wrong about this, who specifically will notice? What will they do about it?" For a grant or two, exploratory diversity would be gained by funding a scholar who answered, "No one would notice" and "Nothing would be done." But in general, science cannot be furthered without funding a competitive, disputatious community focused on the problem. It is my hunch that it is in the frequency of the answer "No one would notice," that the social sciences and physical sciences would differ most. I feel that social science funders need increased recognition that science advances through competitive, disputatious communities of scientists who find it important to keep each other honest and are free to disagree with each other without economic jeopardy. As a part of this, funders should not disapprove of fads among scientists as to hot topics, for these fads provide the crucibles that validate scientific innovations.

7. Fund most those problem areas in which scientists have the means for changing each other's beliefs—that is, for convincing a fellow scientist that he or she was wrong. Where funds can do so, provide the community with the capacity for convincing demonstration. Put this capacity into several inde-

pendent facilities. Give more importance to the capacity for such belief change than to the social importance of the problem, where the goal is the development of valid science.

Some of you will find some of this advice obvious, some obviously wrong. I introduce it to make more concrete the policy relevance of theories of scientific validity, and the value of philosophers of science doing the kind of research and mutual criticism that would test and refine such a theory. For the examples given, both applied and pure, a sociological theory of scientific validity is involved. Even though such a theory is sociological, I believe that philosophers of science are more apt to develop it than are the sociologists of science.

Now I would like to offer a condensed theoretical argument that a naturalistic epistemology for science must be sociological. The stance of the modern evolutionary epistemologist can be epitomized: "Natural Selection would not have left us with eyes that regularly mislead us." Thus, reference to natural selection can be used to "justify" visually supported beliefs in the formula "Knowledge is justified true belief," in the weakened interpretation of "justification" used by all modern epistemologists except skeptics.

This general program of evolutionary epistemology can only with great difficulty, if at all, be extended to the social processes producing scientific belief. I will give two brief explications of the problem. In the evolutionary epistemology program, any "validity," or usefully competent reference, is attributed to the selection processes that weed out, rather than to the generation processes producing the variations. We know of so many selection processes in the generation, publication, teaching, and believing of scientific truth claims that are irrelevant or contrary to improving the competent reference of beliefs that it becomes hard to argue for a dominant role for "Nature Herself" in the selecting. This is in contrast to the case we can make for Her role in the biological evolution of the eye and brain.

A reflexive use of biological evolutionary theory provides a complementary perspective. Both Cartesian and evolutionary providentialists could plausibly say "(God) (Natural Selection) would not have given us untrustworthy eyes." But even if they noted that the social system of science requires great (albeit selective) trust of fellow scientists, neither the old providentialist nor the evolutionary epistemologist would be apt to find it plausible to argue that "(God) (biological Natural Selection) would not have given us untrustworthy fellow scientists." If we can in fact often validly trust fellow scientists, this is because of culturally evolved and fragile social systems, not because of innate honesty and objectivity.

The problem is a result of the fact that for us vertebrates there is genetic competition among the cooperators (Campbell, 1983). For the social insects whose cooperators are almost completely sterile, one can say: "(God) (Natural Selection) would not have given an ant worker untrustworthy scouts." But for us social humans, the belief assertions or public truth claims we make have important utilities for us other than optimally guiding our own behavior. We have selfish (including nepotistic) interests in what others believe and in what others believe we believe, often motivating belief assertions that differ from those that would optimally guide our own behavior vis-à-vis the objects that are nominally the referents of the truth claim. This becomes particularly so when "secondary groups" rather than "primary groups" are involved. That is, the problem may not have been acute for a stage in human social evolution in which inbred tribes of 100 or so were in intense close competition with other similar tribes, with no individual being able to change tribes successfully. But it is acute in all complex social organizations including science.

In developing a sociological theory of validity-enhancing belief-change in science, attention should be paid to the work of sociologists of science. Merton (1973), Zuckerman (1977), and others, such as Joseph Ben David, contribute much of value for such an epistemologically relevant sociology of science. So do the younger participants in the field, even though they would reject such an enterprise as inconsistent with their vow of agnosticism with regard to the truth of the beliefs they study, comparable to the agnosticism of an anthropologist of religion. (See Giere 1984, 1985a; and also Campbell, 1986c, for reviews.)

Although this is not the place to get into my own tentative sociology of validity-enhancing scientific belief-change (Campbell, 1986c), perhaps I owe you a paragraph on what type of a theory leads to the two sample sets of science policy advice offered previously. It comes from taking the ideology of the scientific revolution as a contingent hypothesis about an optimal social system for science. Habermas's (1970a, 1970b) "ideal speech community" made into a contingent conjecture yields something similar. In *The Open Society and Its*

Enemies (1952), Popper devoted some seven pages to such a sociology. In *Personal Knowledge* (1958), Polanyi devoted a long chapter to it, plus several later essays (Polanyi, 1966b, 1969, chaps. 4, 5, 6, 13).

Because my evolutionary epistemology stresses retention fully as much as variation, I have some problem with the deep antitraditional component to that ideology. This I explain away with a sociobiological theory of religion, too arcane to be summarized here (Campbell, 1983, 1984e). But mention of the evolutionary model provides an entree into a pervasive problem for my sociology of science. For example, the required variation is always at the expense of the also required retention. Some compromise is optional. Thus to generate policy advice on this dimension, the theory needs finely tuned parameters. The social norms of science are unique among self-perpetuating groups in that they are "antitribal" (Campbell, 1979c). Yet, sciences need enough ordinary tribal solidarity to preserve and propagate their special traditions, to get their journals published and archived, and to explore thoroughly their theoretical alternatives. Thus were scientists to fully live up to their radically antitribal public norms, the social structure necessary to preserve and improve scientific truth would be lacking. But in spite of such intermediate optima on many dimensions, I do believe that we can achieve epistemological sociologies for science from which meaningful advice on science policy can be derived.

OVERVIEW OF CHAPTER 8

Why is it that the physical sciences arrive at theories that are replicable and widely accepted, whereas the social sciences seem to be floundering with underdeveloped theories and research that provides conflicting findings? In this chapter, Campbell explores the characteristics shared by the physical and social sciences and poses reasons for their differential success. He believed that these differences could be explained in terms of the social, psychological, and ecological requirements for being scientific.

Campbell saw hermeneutics as important to the social sciences. Hermeneutics is the science devoted to the study of specific texts to provide improved interpretations. He lauded several principles emerging from this community of scholars and advocated validity-seeking hermeneutics. He disagreed with hermeneutic nihilism, which encourages new interpretations, because the goal of these interpretations is not to increase validity. The physical sciences were seen as hermeneutic communities. They were more successful than the social sciences because they shared a greater number of facts that were agreed on.

One of the goals of this chapter is to unify the "hard" methodologies borrowed from the physical sciences and the "soft" methodologies borrowed from hermeneutics. Campbell explained how hermeneutic scholars deciphering ancient texts begin with a guess about the overall message and then in repeated cycles continue to decode the text and improve the interpretation. Eventually, the scholar believes he or she understands what the original writer intended, although this understanding is based on many fallible elements. If one accepts the validity of this process, one would deny the empiricist's claims that the validity of our knowledge cannot be greater than the data on which it is based, because that would imply that the data were perfectly reliable and valid and could not be improved. Closer to hermeneutics, the postpositivists believe that reality can only be probed, yet the validity of our knowledge can be enhanced by applying many fallible measures.

Any theory attempting to explain how science increases the validity of its knowledge must include a sociological component. Both the scientific method and the norms of the scientific community have evolved throughout the years. The fact that these have been socially constructed does not necessarily make them invalid, because natural selection could have eliminated other methods of construction that were less effective. There are many components of the sociology of science, called the *strong programme* or *social constructivist,* with which Campbell agreed, and in this chapter, he listed eight of these components. He was ambivalent about the point of whether sociologists of science should be agnostic regarding the truth of the beliefs and the optimality of the belief–change system. Although he believed that the social processes of science that are used to increase the validity of the beliefs were superior to those of religion or politics, he felt that the scientific method needed to be formulated and theory of science further studied and described.

The new relativists rejected Campbell's "epistemologically relevant internalist theory of science" (ERISS) agenda, which sought to ascertain the best way for scientists to convince others to change their beliefs to more accurately reflect the referents. Epistemological relativists would see fit to explore this problem; however, those who are ontological nihilists would see no reason to try to convince others of errors in their beliefs. Campbell reflected that the agnosticism of the new relativists would not encourage them to speculate about unobservables.

For those like Campbell who hold a naturalistic epistemological perspective, the task of studying the best way to foster belief change among scientists is an important one. To meet the challenge of the relativists, it must be demonstrated as plausible that the social processes of science lead to validity-enhancing belief change. Any sociologist of science who believes that nature has even a small role in scientific knowledge would be interested in the sociological and psychological processes of how that occurs. Although any theory of how this comes about would be fallible, it is still of interest to ask what system would maximize the role of the natural world in constructing our knowledge.

Although he accepted the social construction of scientific knowledge, Campbell believed that the winnowing process in science led to acceptance of the few constructions that referred most competently to their presumed ontological referents. Yet sociologists of science have observed that social negotiations, rather than factual proof, are responsible for imposing order on inconsistent or inconclusive findings. This sheds doubt on the reputed objectivity of science. Campbell believed that an expanded sociology of idea winnowing was necessary.

Because Campbell believed it was epistemologically relevant to examine belief change, he offered his views on how mutual trust and honesty among scientists is possible. He began by discussing the disputatious community of truth seekers and how the "organized skepticism" of this group could enhance the validity of the beliefs about the physical world. He pointed out that a version of the fact–theory distinction must be a part of a validity-enhancing sociology of science. Agreement on facts is vital if the science is to be successful, and the social sciences lag the physical sciences in this regard. This is, in part, a result of the referential ecology of the social sciences. Replicability is another feature of this validity-seeking sociology of science, because this aspect ensures the social control that keeps the scientific community honest. This is not highly regarded in the social sciences, in which those attempting replications find their work unpublishable.

Another step in the successful sciences is to relate facts to one another. When simple laws are found to work, more complex theories are posed. In order to do this, these laws must transcend time and place. If theories are to be argued, plausible rival hypotheses must be offered.

In order to rule out these alternative explanations, cross-validation across persons, times, or instances must be carried out. Experimentation and quasi-experimentation are designed to rule out the rival hypotheses that have been raised. Unfortunately, experimental results can never be proven and are always disputable. Even when experiments are flawed, however, they can still be useful by ruling out rival hypotheses. By using methods such as the 1 to 99 doubt–trust ratio and extinction of rival hypotheses, the scientific community attempts to achieve consensus.

Because a scientist's work is evaluated almost exclusively by other members of the profession, the scientist can assume that those critiquing the work share the same underlying set of scientific standards. In the physical sciences where there is less contact with the society in general, a scientist can choose the problem to be studied based on the likelihood of solving it and the availability of the necessary tools. This is a luxury not enjoyed by the social sciences where the problems are urgent and solutions are difficult. Finally, in order for an area of inquiry to proceed, there must be a number of scientists interested in a given problem who feel that progress is being made. In the social sciences, there are many areas that lack this critical mass, begging the question, "If you are wrong, who will notice?"

Based on these features of a validity-enhancing sociology of science, the chapter ends with recommendations for social science policy. Although the social sciences face some problems not shared by the physical sciences, Campbell believed that it was possible for the social sciences to improve the validity of its theories and achieve the successes necessary to keep scientists toiling in the field.

EIGHT

SOCIOLOGY OF SCIENTIFIC VALIDITY

The conclusions of this chapter may seem old-fashioned and scientistic: There are social, psychological, and ecological requirements for being scientific that are shared by successful physical sciences and unsuccessful social sciences. The relative lack of success of the social sciences, as well as possibilities for improvement, are understandable in terms of these requirements. Yet these conclusions come from a point of view that shares most of the assumptions of the avant-garde and humanistic wings of our conference: antipositivism; antifoundationalism; epistemological relativism; historicism (recognition of the historically dialectic indexicality of all scientific concepts); and the acceptance of the paradigm-laden presumptiveness of judgments of scientific progress, the theory-ladenness of scientific "facts," and the discretionary nature of theory choice.

Campbell, D. T. (1986c). Science's social system of validity-enhancing collective belief change and the problems of the social sciences. In D. W. Fiske & R. A. Schweder (Eds.), *Metatheory in social science: Pluralisms and subjectivities* (pp. 108–135). Chicago: University of Chicago. Used with permission.

I am self-consciously using the physical sciences as the model for the social sciences, but as I hope the following discussion of hermeneutics makes clear, I am borrowing from physics the social system of belief change rather than methods or theories. I believe that there are fields in which science can be practiced but in which it could not have autonomously developed. Meteorology may be one, rainmaking by cloud seeding another, long-term ecological effects of industrial chemicals still another. This group probably also includes psychology and the social sciences. Successful science in these areas must be achieved by a cultural colonization in which the social norms and social system of science are successfully transferred. Success experiences are required to sustain such colonies.

COUNTERPOINT ON HERMENEUTICS

The concept of hermeneutics appears throughout this volume. I join those who assert the relevance of hermeneutics for the social sciences. The hermeneutic principles that attract me are those emerging from communities of scholars persisting over several generations and achieving what they regard as improved interpretations of specific texts: principles such as the hermeneutic circle (cycle, spiral), part–whole iteration, the principle of charity (assumption that the text producer is rational and honest), contextual coherence, extension of context, thick description, contrast indexicality, a dialectic of guess-and-criticism, and so on. Habermas identified this tradition as *hermeneutic objectivism* and offers an antifoundationalist version that might better be called *validity-seeking hermeneutics,* merging with his *hermeneutic reconstructionism* (1983, pp. 251–261, n. 8). What I join him in rejecting is the currently fashionable hermeneutic nihilism, in which validity of interpretation is rejected as a goal. Like the nihilists, I applaud the achievement of multiple radical new interpretations, but only because I see these as the inevitably wasteful route to a potential future consensus on a more valid interpretation. The sociology of science that I attempt is hermeneutic in almost all of Habermas's senses, emphasizing intentional communicative acts and intentional interpretative efforts and presuming the rationality and communicative intent of the communicators. More clearly than Habermas, I emphasize that the physical sciences have also been hermeneutic communities, more successful for reasons of referential ecology than for reasons of method. My title for this chapter could refer equally well to "scholarship's" social system of validity-enhancing belief change as to "science's."

Hermeneutics arose out of philology, out of Homeric and classical Greek as well as biblical studies. Although biblical studies currently represent a hermeneutic community successfully achieving what are perceived to be better and better interpretations, there is a sense in which Homeric studies would have been a less misleading basis. Biblical scholars had, in addition to the goal of interpreting the text's original meaning, the additional burden of believing that the texts were both divinely inspired and eternally relevant. Each generation of preachers used the same old texts to speak to the changing religious needs of its day. This is probably an important background motive for Bultmann's historical–ontological relativism (1956), in which the "true" interpretation differed for each generation of interpreters. From this developed a still more "sophisticated" epistemology denying the goal of truth. Of course, Bultmann and his followers (1956) have instead emphasized the changing historical, contextual, presumptive, interpretive frame unique to each generation of interpreters (paradigm-embeddedness à la Kuhn, 1970). But granting this predicament, the goal of understanding the original meaning need not have been given up. This ontological relativism now dominates social scientists' and philosophers' use of hermeneutics. Paradoxically, the community of biblical translation and interpretation has continued to focus on the intended meaning in the original context. They testify to a collective experience of successful progress plausibly based on a cumulative tradition with increasing breadth of context and range of articulated alternative interpretations.

One of the major contributions of the volume edited by Fiske and Schweder (1986) may turn out to be in helping unify the "hard" and "soft" methodologies now being recommended to the social sciences. A useful tactic may be to recognize convergences of efforts in postpositivist methodology from the hard-science camp, on the one hand, and central themes in the hermeneutic tradition on the other. For this purpose, I will extend my discussion of hermeneutics a little further, focusing on the joint rejection of the atomistic–foundationalist component of empiricism. This was the claim that the validity of derived knowledge and theory could never be greater than the individual data on which they were based, imputing incorrigibility to the data in the process. Let me begin with two quotations:

> A scholar is deciphering an archaic text. On first reading he gets only fragmentary hunches which he forces into a guess at the overall direction of the message. Using this, he goes over it again, decoding a bit more, deciding on plausible translations of a few words he has never encountered before. He repeats this hermeneutic cycle or spiral again and again, revising past guesses and making new ones for previously unattempted sections. If he is being successful, and has extensive enough texts to properly probe his translation hypotheses, he arrives at such a remarkable confidence that he can in places decide that the ancient scribe made a clerical error, and that he, the modern, knows better what that ancient intended than what the ancient's written text records. I believe such things happen, and often validly so. It well illustrates this holistic dependence on fallible elements, since any part of the text could have been a clerical error. (Campbell, 1977a, pp. 97–98)

> Both psychology and philosophy are emerging from an epoch in which the quest for punctiform certainty seemed the optimal approach to knowledge. To both Pavlov and Watson, single retinal cell activations and single muscle activations seemed more certainly reidentifiable and specifiable than perceptions of objects or adaptive acts. The effort in epistemology to remove equivocality by founding knowledge on particulate sense data and the spirit of logical atomism point to the same search for certainty in particulars. These are efforts of the past, now increasingly recognized to be untenable, yet the quest for punctiform certainty is still a pervasive part of our intellectual background. A preview of the line of argument as it relates to the nostalgia for certainty through incorrigible particulars may be provided by the following analogy. Imagine the task of identifying "the same" dot of ink in two newspaper prints of the same photograph. The task is impossible if the photographs are examined by exposing only one dot at a time. It becomes more possible the larger the area of each print exposed. Insofar as any certainty in the identification of a single particle is achieved, it is because a prior identification of the whole has been achieved. Rather than the identification of the whole being achieved through the firm establishment of particles, the reverse is the case, the complex being more certainly known than the elements, neither, of course, being known incorrigibly. (Campbell, 1966a, p. 82)

My own most concrete success in a now-seen-to-be-hermeneutic enterprise (Campbell, 1964) comes from work on cultural differences in optical illusions (Segall, Campbell, & Herskovits, 1966). In this study, the comprehension checks were formally similar to the measurement items, differing only in degree. In the language of the title, there was no formal way of "distinguishing differences in perception from failures of communication." Yet using what I now recognize as the hermeneutic principle of charity, assuming a basic human similarity, we not only demonstrated cultural differences in perception but also measured their direction and extent. Had the differences been too large, we could not have distinguished them from failures of communication. Hollis (1967) used a remarkably similar argument for the problem of cross-cultural understanding of ritual language.

RELATIVIST SOCIOLOGY OF SCIENCE

Any theory of how science could produce beliefs of improved "truth," or competence in reference, will have to be sociological in considerable part and will have to be epistemologically relativistic. Both scientific method and the moral cultures of scientific communities will have to be seen as products of cultural evolution, rather than being the products of an exercise in pure logic. Scientific method and the "proper" norms of science will have to be seen as empirically based, empirically winnowed, quasi-scientific hypotheses—that is, they have the status of contingent truths rather than (or in addition to being) analytic truths. Even if there were analytically certain rules for the transmission of validity from beliefs to derived beliefs, they could only be discovered by a biological trial and error of genes controlling brain development or by a comparably presumptive, nonclairvoyant winnowing of individual and socially transmitted hunches. The persisting debates over what it is to be rational or logical support such a view. Recognizing that both the primary and secondary qualities are constructed by the brain–mind does not necessarily render them invalid if natural selection has operated to give us a well-winnowed subset of many competing constructions. Similarly, to see the explicit rules and implicit norms of science as socially constructed processes does not preclude their validity. To hypothesize their validity, however, requires the specification of the winnowing processes operating over them and competing social constructions in such a manner as to plausibly select for validity.

An impressive new movement has been started by the vigorous young school of sociologists of science centered in Britain and Western Europe designated by Bloor (1976) and Barnes (Barnes & Bloor, 1982) as *the strong programme* (see Shapin, 1982, for a convenient review), by Collins (1981b) as the *empirical programme of relativism,* and by Latour and Woolgar (1979) and Knorr-Cetina (1981) as *social constructivist.* Many of the slogans of this movement are ones with which I fully agree.

1. Social–causal analyses of belief adoption, belief retention, and belief change are just as necessary and appropriate in explaining why past scientists adopted beliefs their science now regards as true as for adoptions of beliefs now regarded as false. The new relativists, as I shall call them, rightly criticize Mannheim for exempting the physical sciences and mathematics from his sociology of knowledge and rightly criticize those sociologists of science who have found sociological analysis only appropriate for "deviant" science, exempting the "valid" changes in theory.

2. The Quine–Duhem observation on the equivocality of "factual" (experimental, observational) falsifications or confirmations of theoretical predictions is an unavoidable predicament in all scientific belief change. This equivocality allows for extrascientific beliefs and preferences to influence the inevitably discretionary judgments involved. No certain proof, logical or observational, is ever available or ever socially compelling.

3. In explaining why scientists in the past adopted a belief currently regarded as true, sociologists of science cannot legitimately use this "truth" of the belief as an explanation of why it was adopted. (This argument follows from the preceding point and from the acknowledgment by the scientific community that the best theories are only transiently true.)

4. Political power within the scientific community, based on prestigious professorships, journals, funding committees, placement networks, power in the larger society, and so on, often leads one scientific theory to gain or retain acceptance over equally plausible theories. Similarly, social–ideological commitments (national, political, religious, economic self-interest, etc.) often influence the discretionary choice involved in scientific belief in "facts" or theory, unconsciously if not consciously.

5. Scientists do not live up to the so-called norms of science—for example, neutrality, objectivity, and sharing of all information.

6. When one looks closely at the laboratory culture in which beliefs about scientific facts are produced, the view of the laboratory as revealing nature in any direct way becomes less tenable the closer the inspection. Social persuasion and selection processes are involved that are grossly underrepresented in published articles, scientific textbooks, the popular imagery of successful science, and even practicing scientists' image of the certainty of science in general.

7. Sociology of science should be pursued in union with the sociology of knowledge, not as an independent, unrelated tradition.

8. Sociology of science should focus on scientific belief and belief change, rather than relegating these to the scientists, beyond the purview of the sociologists of science.

I am ambivalent about one claim of the new relativists and disagree with another. The first is that the sociologist and anthropologist of science, like the sociologist and anthropologist of religion, should be agnostic about the truth of the beliefs and the optimality of the belief–change system that is studied. One should not be a believer or a prejudiced disbeliever. History and current practice in both astronomy and astrology should be studied with equal trust and respect. This view is also accepted by some leading historians of science. For the scientific sociology of science, this principle exemplifies the Mertonian scientific norm of neutrality.

My own point of view on this issue is ambivalent (Campbell, 1979c, 1985, 1986b). On one hand, I approve of this point of view for the study of religion. On the other, I must confess to being a believer in the superiority of the social processes of science for validity-enhancing belief change to those of religion, governmental politics, and cultural evolution more generally. But I do not believe that the unique and epistemologically relevant aspects of the social processes of science have been adequately described or that "the scientific method" has been adequately formulated. Furthermore, from such fragments of theory of science as I have, I judge that the development of this social theory of science will be furthered by vigorous disagreements with agnostic sociologists of science. Harry Collins provided a convenient example. In a triumphant summary of the premises, achievements, and conclusions of his school, he asserted that "the natural world has a small or non-existent role in the construction of scientific knowledge" (1981b, p. 3). In his own empirical work, he has documented how far from the ideal of replication and cross-validation the actual practice of science is. Such studies discipline my own efforts at a social theory of scientific competence by challenging the dependence I and my predecessors place on the successful replication of experiments.

Moreover, I feel that sociologists of science who are active disbelievers in science (who endorse the nonexistent role in Collins's statement) can play an important part in disciplining a competent theory of scientific competence if, despite competing perspectives, we remain in detailed communication with each other, our refutational effort accompanied by acceptance of the hermeneutic *principle of charity:* "They too are intelligent dedicated scholars like us, dedicated to not deceiving themselves or others." If some, such as Feyerabend (1975), affect the clown's role, it is only in the service of a more profound honesty, insight, and understanding. If some deny the goal of "truth," it is only because of the higher truth that no such goal is possible.

With almost complete unanimity (for a partial exception, see Gieryn, 1982, 1983), my friends in the new sociology of science reject the task of elaborating the "epistemologically relevant internalist theory of science" ("ERISS," Campbell, 1981) that I have on my agenda. What I see as needed are speculations, mutual criticisms, and theory-probing investigations on what sort of social system of belief communication among scientists would optimally foster belief change to adequately reflect the referents of the belief. This, of course, assumes that an external world exists independent (or partially independent) of such beliefs. Epistemological relativists who stop short of ontological nihilism should be able to speculate and assemble arguments on this issue. Ontological nihilism as a systematic position would seem to undermine curiosity and the motivation to persuade others of the errors in their beliefs. But the new relativists are too agnostic (or too busy and successful in their present undertakings) to undertake the ERISS agenda (Campbell, 1981). Their agnosticism extends to an atheoretical positivist asceticism that allows no speculation about unobservables.

For those of us coming from the naturalistic epistemology tradition, those of us still concerned with a theory of science that shares an agenda with the older philosophy of science, those of us fascinated by past successes of science, the ERISS agenda seems to me unavoidable. We must meet the challenge of the relativists by making it plausible that some version of the social processes of science would lead to validity-enhancing belief change. In doing so, we must take seriously what sociology and anthropology have taught us about belief-transmitting communities of all types and what sociologists of knowledge and science have taught us about the more specialized and dependent subgroups and institutions identified as scientific.

To refer back to the quote from Collins, those sociologists of science who tentatively judge that "the natural world has [at least] a small . . . role in the construction of scientific knowledge" need sociological and psychological explanations about how that partial role comes about. Scientific knowledge is a social construction by scientists, even in those aspects in which the natural world plays a role in the construction. Once one has begun on this sociology, it does not increase the ontological assumptions to ask under what social system of belief communication and target characteristics the role of the natural world would be maximized. Such theory will be fallible, corrigible, presumptive, and contingent, as all science is. Nonetheless, most postpositivists in the theory of science have already made the assumptions required. It is on the basis of fragments of such a theory that I hope to compare the physical and social sciences.

Here is a more specific illustration in a setting already used by the new sociologists of science: MacKenzie and Barnes (1975, 1979) have described how the Batesonian–Mendelians wrested dominance within the biological community in Britain from the Galtonian–Pearsonians around 1910. Their explanation is primarily in terms of ideological class and clique membership in the larger society. They mention many instances of mutual scientific criticism in correspondence, public presentations, and publications. They do not explain fully the sociology of these processes or raise the question of the role of social communication in the social persuasion process. Without denying the relevance in many theory preferences of membership in the larger community's competing cliques, I feel we also need an internalist sociology of scientific belief change. Such a sociology would focus more on the case of Darbishire, assistant and student of the influential Professor Weldon, Pearson's major ally (MacKenzie & Barnes, 1975; see also Provine, 1971). After humiliating cross-examination by Bateson about his data, Darbishire converted to the still relatively powerless Mendelian position against all self-interest (unless an estimate about the eventual "correctness" of theory is credited with providing an element of self-interest to the belief–exchange/belief–change processes). What sociological conditions permitted this humiliation? Why were the upstart Mendelians allowed to participate? Was a partial toleration of minority points of view a factor? What strange sociology lay behind Darbishire's presentation of data that could be interpreted as showing recessive factors and F_2 variability greater than F_1? (For F_2 "segregation" and the phenomena explicable by dominant and recessive traits, the Pearsonian model offered no explanation.) Was it mere stupidity or social norms that kept him from editing his data to eliminate this discrepant outcome? (Probably a rationalizable discarding of only 5 or 10 cases out of his possibly 100 or so F_2s would have done it.) What sociology and psychology led him to be influenced by the unpredicted outcome of his own research? Was the socially evolved ritual of experimentation something more than the sanctification of an already determined belief? Did his social indoctrination within the larger society and within science provide some role for the outcome of the experiment in belief change, in spite of the indubitable availability of Quine–Duhem cop-outs?

The case of another convert (reported on by Provine, 1971) provides an interesting contrast. In the chaotic setting of American universities of the time, T. H. Morgan made an impressive early career as an anti-Mendelian. There were no dominant theorists to whom he needed to defer for the sake of his career. Then he became a vigorous pro-Mendelian, a move that in no way jeopardized his prestige or tenure. Indeed, it turned out to be a very opportune move in terms of salary, influence, and fame. But it was not opportunistic in terms of the contemporary scientific establishment and its beliefs. Given, however, a social system employing experimental replication in its persuasion process, the F_2 variability that he found in his early research may well have convinced him that it was a dependable phenomenon that he could count on others replicating, and thus it could contribute to a self-serving career of intellectual leadership, which he would be denied if he persisted in anti-Mendelian arguments. A more complete constructivist social psychology of science will also eventually include the joys of "discovery" and the gratification of assertions believed to be "true."

Nils Roll-Hansen (1983) provided an important reanalysis of MacKenzie and Barnes's (1975, 1979) presentation of the victory of Mendelism in terms of a rationalist sociology of scientific knowledge. He concluded that Pearson's personal enmities and overemotional commitment to continuous variation and no unobserved entities blinded him to the facts and arguments of the Mendelians that were sufficient to convert the younger biometricians. Pearson's behavior represented a distinct deviation from the current norms of science. My own analysis does not overlap Roll-Hansen's, but is one with which he would probably concur.

Although I welcome Roll-Hansen's contribution, I do not want to identify myself with it. For one reason, I am fully committed to epistemological relativism. For another, my position is "rationalist" only if the competence-enhancing processes of mutation, natural selection, or avoidance conditioning in animals can be regarded as "rational." Rather than focusing the issue on rationalism versus relativism, I prefer to hinge it on hypothetical ontological realism versus positivist antirealism, which in extreme form becomes ontological nihilism. Roll-Hansen, although accepting much of the relativists' contribution, concluded, "It is the general claims about scientific knowledge as a product of 'social construction' that I find unreasonable and lacking in evidence" (1983, p. 511). In contrast, I identify my own position as completely "constructivist" and go on to posit fallibly that to some degree the social winnowing processes characterizing science lead to the selective propagation of those few social constructions that refer more competently to their presumed ontological referents. I prefer to let rest the question of whether belief—change processes that supposedly provide an opportunity for Nature Herself to contribute to the differential retention of some social constructions are to be regarded as "rational."

In Karin Knorr-Cetina's *The Manufacture of Knowledge* (1981), as in Latour and Woolgar's *Laboratory Life: The Social Construction of Scientific Facts* (1979), the production of scientific belief assertions or truth claims is shown by participant observation and ethnographic research to be a process in which order is imposed on a chaotic welter of inconsistent and inconclusive observations through quasi-conspiratorial social negotiations. Thus when life in the scientific laboratory is examined in detail, the factual proof that might be expected never appears. Ambiguity, equivocality, and discretionary judgment pervade. A point at which Nature intrudes and says "yes" or "no" to theory is never encountered. This research experience increases doubts about the reputed objectivity of science.

Just as a study of the microprocesses of vision failed to encounter dependable truth-transmitting links and thus led to a revival of skepticism in seventeenth-century epistemology, so too an examination of the social microprocesses of science leads to a skepticism about how it can be a source of competent belief. For scientific beliefs, validity must come from the contribution of the referent of belief to the selection processes (in addition to, or in spite of, all the other belief-selection processes). The case studies of Latour and Woolgar and also Knorr-Cetina were done in applied biology-laboratory cultures dominated by the experimentalists' distrust of "the theorists": To reconcile their reports with the possibility of Nature Herself participating in the belief selection, one might attend to the hundred or so instances reported (by implication, a sampling from thousands) in which ideas, procedures, and strategies are proposed and rejected because they "won't work." This not-working is often decided by thought and argument and often by laboratory efforts. Some of the rejections no doubt involve blind social conformity to locally preferred belief, but probably only a small proportion can be so dismissed. We need an expanded, still more microprocess, sociology of idea winnowing.

Although mainstream sociology of science in the United States is much less focused on belief change than are the new relativists, a great deal of it is epistemologically relevant. Zuckerman's "Deviant Behavior and Social Control in Science" (1977) directly addressed the problem raised by Hull's sociobiology of scientists (1978)—that is, the social system explanation of how some degree of mutual trust and honesty among scientists is possible. Merton and Gieryn (1982) argued that the threat of the loss of public support by a profession can lead to intraprofessional monitoring for honesty. Merton's analyses (1973) of the norms of science and of the reward systems of science are all relevant. There are also a considerable number of philosophers who have undertaken the beginnings of similar ventures. My own premature fragments offered here are not to be taken as representing this literature (see Campbell, 1986b, for references).

DISPUTATIOUS COMMUNITIES OF TRUTH SEEKERS

The title of this section denotes one sociological feature of scientific belief exchanges. I use it to introduce my version of the ideology of the scientific revolution. However, I have yet to integrate it with the history of the scientific revolution or with Habermas's concept of an ideal speech community, with which it probably has considerable communality (Habermas, 1970a, 1970b; McCarthy, 1973).

As explained in Chapter 1, the ideology of science is antiauthoritarian, antitraditional, antirevelational, and individualistic. Yet scientists stay together in a disputatious, mutually monitoring community described by Merton (1973) as "organized skepticism." Settings in which organized skepticism can be approximated are rare and unstable. Still, it may be a viable sociological thesis about a system of belief change that might improve beliefs about the physical world, both observed and unseen.

In making this assertion, I am invoking the "historically dialectic indexicality" (Campbell, 1982) of ideological statements on the issue of respect for tradition. Variation, selection, and retention are all three necessary for any fit-increasing process, and variation is at the expense of retention (see, e.g., Campbell, 1974a, 1974b). To so stress variation and selection and neglect retention in the official ideology would be adaptive only if, at a particular historical period, retention were grossly overemphasized in the general cultural ideology and practice. At the time of the scientific revolution, retention had gotten entirely out of hand insofar as beliefs about unobservable physical processes and competence in negotiating with the invisible physical world were concerned. An antitraditional counteremphasis was adaptive at that time, as it still is for much of science, as long as it does not jeopardize a tradition-conserving practice within the domains of scientific belief (there the 1-to-99 doubt–trust ratio must hold and must include ordinary beliefs about observables [Campbell, 1977a, 1978]). With such plausible apologies for aspects that in the seventeenth century did not need underscoring, I believe we should seriously consider the ideology of the scientific revolution as a useful, albeit contingent, thesis in an epistemologically relevant sociology of science.

From my perspective, the ideology and norms of science are not clearly distinguished from "scientific method." Scientific method is also to be seen as a product of a cultural-evolutionary process on the part of a belief-transmitting subsociety of many generations. With Feyerabend (1975), I would agree that new criteria of method are developed as new choices provoke new arguments. Like religious commandments, the "rules" may be mutually incompatible in the sense that if any one were to be followed with complete loyalty, it would interfere with compliance with the others. Each is dialectically and historically indexical, a shorthand interpretable only against a background of prior and current norms and practices. Although historically both methods and ideology have fed on concrete successes, it is convenient to regard the ideology and practice of cooperative truth-seeking as coming first and method as a rationalized summary of successful usage in the community. This is more obviously so for the hermeneutic methods, but I believe it also holds for Mill's canons of cause and Fisher's analysis of variance.

VISUAL DEMONSTRATION AND ASSENT TO FACTS

Just as Stegmuller (1976) in his formalization of Kuhn revives a version of theories being checked against facts laden with other theories, so I believe sociology of science will find some version of the fact–explanation or fact–theory distinction essential. In the paradigmatic instance, the "facts" are visually supported beliefs shared by the community and visual demonstrations introduced in a persuasive process. In the early stages of physics, chemistry, and experimental biology, the terms *demonstration* and *experiment* have much the same referent. I believe the persuasive role of demonstrations to fellow scientists (and initially to lay audiences) provides a more important connotation than *experiment* for understanding the social grounding of scientific belief, even though demonstrations are minimally practiced in current experimental science. *Facts* were originally theoretical inferences supported by processes built into the nervous system by both natural selection and learning. At a more mature stage, facts may be microtheories no longer controversial within the scientific community.

REFERENTIAL ECOLOGY

A successful science has to be able to achieve assent on some considerable store of agreed on facts (many implicit) and has to be able to demonstrate new facts frequently. The persistent difficulty of the social sciences in achieving agreement on facts is certainly a major source of its failure to achieve scientific status. Much of this is a referential–ecology predicament that is unavoidable because it is intrinsic to social science topics. Some of the problem, however, is a larger societal ecology-of-support issue. Were social scientists to limit their work to topics on which factual assent could be readily achieved, it might be that society would not support their research, nor students attend lectures limited to their findings, because of their banality. This needs thorough exploration. But some of the fact–assent problems might be alleviated through structural and ideological changes in the social science community, in publication practices, reward systems, funding priorities, and the like.

REPLICABILITY OF FACT

A crucial part of the egalitarian, antiauthoritarian ideology of the seventeenth-century "new science" was the ideal that each member of the scientific community could replicate a demonstration for himself. (Whether or not we end up regarding experimental alchemy as protochemistry, we must recognize that its ideology of secrecy was anathema to scientific exchange.) Each scientist was to be allowed to inspect the apparatus and try out the shared recipe. Collins's sociological studies showing the absence of replication in current physics (1975, 1981a) are to be taken very seriously. Nonetheless, the early study of electricity will show hundreds of Leyden jars, Voltaic piles, and static electricity wheels generating sparks in hundreds of labs, almost none reported in publication. In that sense they are replication or teaching and not "research"; however, they play an overwhelming role in the social persuasion process. A healthy community of truth seekers can flourish where such replication is possible. It becomes precarious where it is not. Those undergraduate chemistry laboratories in which students must fake their results to meet the course requirements are a degenerate form of persuasive demonstration and a threat to the precarious social system of validity-enhancing belief change.

Replication is available in astronomy, in historical studies in which multiple relevant sources and texts are available, and is an ever-present, if rare, potential in public-opinion surveying. Replications can be attempted, but too frequently fail, in the most exciting fringes of experimental social psychology (a referential–ecology problem, at least in part). Perhaps as a result, social psychology has the custom (atypical of successful science) of trusting a single dramatic study in going on to the next experiment without explicit or implicit replication. The effort and cost of replications within a social system that regards them as unpublishable and of low prestige contribute to their absence. The lack of replications greatly weakens the essential social control on opportunistic selection from proliferated small-sample studies, on overediting of data to achieve a publishable paper, or on partial faking of data. Thus experimental social psychology lacks the social control that exists in those sciences in which replication is feasible and regularly succeeds (Campbell, 1979c).

In general, the absence of the norms and practices of replication and of the possibility of replication for historical or cost reasons are major problems for the social sciences. From the standpoint of an epistemologically relevant sociology of science, this absence makes it theoretically predictable that the social disciplines will make little progress. Can planned changes in science policy (in universities, journals, professional associations, and funding agencies) change the situation?

The achievements of disputatious communities in the fields of history and hermeneutics have been at the level of facts, of historically and contextually singular facts, including those about historically specific "meanings" and theory-laden facts involving culturally shared theories of human nature. Moreover, the scrupulous mutual monitoring of such communities in the humanities has often led to an explicit distrust of theory, because theory is seen as a source of distortion and disregard of facts.

THE ECOLOGY OF EXPLANATIONS AND ANTICIPATIONS OF FACTS

It is our ontological predicament that the events and stabilities we come to know lie at the intersection of innumerable forces, restraints, and causal processes, most of them unmapped at any given stage. This is true both of the biological evolution of sensing and predictive machinery and of culture or science. The survival value of perception and memory lies in those ecologies in which the highest order interactions of all of the variables are not significant, in which all things are approximately equal (ceteris paribus). Similarly, the growth of science has required not only the accumulation of facts, but also the achievement of successful approximative theory relating facts to facts. This is most possible in those ecologies in which powerful, oversimplified ceteris paribus laws can be invented to sustain the group's feeling of progress until the shared corpus permits more complex theories with more detailed explanations and increased precision of prediction.

DEGREES OF FREEDOM AND HISTORICAL–CONTEXTUAL UNIQUENESS

Classic theories of physics and chemistry sought laws in which historical date and provincial location could be disregarded. Whether or not physics can ever quite achieve this, the biological and social sciences are much less likely to be able to do so. This difference in referential ecology rightly motivates social science methodologists and metatheorists to be wary of borrowing uncritically a theory of science based solely on the physical sciences. The methods of the humanities appropriately become attractive alternatives to be considered. But it is intrinsic to understanding, interpreting, explaining, predicting, or theory that the principles invoked are to some extent transtemporal and transcontextual. Complete situational and historical uniqueness eliminates not only theory but also any grounds for "understanding" or shared "meaning."

Insofar as the disputatious communities of scholars dispute about theory, they are likely to generate plausible rival hypotheses and to enter into arguments in which mutual persuasion becomes possible only where there exist degrees of freedom sufficient to make possible cross-validation. Such degrees of freedom can come only from attempting generalization across instances, persons, provinces, times, or the like. Of course, we do not want to accept rejection of the "one-shot case study" (as in Campbell & Stanley, 1966), but instead we want to join my later recognition (Campbell, 1975a) of the degrees of freedom available in a case study that come from ability to check some multiple implications of a theory in that setting. Until we have successful cases of mutual persuasion converging on an agreed on theory achieved by such methods, we should continue to regard the problem as serious.

This issue is not at all specific to the humanistic methodologies. It exists in extreme degree for quantitative economics. It seems to me that economists have made the wrong choice in focusing on national economies, where only short runs of 30 or 40 "comparable" years are available, rather than on the economics of, for example, neighborhood laundries, where degrees of freedom for testing hypotheses on not-yet-used samples abound.

ELIMINATING RIVAL HYPOTHESES THROUGH DISCRETIONARY RAMIFICATION EXTINCTION

The social crux of science is the ability to render implausible rival hypotheses. This focus, exemplified in the quasi-experimental tradition (Campbell & Stanley, 1966; Cook & Campbell, 1979), seems to me more central than experimental isolation or experimental control. Insulated laboratory walls, controlled atmospheres, and lead shielding are all secondary and historically specific means for rendering implausible the rival hypotheses that the current generation of disputative colleagues have raised. Randomized assignment to treatments does not prove the hypothesis under test, nor disprove rival hypotheses, but instead renders many rival hypotheses improbable. Unlike experimental isolation in which explicit rival hypotheses guide the design, randomized assignment may seem to rule out a totality of unspecified rival hypotheses. But careful inspection of the degrees-of-freedom problem for multiple hypothesis testing will probably show this conclusion to be unjustified. The narrowing of experimental comparisons by specialized control groups clearly illustrates the role of the current contents of the disputatious dialogue. In early studies of the effects of specific brain-region ablations, the community regarded surgical shock as a plausible rival explanation of some effects, and thus the sham-operation control group was orthodox for a while. But soon the observed and predicted effects were so specific that surgical shock effects became implausible, and the sham-operation control group was dropped. In contrast, in pharmaceutical research, placebo effects remain very plausible, and double-blind drug trials remain orthodox.

The Quine–Duhem equivocality of any experimental result is a very real problem for any community of scholars. It can only be resolved by discretionary judgments of plausibility. Nonetheless, scientific communities often achieve working consensus, often against the interests of the established and powerful. The central mode of argument involved is closer to the hermeneutic methods than to some idealizations of scientific certainty. The strategy of trusting most of the fabric of corrigible benefits while you challenge and revise a few (the 1 to 99 doubt–trust ratio) is central. Within that general strategy, ramification extinction of rival hypotheses is ubiquitous. Each of the Quine–Duhem-type alternatives not only provides an explanation of the fact in focus but also makes many other predictions. When the competing persuaders make them explicit and follow them up, these other implications often turn out to greatly reduce the plausibility of the rivals and lead even the most committed doubters to adopt the new theory. As Moyer (1979) so well described, it was in this way that the British community of astronomers and physicists changed between 1915 and 1925 from overwhelming faith in Newtonian gravitational theory to complete acceptance of general relativity. They were prodded by their 1919 eclipse observations and the ensuing several years of debate. Something similar is described by Clausner and Shimony (1978) for 10 years of testing of Bell's theorem. Each particular experiment was flawed, but through ramification extinction of the alternative explanations, even the hidden-variable theorists who motivated the experiments are convinced that such theories are ruled out by the experimental outcomes.

INSULATION OF THE SOCIAL SYSTEM OF SCIENCE FROM THAT OF THE LARGER SOCIETY

Thomas Kuhn said of the physical sciences that there are no other professional communities in which individual creative work is so exclusively addressed to and evaluated by other members of the profession. The most esoteric of poets or the most abstract of theologians is far more concerned than the scientist with lay approbation of his creative work, though he may be even less concerned with approbation in general. That difference proves consequential. Just because he is working only for an audience of colleagues, an audience that shares his own values and beliefs, the scientist can take a single set of standards for granted. He need not worry about what some other group or school will think and can therefore dispose of one problem and get on to the next more quickly than those who work for a more heterodox group. Even more important, the insulation of the scientific community from society permits the individual scientist to concentrate his attention on problems that he has good reason to believe he will be able to solve. Unlike the engineer, and many doctors, and most theologians, the scientist need not choose problems because they urgently need solution and without regard for the tools available to solve them. In this respect, also, the *contrast between natural scientists and many social scientists proves instructive. The latter often tend, as the former almost never do, to defend their choice of a research problem*—for example, the effects of racial discrimination or the causes of the business cycle—*chiefly in terms of the social importance of achieving a solution. Which group would one then expect to solve problems at a more rapid rate?* (Kuhn, 1970, p. 164; emphasis added)

Because scientists have to live in the larger society and are supported by it in their scientific activity, it becomes probable that science works best on beliefs about which powerful economic, political, and religious authorities are indifferent (Ravetz, 1971). Thus static electricity (rubbing cats' fur on amber) and magnetism were optimal foci of scientific growth. However, it follows from the psychology of discretionary judgment that visual demonstrations vary greatly in clarity and persuasiveness, and those produced in research on magnetic and static electricity were often dramatically convincing. If convincing enough, demonstrations can even overcome political relevance. For example, the successful predictions of lunar and solar eclipses have historically been so compelling that demonstrable competence in this regard led Chinese emperors to replace their well-entrenched court astronomers with powerless Italian astronomers (Sivin, 1980, pp. 25–26). But the combination of perceptually unclear demonstrations with highly important political beliefs, such as is found in the applied social sciences, is on these a priori theoretical considerations unlikely to produce belief change in the direction of increased competence of reference (for a more detailed and less pessimistic analysis, see Campbell, 1984a).

CRITICAL MASS AND SUCCESS EXPERIENCES

There are sociological requirements that must be met for sustaining communities of truth-seeking belief exchangers. A critical mass and the appearance of progress (collective success experiences) are among them. In the natural sciences, fad phenomena in the choice of problems to work on, with a concomitant enthusiasm and intensity of informal communications, are characteristic. For focal problems, these mobilized concentrations of effort supply the critical mass, mutual monitoring, cross-validation, and sometimes sustained perceptions of progress. Without perceived breakthroughs into further problem areas, interest dwindles and experimental energy becomes available for new perceptions of hot problems and promising techniques.

Certainly there are many areas of the social sciences that lack critical mass at the mutual monitoring level. Pressures in problem choice exist in both directions: A strong preference for working in an area with no competitors would certainly dominate were it not for the risk that fellow scientists would pay no attention. Sociology-of-science studies might well ask scholars in various fields about a specific publication. "If you are wrong about this, who will notice? Who will try to check by replicating? Who will publish (or informally publicize) their disagreement? Who will let you know privately about a successful or unsuccessful replication or other data that support or weaken your position?" These studies should focus both on the level of fact and on the level of theory. (This distinction can be made in practice, even though we epistemological relativists recognize all facts as theory, or as theory-laden). Without having such studies available, let me nonetheless hazard some opinions.

There exist mutually monitoring communities in religious hermeneutics for such issues as who borrowed from whom in the New Testament gospels and the proper translation of crucial verses in the Old and New Testaments. It might thus be reasonable for practitioners to claim that cumulative progress had been made.

In anthropological ethnography, no such communities exist, Lewis (1951), Bennett (1946), Holmes (1957), Freeman (1983), Firth (1983), and Brady

(1983) notwithstanding. Instead, one seeks a region as yet unstudied on one's special topic and, once successfully published, may jealously try to prevent others from allegedly needless replication of one's work. The field's shortage of investigators and the genuine collective interest in describing all vanishing cultures before they disappear provide justification. Mutually monitoring communication networks seem better realized in anthropological linguistics and in the theory of the origins of city states as tested by archaeology.

Contrast the ethnomethodology movement within sociology with the behavior-modification movement within psychology. Both are proud, self-conscious deviations from the mainstream of their disciplines. Both have social-solidarity needs that press for the inhibition of internal divisiveness, and hence, for the inhibition of mutual criticism in order to shore up intramovement morale against the neglect or attacks of the dominant paradigm. The behavior modifiers withdraw to their own journals and within them pursue vigorous internal disputation. The ethnomethodologists, on the contrary, produce isolated illustrations of their method and theory but, owing to their lack of numbers and embattled status, never disagree with each other about matters of fact. Insofar as they disagree about matters of theory, they tend toward further sectarianism and reduced communication rather than mutual monitoring. Mutual monitoring fails, not only within this and similar movements, but to too great an extent also in their roles vis-à-vis their parent disciplines. Such movements are of great value for mainstream social science as penetrating criticisms and suggestions for revision. But this effect can only be achieved if both the radical critics and the mainstream scholars remain within a common communication network and listen seriously to each other.

IMPLICATIONS FOR SOCIAL SCIENCE POLICY

If we had an epistemologically relevant sociology of science, and if under the guidance of that theory we had a competent study of current practice in the social sciences instead of the impressionistic sketches I have offered in the last few sections, we would have a basis for recommendations to the social science funding community. The central goal would be to establish and maintain a disputatious scholarly community for each problem area. Perhaps each grant and contract application should answer, for previous and proposed research, questions such as those suggested previously (e.g., "If you are wrong about this, who, if anyone, will notice?"). Given limited funds, this might lead to a decision to fund fewer problem areas more generously. Where promising developments emerge, independent participants should be funded rather than allowing one research team to maintain a monopoly.

This conference and others like it show a great enthusiasm for ethnographic, hermeneutic, participant–observer, ethnomethodological, and phenomenological approaches. This testimony is so impressive that the social science community should respond with generous funding for the application of some of these approaches to some problem areas, so that the kind of community we need is created. Such funding should avoid further extension of isolated, uncontested illustrations. Even though the epistemology of some versions of these movements deny the relevance, there should be replication efforts and sequential studies guided by rival interpretations of prior studies. These approaches have much to offer the validity-seeking social sciences that is not precluded by their relevance for other epistemologies and goals.

Effective communication has been greatly facilitated by journals specifically dedicated to critique and rebuttal, such as *Current Anthropology* and *Behavioral and Brain Sciences.* These are expensive to run and require subsidies. It seems to me of highest priority to create many more of these journals, perhaps with narrower scopes, and some that cross disciplinary boundaries (Campbell, 1969a). Funding priority should be given to studies designed to help choose among well-articulated alternatives emerging from confrontations. Annual problem-centered conference series have in some areas produced similar benefits. Funds for visiting other laboratories working on the same problems are also of value.

Innovation, validation, and cumulative growth in a research tradition are intrinsically incompatible goals, in that too much of one jeopardizes the others. Thus policy advice depends on a judgment of current balance. I judge the social sciences to have a surfeit of brilliant theories, inadequately sifted. Promising lines of growth are often abandoned because of boredom on the part of funders or scholars. But the detailed sociology of our past and current practice on which such a judgment should be based has not been done. My policy recommendations are contingent on the validity of my estimate of the current imbalance.

Funding policy should place greater stress on replication. Grant awards should specify that the central findings of the research being built on be replicated as a part of the extension. These and other replications should be "heteromethod" insofar as theory does not specify method. Their outcomes should be published no matter how confusing. They should be made available for literature review and meta-analysis through subsidized retrieval systems (such as *Selected Documents in Psychology* and *Dissertation Abstracts*). *Psychological Abstracts* and *Sociological Abstracts* should be subsidized so that they can include abstracts of unpublished and semipublished research, including dissertations.

The sociology of science offered in this chapter (see also Campbell, 1969a; Polanyi, 1966a) suggests that underdeveloped areas for science should be colonized first at the fringes that overlap thriving scientific communities. For pure science, this would be my recommendation. For applied social science, the mutual monitoring community of experts must be achieved in other ways. I have spelled out elsewhere the possibilities as I see them (Campbell, 1984a). I reject the one-decision–one-research model of program evaluation. For locally replicable programs, I recommend a contagious cross-validation model. For big-unit programs, I suggest more artificial efforts to achieve replication and to create a consensual validational community—for example, splitting the study into two parallel contracts and legitimizing intrastaff criticism.

OVERVIEW OF CHAPTER 9

Based on his beliefs regarding how scientists come to know, Campbell was asked by a federal agency, the Center for Prevention Research (CPR), to formulate guidelines for monitoring, funding, and reporting requirements that would enhance scientific validity. Specifically, he was asked to consider the research carried out at a number of regional Preventive Intervention Research Centers (PIRCs) funded by the CPR. Campbell recognized that he had to draw on the sociology of scientific knowledge; however, there were no existing principles to point the way. Instead of formal theory, he drew on his knowledge of the ideology of the scientific community and his vast experience in applied social science.

The recommendations offered by Campbell could not be meticulously enforced in the centers because they assumed an ideal situation and unlimited funds. Still, the ideal situation can provide guidance when decisions and compromises must be made. Campbell wanted the funders of social science to be aware of the special needs and pitfalls in applied social research. First, scientifically valid social science is an expensive proposition. Second, the research outcomes of these evaluations are often ambiguous or contradictory. Knowing that research findings are tentative reduces the willingness to provide continued funding for programs with overoptimistic claims and exaggerated results. Finally, problem areas should not be supported simply because they lend themselves to scientific rigor. The importance of the problem should be the deciding criterion. Campbell suggested qualitative methods for those problems that cannot be studied with experimental and quasi-experimental methods. Special attention should be paid to ruling out threats to validity, and where possible, qualitative methods should accompany quantitative research.

The twelve recommendations given by Campbell were intended to create a mutually reinforcing applied scientific community. He advocated funding "thought collectives" such as the center concept, providing ample opportunities for researchers to communicate and brainstorm, supporting numerous studies focused on a particular problem area, and encouraging independent decision making by those studying the problem. He believed the studies should be validated through replication and reanalysis, that the imperfections of the data should be presented with results, and that dissenting opinion reports and whistle-blowing should be encouraged. Finally, he felt that long-term follow-up was essential to verify that the interventions have been effective. Doing such follow-up is difficult because of the time commitment on the part of the researchers and funding sources. Still, CPR should lay the groundwork for future follow-ups.

Campbell was pressed still further to provide guidelines for monitoring the Preventive Intervention Research Centers. He cautioned that monitoring carries with it undesirable side effects; however, he offered some guidelines and discussed the trade-offs if they are enforced.

In the final section, three issues were discussed that were important to the thinking of the federal funders of targeted research. First, formative evaluations should be conducted to keep a record of the decisions and unsuccessful attempts that are a part of setting up a new intervention. The funding source should monitor these formative evaluations, although formal evaluation reports are discouraged because of time pressures. Scientists should be required, however, to compile all documents, such as articles, books, and those records that would aid in replication of the data. The second issue stressed the importance of situation-specific knowledge. This encompasses the experiences, observations, and insights of those involved in the program, including such stakeholders as teachers, parents, social workers, and the recipients themselves. Campbell referred to this as qualitative knowing. This important information should be documented to aid in the interpretation of the quantitative research results and to rule out plausible rival hypotheses. The third and final issue offered guidance in choosing sites and populations to increase the scientific validity of the findings. Stable populations should be chosen for pilot studies even if they do not represent a high-risk target population, because results may be uninterpretable if the population is unstable. If possible, several pilot tests should be conducted on disparate populations to increase the confidence when generalizing to the target group.

As a social scientist, you may be called on to offer guidance regarding evaluations of important social issues. Throughout this chapter, Campbell demonstrates the importance of considering the sociology of scientific validity in recommending policies for governmental programs.

NINE

SOCIOLOGY OF APPLIED SCIENTIFIC VALIDITY

The setting: The Prevention Research Branch, Division of Clinical Research (formerly Center for Prevention Research) of the National Institute of Mental Health funds a number of regional Preventive Intervention Research Centers. I accepted a contract for developing guidelines for monitoring such centers so as to achieve optimal scientific validity. This article presents, in revised form, parts of that report (Campbell & Kimmel, 1985).

There were no prior models of this genre to draw on. Although I brought to the task some 25 years of publishing on social science research methods for nonlaboratory settings (e.g., Campbell, 1957; Campbell & Fiske, 1959; Campbell & Stanley, 1966), such expertise was more appropriate for advising the centers on specific details of research design. What were called for instead were guidelines for science-management issues as they face the National Center,

Campbell, D. T. (1987a). Guidelines for monitoring the scientific competence of preventive intervention research centers. *Knowledge: Creation, Diffusion, Utilization, 8*(3), 389–430.

including monitoring, funding, reporting requirements, problem allocation, and so on.

My own beginning explorations in the sociology of science (Campbell, 1979c, 1986b, 1986c) prepared me to recognize that advice on science management for purposes of optimizing validity would be, de facto, an exercise in applying a "sociology of scientific validity," if such were to be available. Although this awareness was present from the beginning, it would be quite wrong to regard the specific items of advice offered as deriving from already existing principles. Rather than by formal theory, my speculations have been more guided by the implicit ideology of the larger scientific community, insofar as I have shared it, and by vicarious and direct participation in a wide range of roles in the applied social science enterprise.

A SOCIOLOGY OF SCIENTIFIC VALIDITY?

Not only is there no such body of doctrine, but the most active and most exciting sociologists of science deny both the possibility and the desirability of such an enterprise. The several western European groups I have in mind share a focus on the sociological determination of scientific beliefs. As especially emphasized by "the strong program" with its "principle of symmetry" (Barnes, 1974; Barnes & Bloor, 1982; Bloor, 1976), past changes to beliefs presently regarded as correct require a sociological explanation fully as much as earlier adoption of beliefs now regarded as erroneous. (With this point I concur.) Collins (1981b, 1985) identified the movement as "the relativist program" in the sociology of science. If this could be interpreted as "epistemological relativism" (Barnes, 1974), I concur. If it involves a commitment to "ontological relativism" or, because Quine (1969b) has used that phrase for another purpose, to "ontological nihilism," I do dissent. Latour and Woolgar (1979), Knorr-Cetina (1981), and Pickering (1984) emphasized the "social construction" of all scientific beliefs. (With this, too, I concur.)

What I have repeatedly asked these exciting scholars (e.g., Campbell, 1981, 1986c) is to move on to "The second phase of the Strong, Relativist, Social Constructivist Sociology of Science." In this phase, differing social systems of belief retention, belief change, and belief legitimation would be compared as to their effectiveness in improving the validity of beliefs about the invisible physical world (were there to be such). This they have almost uniformly refused to do. Like sociologists of religion, they take a vow of agnosticism about the beliefs they study. They plead no ability to deal with issues of comparative validity, even on a hypothetical, speculative level. But those of us who are willing to give advice on science policy with the goal of maximizing scientific validity have foregone such agnosticism, however tentatively. Note too that many philosophers of science are interested in such an enterprise (Campbell, 1986b, lists some 30).

Jerome Ravetz, in his *Scientific Knowledge and its Social Problems* (1971), is widely recognized in this new sociology of science community as a founder of the modern social-constructivist movement. Fortunately, he lacks the agnosticism of those who have followed, and boldly distinguishes good and shoddy science. His work is especially relevant, because he focuses on the exacerbated problems of the social sciences (in contrast with the physical) and of applied science (in contrast with pure).

RECOMMENDATIONS FOR GUIDING PREVENTIVE INTERVENTION RESEARCH CENTERS

These following recommendations are self-consciously utopian, deliberately suggesting procedures so costly, and so out of tune with the prevailing bureaucratic understanding of these issues, as to be intimidating and destructive if viewed as requirements for either CPR (the National Center for Prevention Research) or PIRCs (Preventive Intervention Research Centers), or the specific research projects undertaken by them. The goal for this draft is not to establish enforceable requirements, but to expand our awareness of what would be desirable (and possible with unlimited funds), the better to inform our specific choices and compromises.

We and the funding communities in private foundations and government labor under the unconscious assumption that social science is inherently less expensive than physical science. Perhaps this is because it is closer to common sense. On the contrary, all argue that scientifically valid applied social science should, on a priori grounds, be more expensive than applied physics, chemistry, or biology of an equal level of scientific precision. This is a result of the complexity of our problems, the unavailability of relevant simplifying laboratories, and the lack of well-tested, dependable theory. One effect of this document might be to increase our courage and persuasiveness in educating our funders to our real predicament, perhaps leading to increases in funding. But the immediate goal is not this, but rather to increase the intelligence of the compromises that we make.

An inevitable side effect of such a document is to increase the modesty of the claims we make for our precariously established findings. This effect we must be ambivalent about. It is hard to exaggerate the pressures from sympathetic members of Congress and administrative staffs on CPR to come up with exciting breakthroughs to justify past funding and to legitimate future budgets. Greater awareness of the inevitable scientific equivocality accompanying the very best of our most promising results will hardly make this task easier. In the long run, there are dangers to continued funding from overoptimistic anticipations and claimed results, but focal initiatives such as Preventive Intervention Research (PIR), whether funded by private foundations or government, do not usually have a long run, and are not apt to survive without short-run "results" to show. One might hope that a document such as this would educate well-intentioned legislators and foundation officers. They, unfortunately, also live only in short runs, stick their necks out in competition with other foundation staffs or legislators, and thus need exciting short-run results to justify continued support of PIR, and for their own organizational survival.

A second preamble may superficially seem to have opposite implications. Applied social science efforts such as PIR are obligated to stick with the problem instead of shifting to tasks that are currently more amenable to full scientific rigor. This obligation is in contrast to the National Science Foundation's (NSF) support of pure science efforts in sociology, anthropology, political science, psychiatry, psychology, and so forth, for which it might very well be appropriate to adopt a policy of supporting only those inquiries that can be done now rigorously.

The pure-science ideology with which we have all been trained, plus the necessity to get research proposals approved by review committees whose main common denominator is expertise in scientific method, plus staff career needs for research products publishable in respected scientific journals, all combine to push those who do targeted applied research (such as PIR) toward abandoning targeted problems and substituting scientifically feasible research of dubious target relevance. A document such as this must take special pains to avoid adding to this bias.

CPR should support research that is as scientifically valid as possible without changing its problem target. But where the problem target is *not* amenable to quantitative experimental research, it should support target-appropriate research using "clinical," "humanistic," "ethnographic," or "qualitative" approaches. In supporting such qualitative research, there is much that can be done to optimize validity, including self-conscious attention to all of the threats to validity that would be shared by hypothetical experimental and quasi-experimental studies in the same setting.

CREATING A MUTUALLY REINFORCING APPLIED SCIENTIFIC COMMUNITY FOCUSED ON PREVENTIVE-INTERVENTION RESEARCH

When the applied social science movement got underway, we and our patrons in government thought that one well-trained and honest scientist (or team) producing one research report resulted in a valid scientific solution to a given problem. The logical-positivist theory of science did little or nothing to disabuse us of this perception, and it still dominates governmental policy in applied social science today.

Modern theory of science, in contrast, sees scientific validity in the physical sciences as a product of a mutually reinforcing (rewarding and disciplining) scientific community. The validity of scientific truth claims does not come from the innate or indoctrinated honesty and competence of a single scientist. It comes from competitive replication and criticism, from fear of humiliation because of failed replication efforts, from competition for discovery and eminence so organized as to disclose (rather than cover up) error, incompetence, and fraud.

Central to this social system is a mutually monitoring scholarly community that stays in close communication on a shared puzzle or focus, while still vigorously criticizing each other and competing for priority in valid discovery.

The rules within this community exclude using social status, authority, or power as persuaders. Instead, persuasive efforts are to be limited to equal-status demonstration, logical and empirical. Merton (1973) has aptly spoken of the "organized skepticism" of science. Such a community is sociologically precarious, however, because of the natural tendency to splinter into like-minded groups that cease communicating where they differ.

From this perspective, the duty of CPR is not only to distribute targeted funds for individually worthy PIR research projects, but also to do so in a way that creates and sustains a mutually monitoring applied-science community focused on PIR research. The responsibility is particularly great because PIR is not an area in which science could have first emerged. If PIR is to be done scientifically, it must be as a colonization effort from successful science. Note that it is the social system of belief change in the physical sciences that is being borrowed, not specific physical theories or methods. Note that this social system turns out to be very similar to that of successful validity-seeking hermeneutic traditions (Campbell, 1986c). From the standpoint of this sociology of validity-enhancing belief change, if quantification and experimental design come to play a role in PIR research, it will be because the participants find them persuasive and find that they facilitate progressive belief change. They will not be used simply because they are used by physicists, or because they are essential prerequisites to improving the validity of belief.

Much that the CPR is doing already has the goal of creating a scientific community focused on PIR. The establishment of numerous stable centers (PIRCs), each with a group of scientists sharing a common focus, the organization of periodic meetings among PIRC staffs, the circulation of project proposals and research reports, all represent important and farsighted investments in creating a mutually stimulating, rewarding, and disciplining scientific community. This represents a wise (albeit expensive) investment in social structure, over and above the funding of the specific research. The general tenor of this chapter is to recommend an increase in such investment, even at the cost of funds for research. Such investment is in continual jeopardy because of the absence of tangible products and because of the competition from other worthy uses for scarce resources. It is to be hoped that a document such as this can provide a rationale for protecting such investment. So, too, the more specific recommendations that follow will often include policies already in practice, and warn against general tendencies in applied social-research funding, which CPR is already avoiding, but which may exist as continual pressures for change in a less than optimal direction.

1. Funding Through Centers

The CPR and its legislative mandate now show full commitment to the research center concept. This decision is no doubt under some degree of recurrent challenge in favor of investing these funds in more individual applications.

Our sociology of applied scientific validity strongly supports the center concept in essentially its present form. This support is scattered throughout this entire section, yet its importance is underscored by making center support our first recommendation. The PIRCs provide the very essential beginnings of scientific "thought collectives" (Fleck, 1979) devoted to the problem.

2. Funding Annual PIR-Research Conferences

These already exist, with two or three representatives of each PIRC attending. These might be expanded to involve all active scientists, including those of graduate student status. Perhaps the location of these conferences should rotate from PIRC to PIRC, and allow time and occasion for observing treatment and measurement procedures, special equipment, and the like.

3. Avoiding Problem–Method Monopolies, Encouraging Overlapping Research Agendas

Given shortages of funds and scientific personnel, and given the background belief that an individual scientist produces scientific validity, the natural tendency will be to delegate specific problems and approaches to specific centers, teams, or scholars. Under the social system perspective, this leads to the absence of an informed scientific peer group to applaud discoveries, corroborate and correct them, criticize and build on them.

A dramatic difference exists on this score between the social sciences and physical sciences in their "pure science" forms. Were one to ask grant recipients,

"If you make an important discovery in this project, which of your fellow scientists would recognize and applaud? If you publish in error, who would discover this? Who will build on, replicate, confirm, or challenge your findings?" it is probable that almost all physical science awardees could name several, but many social science awardees, pure or applied, should (if honest) respond "no one." CPR should operate to correct this.

If we searched out the typical location of those key (stimulating/rewarding/monitoring/shared-expertise) peers, it might very well turn out that they are rarely in one's own department, center, or university, but rather at some distance. It seems a general sociological dynamic that within a single locale, each scholar tends to articulate a specialty minimizing close overlap, mutual monitoring, or competition. This may be a sociological tendency worth encouraging, because it seems so ubiquitous in successful science, wasteful as it may seem in terms of duplicated laboratory equipment, costs in conferences, long-distance telephone calls, and so on. Tentatively, our recommendation is to tolerate or encourage overlap between centers and diversity within.

The individual scientist has an intrinsic ambivalence here. On the one hand, it is rewarding to be the world's most eminent explorer and expert in a given area, to have a specialty that one is king of. This creates a motive to move into an unoccupied novel field that has obvious importance. On the other hand, the scientist also has an opposing motive to move into "exciting" fields in which truly competent peers will be available to provide the reward of well-informed acclaim. In the long run, a monopoly is not personally rewarding if there are no properly qualified peers who inform themselves in detail on one's work and then applaud. Science planners and funders often seek to curb scientist's "fads" to prevent the neglect of important areas not currently fashionable. However, for the applied social science front, the tentative recommendation is to tolerate or foster such fads if they can be made to operate as a discovery and validational process. CPR should not allow one-person or one-lab monopolies to remain past exploratory stages. If the area is really promising, or so important as to justify many failed explorations, then rivals should be deliberately funded.

Any such recommendation implies a judgment on where we are between opposed poles. Do we need more novel ideas? Or do we need to winnow out among the many intelligent and creative ideas those that are most valid and worthy of building on? Contrasting the unsuccessful social sciences, pure and applied, with the successful physical sciences, leads this chapter to favor emphasizing the latter. This has the major policy science implication that we should be funding fewer topics more densely. The applied social sciences are spread out too thin, and thus lack the critical mass to sustain mutually reinforcing, validity-enhancing scholarly traditions focused on specific problems.

It follows that CPR should also encourage PIR funding by other government agencies and private foundations. Where other PIR funders exist, there should be no implicit or explicit agreements to divide up the terrain. On the contrary, overlap in choice of problem and general approach should be encouraged, but this should be combined with independent decisions on details of treatment and measurement. Contrary to a consensus among respected methodologists, I recommend against using the same specific questionnaires and tests, unless these are included as add-ons for comparability in a multiple-measures strategy.

Diversity of decision making should extend to "gatekeeper" roles also. Rather than using the same set of reviewers for each round of funding decisions on proposals, CPR should use separate sets each time insofar as this is possible. Although it is inevitable that proposal reviewers be drawn in part from prior recipients of CPR awards, competent reviewers should also be sought from among unfunded proposal submitters and their intended participants and consultants. Insofar as possible, CPR should also avoid employing the same reviewers other PIR funders regularly use. But in seeking diversity of decision makers, CPR central staff should use it in such a manner as to increase, rather than decrease, the development of overlapping problems and expertise in the funded community.

4. Permitting Independent Grant Applications From Several Scientists in Any One PIRC

Although the start-up grants to any one PIRC will typically come in as a single grant, once it is established, individual scientists should be permitted to apply separately to CPR for specific study grants. These grants should earmark a portion of the funds for center coordination, should be PIR relevant, but otherwise should be emancipated from the organizational authority structure of the PIRC. Arguing for this are the considerations mentioned under 3, the importance of multiple independent decision makers, and peer (as opposed to authoritative) persuasion processes in a social system of validity-enhancing belief change (Campbell, 1986b, 1986c). Also arguing for it is the reputedly dilute participation and poorly thought-out decisions in large research labs in which a single PI (principal investigator) makes the crucial decisions for dozens of post-docs. Such waste would be further reduced by eliminating the pretense of leadership that leads a laboratory director to appear as a coauthor on all publications from his or her center.

5. Splitting Large Studies Into Two or More Parallel Studies

Even when interventions are being tried out that require massive integrated efforts, it is still recommended that they be split into at least two studies The scientific target should be the same, but all ad hoc decisions on implementation and method (down to specific questionnaire wordings) should be made independently. They should be funded to keep them mutually informed at all stages, and to facilitate reanalyses of each other's data, but with no pressure toward methodological identity. Such a policy increases the size and the administrative autonomy of a mutually monitoring scientific community.

6. Funding Cross-Validational Research for the Implementation of "Established" Procedures

In the physical sciences, engineering applications and product use provide implicit cross-validation of the scientific principles involved. For the applied social sciences, it is in the operational implementation and dissemination stage that the required replication and cross-validation is most likely to be available. Yet such cross-validation is usually missing. For mechanical products, failures to operate properly are readily observed in engineering application, manufacture, and consumer use. For social interventions, by contrast, it takes formal research design to confirm beneficial effects, discover malfunctions, and so on, and this confirmation in application is usually lacking.

The CPR and the PIRCs will come up with preventive interventions that seem so promising as to warrant widespread implementation. Funding for such implementations will not be through CPR, but rather from a variety of sources, ranging from local education and health funds to federal legislation. CPR's role in such implementation should include not only providing treatment-design details, but also funding an evaluation component, which should be implemented until there are dozens of cross-validating results and until the factors leading to effective application can be mapped.

7. Facilitating Reanalysis and Meta-Analysis

Where replication is unfeasible because of costs or changed circumstances, independent reanalyses of research data provides part of the competitive cross-validational process of science, and one that has been particularly important in the program-evaluation area. Still more important are the many meta-analyses of dozens of studies on a given topic, such as effects of psychotherapy, effects of desegregation, and so on that often require access to the raw data to optimize comparability.

Probably the CPR should receive and archive a complete set of data tapes with full documentation from PIRC research projects, in conjunction with the submission of each research report. The PIRCs themselves should also keep these data in reanalyzable form, and transfer them to some other appropriate archive if and when disbanded. Because of the importance of long-term follow-up, names and social-security numbers of all intervention recipients and controls should be archived independently, but in such a way as to permit subsequent linkages under conditions that preserve confidentiality (Boruch & Cecil, 1979; Campbell & Cecil, 1982; Campbell et al., 1977).

8. Facilitating Self-Criticism: The "Redesign Component" in Research Reports

The pressures for excellence coming from dissertation committees, journal editors, and research funders often work to lead scientists to cover over known weaknesses in research reports, thus reducing their dependability. Among clinical and field-research psychologists, our training in "pure" experimental psychology is apt to intimidate us into sweeping all imperfections under the rug, and make us feel personally guilty for these imperfections. Instead, our attitude should be as follows.

We are pioneering beyond the sheltered laboratory, doing the very best that can be done without abandoning our task. The imperfections and the research disasters are attributes intrinsic to our problem area, and will be encountered by those who follow us, too. They are not at all signs of our incompetence or venality as scientists. We, the pioneers, should blaze the path for those who follow by fully describing our frustrations and by speculating about their implications and how to control them.

The CPR can do much to further this pioneer morale and the self-critical disclosure of imperfections. It can try to avoid rewarding those whose research seems most perfectly executed only because of failure to disclose imperfections. Some specific institutional practices might be devised to encourage this. One might be to require an appendix to each research report dedicated to describing how the study would be redesigned if one had it to do over again, with and without a larger budget, and warning of mistakes that others should avoid.

One source of shoddy research comes from underbudgeting, overpromising, and underestimating the costs in time and money. The NSF's express policy allowing the researcher freedom to change problem scope and topic after the award is made sets a good management model, but one that a problem-targeted agency may find hard to follow. CPR should use its experience to expand awards beyond what is requested in instances in which the awards are inadequate, or to suggest reductions in scope without reduction of budget. One institutional innovation that would assist in making such judgments would be to request an approximate accounting in time and money of the research in each specific report, and figures on the per-unit cost of repeated items, such as interviews, schoolroom data collections, transcribing of tapes, and standardized data-analysis routines. (Although this adds a substantial additional burden, this might also greatly improve the scientist's own budgeting and management of funds.)

9. Legitimating and Facilitating Supplementary and Dissenting-Opinion Research Reports From the Research Staff

This is the most radical of the suggestions, but one that would probably seem obviously proper were it already in place. The goal is to give to all of the professional research staff the right to the data for eventual reanalysis and publication, comparable to that available to external secondary analysts. The benefits expected are threefold:

1. It is expected to increase the multiple perspectives triangulating on the same problem and body of data made available to the PIR community. Note that because of their informal participatory knowledge, former members of the research staff are optimal secondary analysts, because much of the valuable background knowledge needed for interpretation is not recorded on the data tapes or elsewhere. Moreover, during the project, there have often been high-quality discussions of alternative theoretical interpretations and differential weightings of evidence that get crowded out of a unified project report.

2. To make fuller use of the data collected, including use for other purposes than originally intended, it is a general experience that the data from large projects ends up grossly underanalyzed. A part of the problem is that project budgets still grossly underestimate data-analysis time, and also that analysts' time must be diverted to writing up new proposals to keep the center operating. Granting all professional staff members explicit co-ownership of data sets that they have participated in collecting would produce a gain in completeness of utilization far outweighing the occasional unnecessary replication of findings in the published literature. (Norms of offering the other professional participants both coauthorship and the privilege of individually authored footnote insertion would probably avoid increasing replicate publishing beyond its present level.) I have experimented with the principle of multiple "ownership" of data in two cross-cultural studies, both regarded as successful, even though in one a collaborator scooped the integrated project report by seven years (Campbell & LeVine, 1970, pp. 383–384). For the second of these, a formal signed agreement was employed. (An example of such a document appears in the original publication of this article.)

3. The most important anticipated effect would be to increase the caution, complexity, and self-critical awareness of alternative interpretations presented in the official research reports and publications of the research projects. Making all aware of the potentialities of competitive reanalysis would in many instances curb oversimplified and overoptimistic reporting and increase the dependability of the official reports. Moreover, to some small degree, this policy would

reduce the status differences created by tenured position and employer role, and thus help to establish a mutually disciplining scientific community within each PIRC.

Another background for this suggestion is the great value that whistle-blowing has had for the validity of physical and biological research results when these have been done under conditions of extreme policy relevance. (I am thinking of research on the dangers of chemicals to manufacturing workers and food consumers, the dangers to and effects on humans and sheep of irradiation from nuclear experiments and power generators.) Although such whistle-blowing occurs, it is still experienced as a guilt-producing team disloyalty, both by the whistle-blower and coworkers, who may react with ostracism. It would improve the scientific and political validity of applied physics, chemistry, and biology if whistle-blowing were legitimated by reconceptualizing it as the right and duty to generate dissenting-opinion research reports, and if all laboratory staff were provided official access to all data for this purpose. Insofar as our research results are inherently more ambiguous, even more do we need this in applied social science (Campbell, 1984a, p. 40).

10. A Journal: Planning Preventive Intervention Research

A published record of research proposals and critiques, with continuing discussion, seems desirable. Many scholars submitting research proposals that do not get funded are still proud enough of them to welcome their publication in such a journal. A few of the referees' comments are worth expanding into signed, published critiques accompanying the proposals. Later commentary from readers could be printed. Other subjects on which articles could be solicited include research-design lessons from implemented research proposals, special essays on theories needing testing, ideal approaches too costly for present implementation, and recurrent methodological problems with alternate solutions.

There are several aspects of the present situation in government-initiated targeted programs that would benefit from such a publication. First, because there are usually shortages of funds, many well-thought-out and carefully researched proposals are destined for wasteful oblivion. Yet just because the programs are very expensive, public scientific debate on research design is desirable, so that the final research designs later chosen may profit from the pooled wisdom of the entire PIR community.

The pattern of government funding (full-strength startup; short lead time for RFPs [request for proposals]; high turnover in federal agency staff; and high turnover in responding research participants) can lead to the funding of inadequately thought-out proposals and loss of contact with previous thinking and experience. Such a journal would be high-priority reading in designing new research.

Unfunded scientists, proud of the important new concepts in their proposals, sometimes have paranoid thoughts when they later see the same concepts appearing in projects funded in subsequent years, and wonder if some of the dozen or so advisory-panel reviewers might unconsciously have borrowed them. Such a journal would enable later proposal writers to give credit where credit is due. The duty of the later proposals is to be the best possible, not to be original per se, and such pooling of expertise and inventiveness in the few funded PIR researches is optimal for science. Such a journal would both further such excellence and increase public recognition of the sources employed.

11. Sociology of Interdisciplinary Research and Tenured Careers in PIR

It seems desirable that CPR consider sociology of scientific validity problems even when it will only be occasionally or indirectly that CPR decisions can affect them. PIR research is intrinsically interdisciplinary, in that psychiatry, psychology, sociology, education, epidemiology, and public health expertise may be involved. Properly done, PIR requires scientists with interdisciplinary competence, able to use the relevant aspects of several disciplines, or the relevant aspects of several institutionalized and mutually isolated specialties within a single discipline, such as the experimental social psychology of persuasion and the clinical psychology of behavior modification.

Working against interdisciplinary career specialization are most of the social institutions of science, such as university departmental organization, national professional associations, government research funding, and any other institution that organizes competence evaluation and budgeting by existing well-established specialties. These well-established career channels may have been assembled by historical accident and may be essentially no more coherent in focus than PIR, but they do offer "tenure tracks" and other well-blazed career lines.

The route to PIR specialization cannot be achieved by demanding breadth—that is, competence in all of the relevant established fields. Instead, like intraspecialty expertise, it must be able to achieve a profound specialization. Thus PIR as a "novel narrowness" is what is needed. Established disciplinary institutions work against the creation or survival of careers with novel narrowness, just as they do against careers attempting multidisciplinary breadth.

The field of PIR needs specialists whose regular scientific reading pattern and own publications overlap parts of two or more disciplines. It needs careers accumulating unwritten scientific experience and wisdom, to ensure the competence of PIR-research planning, execution, and interpretation. It also needs such careers to make possible long-term follow-ups, as discussed later.

In making PIR research awards, and in funding major PIRCs, the scientific duty of CPR is to the problem area, even if this results in decisions that seem unfair to individual scientists. Thus, for example, universities offering tenure in interdisciplinary centers or specialties, or using specialty-relevant tenure and promotion committees, should be favored. Thus, too, already-tenured faculty are to be favored over untenured ones. But even tenured faculty should be queried about the reward structure of their local and national environment for PIR research.

12. Guiding Long-Term Follow-Up

Intrinsic to the preventive-intervention agenda is the desirability of long-term follow-up to verify that the intervention has led to lower incidence of mental illness or antisocial behavior years later. This implicit requirement is a very difficult one for all aspects of the CPR–PIR responsibilities. Although long-term follow-ups are needed for the ideal outcome measures, they could jeopardize the whole enterprise if they were the only ones employed. They are too delayed to support a mutually reinforcing scientific community. For example, if 10 years delay were required, it would be 15 years from initiating the research to outcome reports in an initial study. A cross-validating replication would not take place until another 15 years had passed. The time spans for the first study are longer than the typical positional and topic commitments of 90% of academic scholars, to say nothing of their governmental funders.

Thus it is essential for PIR projects to have immediate-outcome measures, focused on proximal indicators of those traits that theory specifies as mediating the long-term preventive benefits. But because the theories involved are not themselves well tested, and can only be tested by long-term follow-ups, it is a major part of CPR's responsibility to prepare the basis for later follow-ups.

STRATEGIES FOR ARCHIVING CPR CONTINUITY MAKING POSSIBLE LONG-TERM FOLLOW-UP

We have already seen that the need that new members of Congress and new administrators have to introduce innovations that can be credited to them works against CPR longevity. We also can note that the need for news of prompt "breakthroughs" on the part of those legislators and administrators who took political risks to establish and maintain CPR, makes even these founding supporters reluctant to put emphasis on pilot studies that will not pay off until 15 years from now. An adequate hypothetically normative sociology of scientific validity would address this problem and provide theoretical understandings that could suggest remedies. We are not at that stage yet, but should begin brainstorming on the problem.

We might try educating our sponsors to the problem and to the destructive effects of some understandable reactions to their predicaments. Once pointed out, and with the compensatory education results as illustrations, the necessity for long-term follow-up in PIR should be obvious. In our dealings with our patrons (and in their dealings with their constituencies) it may help to emphasize the importance of the problem and how that importance justifies our trying out many approaches to see what works, rather than falling into the tempting "overadvocacy trap" (Campbell, 1969c).

UNDESIRABLE SIDE EFFECTS FROM MONITORING

This section has no subheads, only this one entry. It is given this special status to stress its importance for the sociology of applied scientific validity.

CPR has the official mandate not only to initiate, coordinate, and inspire the PIRCs, but also to supervise them and monitor their effectiveness. But just as journal editors and other gatekeepers can exert pressures that decrease the validity of research reports, so too could CPR carry out its supervision and monitoring efforts in a manner that leads to covering up flaws rather than reporting them for the benefit of subsequent PIR research.

A related point: There have long been scattered reports of negative side effects from accountability efforts (Blau, 1955; Ridgeway, 1956), the lawlike nature of these feedback mechanisms, and their ubiquity. These side effects have been discussed in the evaluation and scientific management literature (Campbell, 1979a, pp. 84–86; Dirsmith & Jablonsky, 1978; Ginsberg, 1984; Prakhash & Rappaport, 1977).

The literature on the evaluation of scientific laboratories and research and development centers can be very misleading in this regard. Take the recent conspicuous case of the negative British review of their major astronomical observatory. In this, indicators such as number of publications and number of citations of these publications are used. Similarly, in studies of R&D laboratories in the Pelz and Andrews's (1966) tradition, the number of patents awarded has been used. A lesson from the "dysfunctional side effects" literature can be applied as follows: Such indicators may have considerable validity when collected from time periods before they were used as evaluation criteria. But once they are known to be employed for managerial decision making and control, they tend both to lose their validity and distort the processes they are designed to describe.

If, for example, great weight is put on number of publications, this will increase the proportion of effort devoted to submitting and resubmitting articles, to fragmenting research into more articles with disregard for repetitiousness, to lowering standards of importance for articles submitted, and to selecting research topics for their likelihood of short-run publishability, rather than for their genuine importance for PIR. Citation rates might seem more immune to inflation, but as gathered by the Social Science Citation Service, actual rates per article per year are so low, with both mode and median at zero, that they can be manipulated. If cliques conspire in their publications to add in citations to their fellow clique members' articles, substantial gains can be achieved in a 5-year period. Style of article can help, with reviews of the literature being more cited than the constituent articles they review, and so on.

The first draft of these recommendations was judged incomplete because it failed to specify explicit criteria, the accomplishment of which could be accurately monitored. Although our project monitor liked our product, and agreed with our reasons for evading such specifics, she had no choice but to demand them. In the end, we compromised on a list of 22 desiderata, which began as follows.

> This is a criterion-oriented abstract based on the attached report, *Guiding Preventive Intervention Research Centers for Research Validity.* The criteria presented here are to be interpreted in terms of the full document.
>
> Close analysis of any such list of criteria will show that any one criterion, if made predominant, jeopardizes attention to the others. Given shortages of funds and staff time, all criteria are in competition. Moreover, the addition of nonscientific reporting requirements (such as required evidence of criterion compliance not of use in scientific communication) adds further time demands.
>
> The basic document places great stress on such tradeoffs and undesirable side effects because it draws from the sociology of scientific validity in addition to, or instead of, borrowing managerial principles originally developed for profit production, translating these into rules for maximizing report productivity or report persuasiveness.
>
> 1. *Discovery of effective new mental health preventive interventions.* All monitors of PIR research, readers of research reports or site visitors, will be alert to, and value highly, such breakthroughs. In this sense, criterion 1 does not really need listing. Moreover, making such breakthroughs a requirement for each and every PIRC will tend to produce preliminary claims that turn out to be transient flashes-in-the-pan, failing to hold up when rigorously cross-validated. Although effective discoveries will appropriately lead to high marks and increased funding, CPR should give full continued support to PIRCs that are working with high scientific rigor and ingenuity on important preventive intervention hypotheses, even though the research outcomes reported are discouraging.
>
> 2. *Volume of research reports.* All other things being equal, the more research reports produced, the better. But if emphasis is put on volume, then

"other things"will not be equal, and the emphasis on productivity will have undesirable side effects, jeopardizing other criteria.

3. *Relevance of research to preventive intervention in mental health.* Because of pressures to meet scientific experimental standards, the need for short-term research results, the need for publishable articles in scientific journals, and so forth, this criterion needs continual attention.

4. *Publications in refereed journals.* The emphasis throughout is on the need for a mutually reinforcing (rewarding and disciplining) scientific community. Publication in refereed journals is traditionally the major formal evidence of approval by such a community. But the rigor of such refereeing is reported to be less in the very journals specializing in preventive intervention research. On the other hand, the more strenuously refereed journals are less apt to be directly relevant to preventive intervention.

VALIDITY CONCEPTIONS FOR QUASI-EXPERIMENTAL DESIGN ON TARGETED PROBLEMS

Half of the initial proposal was devoted to methodological advice for the conduct of PIR research, addressed to investigators in the PIRCs designing specific studies. Of these 12 subsections, the following three seemed equally relevant to the thinking of the national CPR in its overall goal setting, funding decision making, and in guiding and monitoring the PIRCs. These three have been included. Although less clearly relying on implicit assumptions about the sociology of scientific validity, such considerations are nonetheless scattered throughout.

1. Formative Evaluation in Project Development and PIRC Monitoring for Scientific Validity

The concept of *formative evaluation* stands in important contrast to the more formal *impact evaluation.* We need to be reminded of the many informal tryouts and decisions that go into putting a treatment program together, testing out an application setting for feasibility, and developing measures. These decisions abound in causal inferences, and as such, one could conceive of doing a formal experiment for each, but this would be overwhelmingly unfeasible. Instead, formative evaluation, debugging, and designing at all stages must be done using the informal evidentiary and causal-inference processes of everyday life. These formative-evaluation decision processes can be done poorly or well. The beginning of a methodology of formative evaluation consists of rules of thumb in this regard. This process is so important that it is worthy of a well-summarized conference to which PIRC staffs bring their experiences, recommended tactics, and unsolved problems. In advance of that, here are some tentative principles.

A first recommendation is that the process be made more self-conscious, with fuller awareness of the causal inferences being made and the alternatives considered. If each participating scientist were to keep a log of the choice considered and their pros and cons (a formative decision log), this would serve increased self-consciousness. Even more important, it would provide a record that would be invaluable for guiding program revision years later. Such a log could include notes on decisions made judgmentally that might someday be worth formal experimental testing.

Another recommendation is that these decisions be discussed within each PIRC staff and its team of regular consultants. Major advantages are those of brainstorming: to increase the range of alternatives being considered, and the range of possible implications, advantages, and disadvantages of each. From this plethora of considerations, a choice must be made. It seems favorable to scientific validity that it be made as independently as possible of social or organizational status per se, and of peer-conformity pressures per se. I recommend keeping the collective expansion of considerations formally separate from the process of decision making. For judgment pooling that optimizes error elimination, averaging choices that were arrived at independently is probably better than a group meeting in which independence and multiplicity is lost through the persuasive power of one participant or through a bandwagon conformity process. Actual group meetings for brainstorming on alternatives may or may not produce a wider range of relevant considerations than a series of individual consultations. We make no recommendation on this score. I do, however, recommend that the decision choices to be made in the formative process be shared only by those persons whose anticipated responsibility is such as to merit coauthorship. (A principal investigator should of course have a veto, whether or not he or she intends to be a coauthor on a project.)

This can be summarized in the modest and common-sense exhortation, "Make these formative decisions carefully: You are going to have to live with them for a long time." Nonetheless, a burdensome degree of decision-log keeping, consultation, and staff meetings have been described. In many settings even this is more than will be possible.

The PIRCs will inevitably be in the Catch-22 situation of pressures for immediate implementation/results and pressures for perfectly validated plans and choices. This chapter should not add to the impossibility of this impossible situation. Although its overall tenor is designed to provide arguments against the wasteful hurry-up pressures, these pressures represent bureaucratic necessities coming with the funding itself. It would be destructive advice to recommend ignoring them.

This evaluation planning is addressed primarily to CPR, and only secondarily to the PIRCs. CPR is mandated to monitor the formative evaluation processes, as well as the research results. What specifically should this monitoring consist of? I recommend that no regular formative evaluation reports be required. In general, the PIRCs should be asked to do no report writing over and above the texts of articles and books submitted for publication, lectures delivered at scientific meetings, and the more detailed appendixes to these reports necessary to guide replication and reanalysis of data. These documents should all be shared with CPR in final form, and be available on request in preliminary form. The indispensable annual report should have as its required component only a listing of these various documents, including in-house planning documents, treatment descriptions, copies of questionnaires, and so on.

CPR staff should of course be optionally available as consultants in the formative brainstorming stage, but care should be taken that their role as funder does not lead their suggestions to be overvalued in the formative decision making. The CPR staff should have the right to examine PIRC staffers' individual formative decision logs on demand. But a requirement that these be written up and submitted on a regular basis is judged to be nonproductive for scientific validity, and a use of time and writing energy better spent on scientific products.

2. Situation-Specific Knowledge and Plausible Rival Hypotheses

PIR studies now being undertaken by the PIRCs involve random assignment of treatment and no-treatment (or other contrast) conditions to experimental and control participants, and in that sense are "true" rather than "quasi" experiments. Nonetheless, it is important to consider them in the light of "threats to validity" and "plausible rival hypotheses" as employed in the quasi-experimental tradition. Indeed, "eliminating plausible rival hypotheses" is fundamental to scientific inference, encompassing nonexperimental astronomy, physical features of laboratory control in the "experimental-isolation" tradition, as well as random assignment to treatments. The airtight chambers, temperature and humidity controls, lead shielding, and so forth, are all designed to eliminate those change agents likely to affect the measures, other than the focal experimental variable. These change agents are the "plausible rival hypotheses" of the quasi-experimental tradition.

In random-assignment experiments, the variables blocked on prior to randomization are again usually plausible rival change agents affecting the dependent variable. In laboratory controls, in the choice of blocking variables, and in quasi-experimental designs, the plausible-rival hypotheses have to be specified in order to be controlled. In random assignment to treatment, on the other hand, there is a sense in which a host of unspecified possible change agents are "controlled for," rendered implausible, by the way in which randomization artificially breaks up the causal threads that normally make exposure to treatments a selective, symptomatic process. But in practice, this advantage of randomization can easily be exaggerated. Even when everything goes well in a randomized experiment, there are inevitably many rival hypotheses that can be generated to explain away the measured outcome.

The long lists of threats to validity such as those provided in Cook and Campbell (1979) are some of the general classes of recurring threats. The specifics must come from the particular experiment, and it is in terms of their specificity that the "plausibility" must be judged. These will include such ordinary environmental intrusions as when schools or hospitals close because of snow or strikes, and management disasters such as when a charismatic student ringleader leads a humorous mockery of answering questionnaires or tests.

In physical science laboratory work, traditionally, the laboratory notebook has provided a record of what goes wrong: calibration problems, runs discarded and for what reason, and so on. These have been complete enough, and well-enough archived, so that years later historians of science have been able to use them even to "correct" the scientist's published report (e.g., Holton, 1978, and Franklin, 1981, on Millikan's oil-drop experiments asserting the quantum nature of the electrical charge). This laboratory notebook or log is not usually a part of the social scientist's apprenticeship training (except in anthropology). Much more is needed, and PIR studies should reinstitute this custom.

Applied social science of the black-box, input–output, summative, external types has been justly criticized for lack of knowledge of what went on between input and output. An example is analyzing pretests and posttests gullibly unaware that no differential treatment had actually been delivered or that the measuring processes had gone awry. Indeed, there are instances in which there were three different external teams, none of them knowing the conditions of treatment and measurement; one would collect the pretests, one the posttests, and a third would do the analyses. Such external quantitative studies are a far cry from the physical-science model. Somehow in borrowing the quantitative emphasis from physical sciences by way of philosophy of science, quantitative–experimental social science has made the mistake of assuming that quantitative approaches replaced qualitative knowing. Instead, valid quantification depends on a qualitative monitoring of the experimental and measurement setting: Quantitative analysis goes beyond qualitative knowing only by depending on it, not by replacing it (Campbell, 1978).

All treatment-delivery and data-collection staffs should keep laboratory notes, at least at the level of nursing notes in a hospital. Often the intervention and measurement teams will be "guests," as with teachers whose classes are being trained and tested. These "well-placed observers," such as teachers, should be encouraged to provide notes on anything that they believe might affect the results. When later inconsistencies in the findings and other puzzles of interpretation arise, these notes can be examined both to generate plausible rival hypotheses and to render them implausible and thus often "control" them.

So important is this qualitative background, using the everyday modes of knowing, observing, gossiping, and asking, that *the quantitative applied scientific validity of a large project will be increased by funding a project ethnographer or historian.* This should not replace the laboratory notes kept by individual members of the research team.

The "well-placed observers" include not only the parents, teachers, nurses, and social workers, but also especially the intervention recipients themselves. Although these groups cannot be expected to keep laboratory notes, it would seem desirable to bring them together in small groups when the major results are in, and ask them to speculate on what caused these outcomes. These are the persons in the best position both to spot and discount plausible rival hypotheses.

This suggestion is similar to Orne's (1969) proposal that researchers use quasi-controls for detecting demand characteristics in human-participant research. Quasi-controls are participants who serve to suggest alternative explanations that the research design does not exclude. Such participants could be used in a postexperimental discussion directed toward clarifying their perceptions of the research situation, or during a preinquiry in which the research procedure is explained to a group of persons from the same population as the eventual participants and who are asked to imagine that they are participants themselves. One must be cautious in drawing inferences to the actual situation in real life from results obtained using these techniques. But, as Orne suggested, when adroitly used, quasi-controls become essential tools to clarify the results of actual investigations.

3. Selecting Sites and Populations for Clarity of Scientific Inference

In keeping with the principle of local molar validity, which was discussed in Chapter 4, and the priority given to internal validity over external validity, we recommend that sites for pilot studies be selected to optimize clarity of scientific inference. Thus, other things being equal, a middle-class school with excellent records and low residential turnover is a better laboratory than an inner-city school with 50% or more pupil turnover per year. It is better to know clearly that the intervention is effective in some portion of the target population, even though atypical, than to do the pilot testing in a more typical arena within which its effectiveness cannot be ascertained. Eventual implementation for such high-risk areas may have to be made on trust, extrapolating from areas in which effectiveness of intervention has been confirmed.

Such trade-offs must be discussed. Certainly this advice stands in considerable tension with the general exhortation that PIR stick with the problem, rather than allow the requirements of pure science precision to lead to a change in topic. Clearly, I am recommending that principle more consistently for the preventive-intervention treatment package than I am for the selection of target populations. Certainly, pilot experiments should be on groups within the target population even if these are not the highest or typical risk level targeted.

Employing the *principle of proximal similarity*—doing pilot experiments at several different points on the dimension of known selection bias—one can extrapolate with more confidence into the center of the target population than one could with but one bias level (Campbell, 1969c, including citations to Naroll).

I recommend looking for "optimal laboratories" for PIR first, and then designing interventions and measures that can be fitted into them and can make use of records being kept anyway. HMO memberships may be ideal, for example. Other health-insurance programs with mental-health services may be almost as ideal.

Encompassing administrative record systems is a possible abstraction appropriate to many ideal laboratories. In site selection (even where an HMO, for example, provided the basic population), access to coroner's records (for suicides and drug overdoses), police records, hospital records, and so on, should be negotiated in advance. Such encompassing record systems (potentially including the Social Security records on Medicaid, Medicare, use of unemployment insurance, and earnings) make it possible to follow up on attrition cases. Social Security Administration records have been used in this way, with confidentiality preserved (Boruch & Cecil, 1979, 1982; Campbell et al., 1977).

WINDUP

This has been an uneven, disconnected, and incomplete listing of suggestions coming from quite different corners and levels of a framework still to be made explicit and coherent.

What I hope is that its scattered illustrations will convince others of the need for a general sociology of scientific validity, from which could come policy recommendations for an applied social science such as preventive-intervention research.

Author's Note: This article is for the most part drawn from the final report of 5/20/85 on order no. 83M054244901D, issued by DHHS, PHS, DMM, Room 5-101, 5600 Fishers Lane, Rockville, MD 20857, on 7/20/83. Contractor: Professor Donald T. Campbell, Social Relations, Lehigh University (40), Bethlehem, PA 18015. Project Monitor: Dr. Doreen Spilton Koretz, Center for Prevention Research, Division of Prevention and Special Mental Health Programs, National Institute of Mental Health, PHS. The opinions expressed here are those of the author, and are not to be interpreted as government policy.

OVERVIEW OF CHAPTER 10

At a symposium on the sociology of scientific knowledge (SSK), Campbell defended the superior objectivity of science; however, there was a limit to how far he could take this claim. He lauded the progress made in the first stage of the sociology of scientific knowledge that is symmetrical—in other words, it explains how scientists came to believe both those truths now believed to be valid as well as those now regarded as wrong. Still, he called for a second phase that asks the question, "If there is a physical world independent of our perceiving–believing processes, what social system would ensure that our beliefs about it are the most valid?"

To advance his argument, he discussed two features of the sociology of scientific knowledge, namely, social construction of scientific beliefs and epistemological relativism. The first feature refers to the fact that all scientific beliefs are the result of social processes in the scientific community and involve trust and persuasion. The "facts" that are agreed on are the result of social negotiation. And Campbell accepted the second feature, epistemological relativism, because he believed that "direct reality" is not a matter of objective verification. On the contrary, a scientist views this external reality through the prism of his or her unproven and unconscious presumptions.

because the inductive process used by scientists is incomplete. There is a flaw in the logical argument because there are other possible explanations for the occurrence of a particular phenomenon in addition to the original hypothesis. Yet the argument is an important one because it can be used to rule out hypotheses that do not support the original assumptions. The uncertainty even extends to the "facts" used to confirm a particular hypothesis. In studying the mechanism of vision, Descartes recognized the possibility that under certain circumstances false transmissions can occur. This is further clouded by the fact that we depend on the reports of others who may have motives driven by self-interest other than reporting valid knowledge. And finally, one might personally misunderstand the facts being reported. All these areas of uncertainty point to epistemological relativism. This essential first phase of the sociology of scientific knowledge implies a skepticism about the absolute truth of any of our knowledge, as indeed it should.

The second stage of the sociology of scientific knowledge asks which social system would ensure that our knowledge is most valid. This phase would be based on conjecture. Because proof is not possible, the best that can be achieved is plausibility. The two movements in epistemology, the coherentists and the naturalists, both neglect the social processes that would ensure that the accepted scientific beliefs were most competent. Certain epistemologists, however, were beginning to consider these issues. Knorr-Cetina maintained that socially constructed beliefs were not necessarily invalid. By stating this, she allowed the possibility that the physical world plays a role in the construction of these beliefs. Pickering acknowledged that the "material world" played a part in the construction of scientific beliefs through resistances that rule out some beliefs. Although the coherence between the material world and the beliefs about it are socially constructed, the beliefs are selected by the degree of their coherence with the material world.

Campbell maintained that knowledge was based on the community's judgment of the plausibility of the explanation rather than on proof. He believed that selection, akin to the selection that operates in the physical world, offered a plausible explanation for how some social systems improve their beliefs. That is, the physical objects play a role in selecting the beliefs we hold about them, although they are not the only systems improve their beliefs. That is, the physical objects play a role in selecting the beliefs we hold about them, although they are not the only selector. This phenomenon is referred to as "coselection by referent." The literature on the research achievements of the first phase of the sociology of scientific knowledge is replete with examples of other influences on the consensus of scientific beliefs. Though he admitted that the role of the referent was a small one, Campbell still wished to explore the social customs and structures of science that would maximize that role. The social system in which the referent plays the optimum role would provide a hypothetically normative theory of scientific validity. That is, it would provide a conjecture regarding "best practices" for the community of scientists if they were to improve the validity of their beliefs. It was proposed that a social system that emphasizes freedom from religious and political authority, features experimentation, and encourages individuals to replicate experiments would increase the influence of the referent in the selection of beliefs about the physical world.

Case studies from sociologists of scientific knowledge give examples of many proposals and hypotheses that were abandoned. The reasons for abandonment may be purely social or may be guided directly or indirectly by "the way the world is." Campbell alluded to the possibility that there was some bias on the part of sociologists of scientific knowledge to minimize the role of the referent and emphasize the social component in the selection of scientific beliefs—although science ethnographers may have had the opposite bias.

In this chapter, Campbell made several important points. First, sociologists of scientific knowledge maintain a skepticism about the absolute truth of knowledge. We must recognize we can only presume that the process of scientific inquiry increases the validity of the knowledge; this can never be proven. The goal now is to demonstrate the plausibility that the physical world plays a part in selecting scientific beliefs, and to examine the social system that would maximize that role.

TEN

"SOCIAL CONSTRUCTION" IS COMPATIBLE WITH "VALIDITY" IN SCIENCE

Are the achievements of the sociology of scientific knowledge (SSK) compatible with the view that the beliefs of scientists have increased in empirical adequacy or validity? We are beginning to see integrative efforts in this direction (e.g., Franklin, 1990; Galison, 1987; Giere, 1988; Gooding, 1990; Kim, 1989). But there are many ways of conceptualizing such a unification, and the theory-of-science community has need for more alternatives before consensus on this point is achieved. The present sketch attempts to offer one such, characterized by a greater sympathy with SSK, a strong identification with the skeptical aspects of traditional mainstream epistemology, and a selectionist (a.k.a., evolutionist) version of naturalistic epistemology.

Campbell, D. T. (1993). Plausible coselection of belief by referent: All the "objectivity" that is possible. *Perspectives on Science, 1*(1), 88–108.

This chapter retains the form and focus of the original setting, the October 1990 meetings of the Society for the Social Studies of Science, and within that a symposium with an SSK agenda. For this symposium, I assumed the role of vigorous defender of the superior objectivity of science. My limp title shows the limit of how far I can go in this direction. I hope that through this debate, we can advance to a new synthesis, which I designate the second phase of the symmetrical, social constructivist, relativist sociology of scientific knowledge. This is a comparative phase that might, for example, compare the social structures and ideologies of astronomy and astrology.

By *symmetrical* I refer to a key feature of Barnes and Bloor's (Barnes, 1974; Barnes & Bloor, 1982; Bloor, 1976) "strong programme." A sociological analysis of believing is needed to explain both shifts to scientific beliefs that the relevant scientific community currently regards as valid and the adoption or retention of beliefs that are now regarded as wrong. In both cases, a complete sociological explanation should attend to the extrascientific society and culture, as well as to the intrascientific subculture, with its own customs and norms. (Although such a complete explanation is part of their program, Barnes's and Bloor's own research has tended to neglect the sociology of persuasion within the scientific community.)

The central beliefs of scientific communities are socially constructed (Knorr-Cetina, 1981; Latour & Woolgar, 1979; Pickering, 1984). The image of a solitary scientist discovering nature "directly" is wrong on both counts. The laboratory results are never "direct viewings." The context of beliefs within which the meter readings are interpreted are based on trust of the reported observations of other scientists, rather than on a solitary scientist's own perceptions. The trusting and persuading are social processes. By the time the laboratory group or the larger field of participants have agreed on a "fact," this has been achieved by a social negotiation process that has selectively discarded and augmented the "meter readings."

The epistemological *relativism* of SSK (Barnes, 1974; Collins, 1981b) derives from the fact that our epistemological position is profoundly indirect and presumptive. Vision, for example, is based on inferences based on contrasts in superficial reflections. Its anatomy and physiology show nothing like "direct" transmission of "knowledge" of "external" objects, but instead highly presumptive unconscious constructions. Those ordinary perceivers who posit "real objects" can do so only by unconscious presumptions, not by proof. When we come to the "real objects" posited by scientists, whether visible or invisible, macroscopic or microscopic, these are even more presumptive. We epistemological relativists reject all claims of direct realism, or, as I like to call it, "clairvoyant realism." My own "hypothetical realism" (Campbell, 1959) has all along been closer to current "antirealisms" than to a direct realism. From the first, I have accepted the analyses of the skeptics and have made clear that I went beyond this only by unproven assumptions. For those unwilling to base belief on unproven assumptions, epistemological relativism becomes ontological nihilism.

My own agenda has been to relate the philosophers' epistemological problems to evolutionary theory, and to that more abstract model of discovery and adaption shared by trial-and-error learning, natural selection, cultural evolution, acquired immunity, radar, sonar, echolocation, and vision: *selection theory* for short. Pursuant to that agenda, I would like to relay graphically in Figure 10.1 what I take to be the consensus position of modern epistemologists and philosophers of science. It is a perspective that provides philosophical warrant for the symmetrical, relativist, social constructivist, sociology of scientific beliefs.

One of the scandals of induction can be expressed by noting that science makes use of an invalid logical argument, making the error of the "undistributed middle term" or of "affirming the consequent." But although invalid, the argument is not necessarily useless.

The logical argument of science has this form:

> If Newton's theory *A* is true, then it should be observed that the tides have period *B*, the path of Mars shape *C*, the trajectory of a cannonball form *D*.
>
> *Observation confirms B, C,* and *D* (as judged by the scientific consensus of the day). Therefore Newton's theory *A* is "true."

We can see the fallacy of this argument by viewing Figure 10.1. The invalidity comes from the existence of the crosshatched area—that is, other possible explanations for *B, C,* and *D* being observed. But the syllogism is not totally useless. If observations inconsistent with *B, C,* and *D* are agreed on by the consensus of participating scientists, these impugn the truth of Newton's

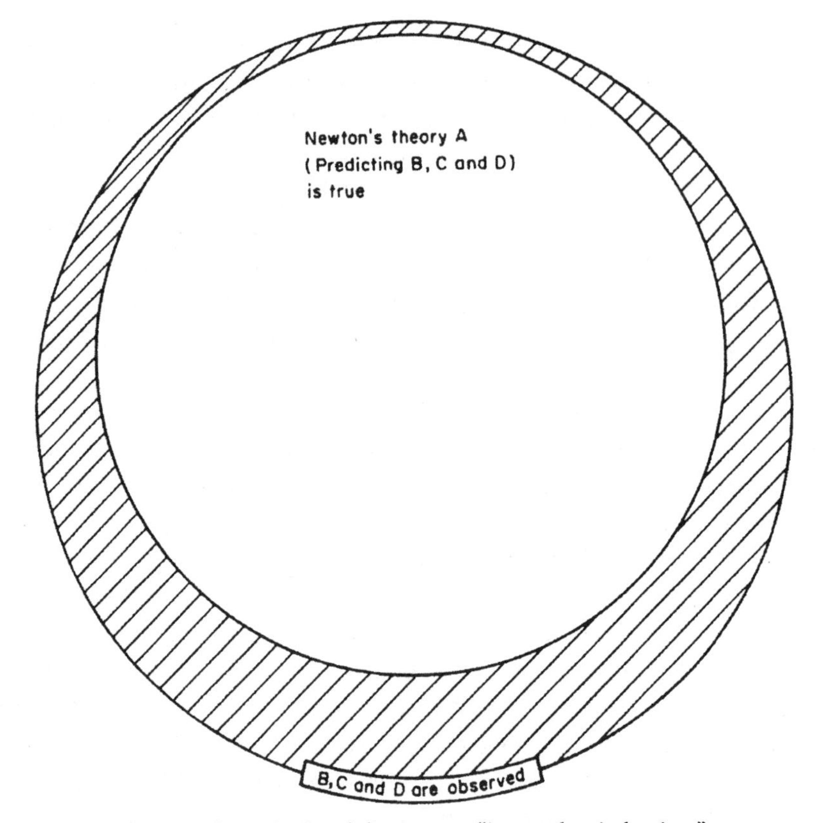

Figure 10.1. Newton's gravitational theory as an "incomplete induction."

theory *A*. The argument is thus relevant to a winnowing process, in which predictions and social consensus on observations serve to weed out the more inadequate theories. Furthermore, if the predictions seem confirmed by the consensus of current experimentalists, the theory remains one of the possibly true explanations.

All inductive achievements are "incomplete inductions" (Campbell, 1990), as illustrated in Figure 10.1. It is now generally recognized that this incompleteness also applies to the so called "facts" that test or "falsify" theories. Any "well-established" scientific fact that falsifies a theory is a socially negotiated consensus for which a diagram such as Figure 10.1 could be drawn, with a fringe area of plausible rival interpretations.

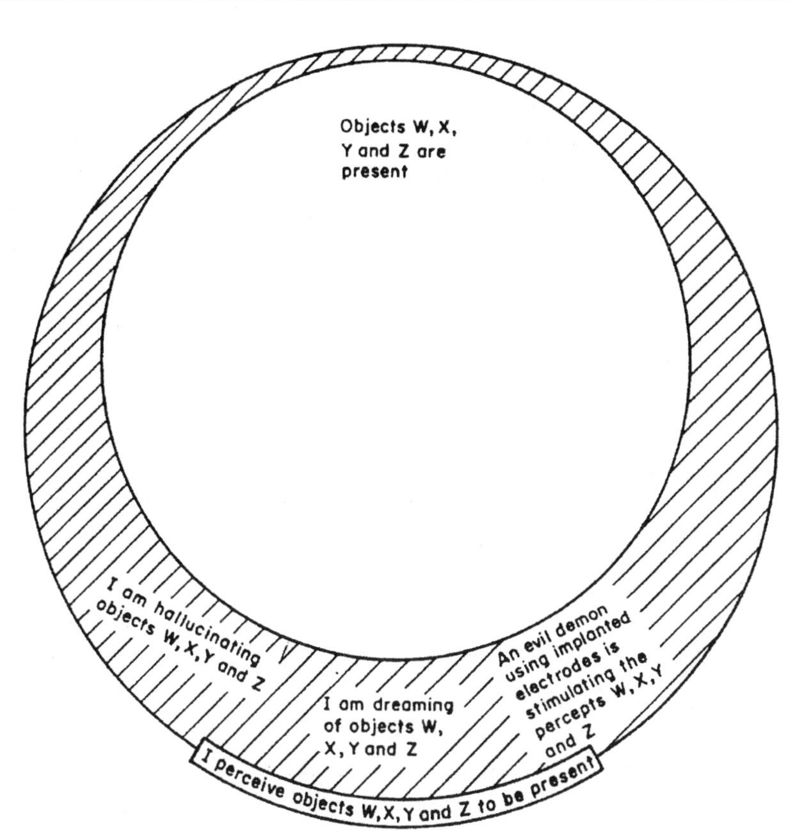

Figure 10.2. Visual perception as "incomplete induction" (with apologies to Descartes).

The quasi-Euler diagram is also useful in presenting Descartes's skepticism about sense perception, as in Figure 10.2. Descartes was perhaps the best neurophysiologist and physicist of vision of his day. In this role he took a reflexively realistic stance in taking the machinery of perception to be made up of real objects and events in the world, comparable to the ordinary objects of perception. He studied the physics of light rays and their propagation through pinholes and lenses. He posited a subsequent message transmission through nerves to the brain (hydraulically, by fluids in neural tubes). All of these mechanical links increased his awareness of the possibility of malfunction. For example, pseudotransmissions might be initiated at an intermediate link rather than by the "perceived object" itself, or other factors might intrude into this mechanical sequence from tangential causal chains. All this increased the plausibility of the skeptic's argument from illusion. Trust in perceptions produced by such a mechanism required faith in powers or processes that would keep the vulnerable causal chain insulated and free of defect. Lacking selection theory, Descartes chose God. Those modern evolutionary epistemologists who invoke biological evolution (e.g., Goldman, 1986; Quine, 1969b; Rescher, 1977; and the many others cited in Campbell, 1974a, and Cziko & Campbell, 1990) use dear old Mother Natural Selection to support a parallel trust in vision, albeit a more qualified trust, not providing incorrigibility. But many selectionists at the level of the evolution of the visual system tend toward a complacent foundationalism with regard to the momentary operation of vision.

Descartes got to his skepticism about vision from what he took to be the illusory vividness of his own dreams, from an up-to-date knowledge of the physics, anatomy, and physiology of vision, and from a pathological need for certainty. But his analysis has been a part of the great tradition of perceptual skepticism back to the pre-Socratics. Plato's parable of the cave (bk. 7 of *The Republic)* has that theme. The "strange prisoners" are "like ourselves." "They see only . . . shadows. . . . To them, the truth would be literally nothing but the shadows of the images." In this allegory, "the prison house is the world of sight." Note how compatible this is with our modern physics and physiology of perception, in which the brain reifies objects from patterns of light indirectly and superficially reflected from them.

From the epistemology exemplified by Figures 10.1 and 10.2, all knowing can be epitomized as guessing what is casting the shadows—the shadows on our retina or the shadows on our laboratory meters. Figures 10.3 and 10.4 provide an advertiser's illustration of the equivocality of shadows. But the more fine-grained detail of a photograph (or of a "direct" perception) differs from the silhouette shadow only in degree, not in fundamental epistemology. Psychoepistemologically, the "guesses" of direct perception are unconsciously automated, and the conscious experiential "givens" are of external objects as though directly, unmediatedly, known. But this did not mislead Plato or Descartes, and it should not mislead us as epistemologists. In anticipation of the title theme, note that any transient belief that the shadow caster was a dark-alley gunman, or the belief that the photo was of a tennis player, is only

Figure 10.3. Shadow of a menacing gunman as an "incomplete induction."
SOURCE: Copyright Information Resources, Inc. Used with permission.

*co*selected by the shapes of shadow and photo: Essential also to their formation are the culture and experiences providing the repertoire of possibilities, one of which was triggered by shadow form and photo contours.

As social animals, we acquire confident beliefs through the reports of others. The layers of equivocality are then more numerous, as shown in Figure 10.5. As the process is diagramed, I may end up confidently believing that, in the next room, out of my sight, the cat is on the mat. This belief is compatible with the cat's really being on the mat (the inner, clear, circle). But the reporter who supposedly is in a position to see the cat in the next room may have hallucinated, and the cat was not really on the mat. The inner two circles are as

Figure 10.4. A plausible rival interpretation of the shadow of Figure 10.3.
SOURCE: Copyright Information Resources, Inc. Used with permission.

Figure 10.2 (Descartes). The reporter speaks out, "The cat is on the mat." Now this may be a result of the cat's being on the mat, and being perceived as on the mat, or it may be a result of idiosyncratic semantics for "cat," "mat," and "on," or it may be a result of the reporter's wanting me to believe that the cat is on the mat, whether or not it actually is.

For example, the reporter may be my child, who sides with the cat's preference for sitting on the sofa and does not want me to go in and discipline the cat. This source of equivocality is one peculiar to social vertebrates, not shared by the social insects. For us social vertebrates, in our public truth claims, there are always two motives: (a) to report validly to the best of our own knowledge and (b) to influence the decisions we expect to be made on the basis

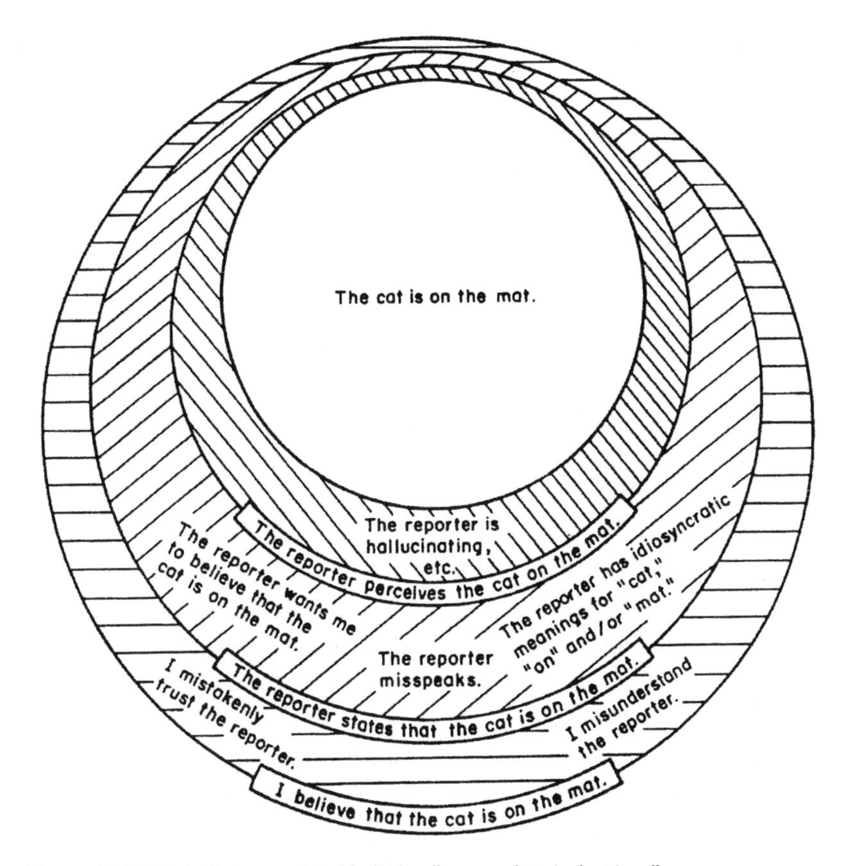

Figure 10.5. Verbally transmitted belief as "incomplete induction."

of our report in a direction deemed favorable to us. These two motives are often in conflict. The second is often the stronger, particularly if our very lives and livelihoods depend on those decisions. Because 99% of the beliefs of a scientist are solely dependent on the observations of others, this makes social control of the validity of reporting central to an epistemology of science (as Hull, 1978, has noted) and to ordinary knowing of social animals.

The crosshatched areas of Figures 10.1, 10.2, and 10.5 can never be entirely eliminated. Beliefs, and the best of current scientific theories, will always be underdetermined, underjustified. This ubiquitous "inductive incompleteness"

(Campbell, 1990) leaves ample room for the influence of social and personal interests seemingly tangential to scientific inquiry. The research achievements of the symmetrical, social constructivist, relativist programs in the sociology of scientific knowledge amply document such influences. This is the triumphant first phase that all on the symposium applauded.

The first phase of the sociology of scientific knowledge has been more epistemologically relevant than previous sociology of science in the sense that it has focused on the social determinants of scientific beliefs, rather than just on the social organization, communication patterns, and publicly affirmed norms of science. Although its goal has been descriptive, insofar as it has supported epistemological positions, skepticism has been implied, as indeed it should if knowledge be defined as fully justified, true and known to be true, belief.

The second phase that I call for would be a comparative sociology. If there were to be a physical world independent of our perceiving–believing processes, what kinds of social systems of belief revision and belief retention would be most likely to improve the competence-of-reference of beliefs to their presumed referents? Particularly challenging are scientific beliefs about invisible physical events, processes, and entities, such as winds, sunbeams, electricity, magnetism, gravity, atoms, and quarks. Such a second phase will have to be conjectural, like science itself. It will never achieve certainty, or proof. The best it can achieve is plausibility, and that plausibility will have to be based on trusting the great bulk of one's ordinary beliefs.

There are two such enterprises going on in epistemology today. Mainstream Anglo-American philosophy since G. E. Moore has proposed starting the analysis with the assumption that we have a great deal of knowledge (rather than total doubt), thus knowingly committing the other scandal of induction, using presumed knowledge to explain how knowledge is possible. They employ arguments to the best explanation, causal scenarios as to how the referent caused the belief, arguments for trusting modes of belief formation that have seemed reliable in the past, and so on. Justification no longer is expected to provide incorrigibility, indefensibility, or a guarantee of certain truth. "Coherentism" (Lehrer, 1974) has taken the place of perceptual foundationalism. Indeed, Pollock (1974) and Lehrer (1989) would now revise the traditional definition to read "Knowledge is undefeated justified true belief," and would probably acknowledge that "not yet defeated by apparently falsifying observations" is an appropriate expansion, and that the "truth" of the belief that is required to sanction the use of the term "knowledge" can never be independently ascertained.

Unlike this mainstream position, "naturalist epistemology" (Kornblith, 1985; Quine, 1969b) is still perceived as very much a minority point of view. This is probably a result of misperceptions. Certainly it is no more flagrant in committing the scandal of using presumed successful inductions to explain and justify the inductive process. Naturalistic epistemology explicitly assumes theories of biological evolution and neurophysiology to explain how we can usually trust beliefs generated by our eyes and by our other innate inference processes (and why in settings different from the ecology of evolution they can go awry). Naturalistic epistemology tentatively also trusts modern physics to describe the ontology of the physical objects the biological perceivers are forming beliefs about.

For the most part these two movements make little distinction between the epistemology of individual perceivers and the epistemology of scientific beliefs. Concomitantly, the essentially social process aspects of science are neglected by both. The second phase of the sociology of scientific knowledge would fit in here. It would be a speculative, presumptive enterprise. It might, for example, compare the social systems of astronomy and astrology, asking, were there to be a real world of independent referents, which social system of rituals, ideology, mutual persuasion, and belief change would be more likely to improve the competence of reference.

Within the movements in the sociology of scientific knowledge there are signs of readiness for such a second phase. Karin Knorr-Cetina (Knorr-Cetina & Mulkay, 1983, p. 6) has taken pains to make it clear that "epistemic relativism" does not entail "judgmental relativism"—that is, that *socially constructed* does not entail *invalid.* Thus she has conceptually opened up the possibility of a theory of the social construction of scientific knowledge claims in which the hypothesized social system of knowledge-claim retention and revision allows some role, however small, for "physical reality," for example, in the social construction of physicists' beliefs. In a later article, Knorr-Cetina (1987) has made a substantial contribution to this agenda with an extended discussion of the relation between microsociological constructivist sociology of science and evolutionary theories of scientific conceptual change (i.e., "selection theory"). She described the individual ideational discovery as an "editorial process," with the pool of stored-up "variations" available to the "ethno-epistemological work" of negotiating a new consensus. "Consensus formation . . . [is] the selective affirmation of knowledge claims" (p. 194), being a very *local contextual* activity, typically negotiated in laboratory settings. "The process in which conceptual variations are produced is *at the same time* the process in which previous variations are either selectively reproduced or 'weeded out'" (Knorr-Cetina, 1987, p. 194). I am in essential agreement with this important essay.

Pickering (1989) posited that the material world not only provide a "plasticity" of alternate experimental arrangements, but also, in the limitations of its plasticity, provide "resistances" that constrain the socially constructed coherences. To sample from his own words: "The nonuniqueness of coherence, it has been suggested, excludes a constitutive role for interaction with the material world in the articulation of scientific knowledge. But I believe this latter conclusion to be mistaken. In what follows, I seek to articulate a more adequate alternative to correspondence realism while avoiding antirealism. I argue that coherences between material procedures and conceptual models should be seen as made things, as actors' achievements, and not as arising naturally and uniquely from the material world itself. But I maintain a noncorrespondence realist perspective on the making of coherence. The thrust of my analysis is that, if one examines the evolution of passages of experimental practice in time, it becomes clear that scientific knowledge is articulated in accommodation to resistances arising in the material world. There is a direct and analyzable relation between scientific knowledge and the material world, though it is one of made coherence, not natural correspondence."

Latour (1987) also spoke of resistances in "trials of strength." Although he portrayed dominant consensuses in science as products of political–military power struggles, he did not assert that those scientists with the greatest political power at any one time are able to dictate the scientific consensus then being formed. Quixotically, Latour characterized microbes and coral reefs (etc.) also as political actors with independent power, whom the competing scientists attempt to co-opt as allies.

I also find passages in Barnes's *T. S. Kuhn and Social Science* (1982), in Bloor's *Wittgenstein* (1983), and in both Barnes and Bloor as portrayed by

Collins (1985, pp. 172–174; see also Bloor, 1984, 1989) that permit such a move (Campbell, 1989, esp. p. 155).

PLAUSIBLE COSELECTION OF BELIEF BY REFERENT

It is high time that I got to my main and terminal theme. Whatever "objectivity" is achieved is based on socially shared plausibility judgments rather than proof. In the informality of my oral presentation, I asked the audience to take this example: "Your belief that I am now up here speaking to you is most plausibly explained by my actually being here (for those of you who have such beliefs). I would ask that we all pinch ourselves right now to see whether we are dreaming, except that I have had what I later decided were vivid dreams in which performance of that test indicated I was awake." Of course, many of the plausibility judgments involved in believing in science will be much less compellingly plausible than that, such as belief in electricity as a cause of lights, sparks, and electromotive motions.

My own plausible scenario for how some social systems might improve their shared beliefs about the physical world is epitomized by "selection." I have for 35 years indulged in an "evolutionary epistemology" full of analogues to natural selection in biological evolution (Campbell, 1988b). The conjectured "fit" of animal form to environmental opportunity is conjectured as being a result of the fact that animal form has been selected, in part, by that same environment. The fit of radar screen image to airplane in the sky (if there be one) is a result of the fact that the radar image has been, indirectly and partially, selected by the airplane. (Inductive incompleteness abounds: It could have been a flock of birds or aluminum confetti.) The fit of one's belief in visibly present objects is plausibly explained by the scenario that the belief has been selected in part by the presence of those objects as reflectors of ambient light (Campbell, 1956). (Such objects operate only as coselectors, not sole selectors, in that visual research shows that biased readinesses to perceive are also essential and that these expectations can cause misperceptions.)

When my analysis is extended to changes in consensus in scientific beliefs, assertions of improved competence of reference require a plausible scenario to the effect that the purported referents of the belief have played some role in the social selection and social construction processes involved. The research achievements of the symmetrical, relativist, social constructivist sociology of scientific knowledge make it clear that many other interests and restraints have also been involved. In a 1981 triumphal summary of this literature, Collins claimed empirical support for the conclusion that "the natural world has a small or non-existent role in the construction of scientific knowledge" (Collins, 1981b, p. 3). My admiration for the achievements of science lead me to reject the "nonexistent" extreme. It is that "small role" that I want to provide plausible scenarios for. In the comparative sociology of believing communities of the second stage of the sociology of scientific knowledge, I look for plausible scenarios as to which types of social customs and structures would maximize that small role and which would minimize it.

The slogan "coselection by referent" represents a significant modification of the evolutionary epistemology program as applied to science. Popper (1935), Toulmin (1967, 1972, 1981), Richards (1981, 1987), Hull (1988a, 1988b), and others have used selectionist models of the history of science to explain historical increases in the presumed validity of scientific theories. But they have left the details of the selection process unspecified, as though all of the social processes of proposing and weeding out among scientific beliefs and laboratory apparatuses were ones that plausibly increased validity. This is obviously falsified by the detailed studies of belief winnowing that the sociologists of scientific knowledge have provided.

David Hull's (1988b) *Science as a Process: An Evolutionary Account of the Social and Conceptual Development of Science* is a troubling case in point. As second-phase sociology of science, it is admirable, with descriptive theories about how seeking esteem from fellow scientists and mutual criticism lead to increased validity. But it is not his selectionist analogues to biological evolution that achieve this. As John Maynard-Smith has stated in the book's most famous review to date, "[Hull's] explanation for such cooperation is that the replicators (genes) in the cells are identical . . . in different members of the group and will be transmitted to future generations only insofar as the group as a whole . . . is successful. Now an analogous argument might explain the loyal cooperation of members of a tightly knit research group, but would equally well explain the cooperation of the members of a religious sect or of a group bound together by a common political or artistic program" (Maynard-Smith, 1988, p. 1182).

Independently, I have made a similar observation: "The resulting selection (whether of replicators or interactors) has no formal implications for increasing validity in scientific conceptual evolution. Differential extinction and proliferation occurs in religious cults as well, without producing increased validity in beliefs held about the invisible causes of visible phenomena" (Campbell, 1988a).

In an 1895 sympathetic critique of evolutionary epistemology, Georg Simmel found that the concepts of "useful" and "true" become indistinguishable. In a parallel application of the natural selection analogue to science, Steve Fuller (1988, p. 100; also Campbell, 1988c) denied that "the cognitive value of a [socially] reproduced perspective [can] be clearly distinguished from its social or 'survival' value." That is, in evolutionary epistemology, scientific truth is to be redefined as those beliefs that lead to the tenure, promotion, and fame of a science professor and his students. Without further specification of how the referents of scientific beliefs participate in their selection, this is where evolutionary epistemologies of science leave us.

In biological evolutionary epistemology, one can pass the buck for our trust of visually generated beliefs to the adaptive bias in natural selection, without specifying or distinguishing among the many components in the selection process. In parallel with Descartes, the believer can say "[God] [natural selection] would not have given us eyes that regularly deceived us." But unlike the ants, termites, and bees, most of we vertebrates cannot say "[God] [natural and cultural selection] would not have given us social fellows [fellow scientists] who would deceive us" (Campbell, 1987b; Hull, 1978).

In this respect and in many others, evolutionary analogues applied to the processes of science fail to provide epistemological comfort for scientific beliefs. The epistemological problem or scientific beliefs is not to explain the social and political success of scientists or their institutions. Rather, the puzzle to be solved is the fit of the beliefs of physicists to the supposed physical world referred to. The new selectionist dogma requires that such fit be explained by "the physical world's" role in selecting such beliefs. Any close analysis of the proposal and abandonment of beliefs in science will show that there are many other systematic selection processes involved that are irrelevant, or negative, or only indirectly relevant (Campbell, 1979c) to improving such fit. An evolutionary model that invokes all of the belief selectors in the social system of science indiscriminately is failing to be epistemologically relevant in not providing a plausible scenario about how the beliefs of physicists might be improving their competence of reference.

COSELECTION BY REFERENT

The "co" is meant to emphasize that, in any possible ideal social system for science, referents will never be more than coselectors, one selective force among many, but the essential one for progress in the competence of reference of scientific beliefs. My call for the second phase in the sociology of scientific knowledge is a call for discourse on the possibility that some social systems differ from others in the likelihood that the referents are participating in the social process of belief selection. A social process theory for optimizing coselection of belief by referent can be used as a hypothetically normative theory of scientific validity (Campbell, 1979c, 1986c).

For example, a social ideology valuing the advance prediction of astronomical events might plausibly increase the likelihood that the way the astronomical world was (were there such a world independent of current beliefs) would play a role in selection among competing beliefs. Still, this selection is bound to be fallible. It was a daring rhetorical ploy for the thirteenth-century Jesuit astronomers in the court of the Chinese emperor to claim to be able to predict eclipses of the moon (Sivin, 1980). It became most persuasive rhetoric when the emperor and the Chinese astrologers and astronomers reported perceiving such eclipses approximately when the Jesuits predicted them. Socialization to honesty among in-group peers is needed to explain why the Chinese astronomers did not just deny that the predicted eclipses had occurred, as their social interest predisposed. Within a few decades, these Chinese astronomers had adopted the Ptolemaic system, thus plausibly increasing the fit of their beliefs to their purported referents. Later, the Ptolemaic theory was replaced by a Copernican–Keplerian–Newtonian astronomical theory. A social ideology emphasizing independence from political or religious authority, the rituals of experimentation, with the ideology of each believer's being free to replicate the experiment, and so on, might increase the likelihood that "the way the world is" could influence belief selection (Campbell, 1986c). Even the now obviously false belief in "facts that speak for themselves" (Campbell, 1979c) might increase the likelihood that the way the world is participates in scientific belief selection.

Let us return now to the sociologists of scientific knowledge. Their case studies of the social construction of scientific consensuses report on the proposal and abandonment of many hypotheses about process and instrumentation. They offer microprocess studies of belief selection appropriate to selectionist accounts. Thinking of Latour and Woolgar (1979), Knorr-Cetina (1981), and Pickering (1984), for example, we probably have a hundred or so instances. These could be tentatively classified as to the type of selection involved. Some of these episodes will be classified as purely social: An idea is not followed up because it would offend the laboratory head, or because it would give comfort to a rival research group, or because of lack of funding.

Other ideas are reported as being tried out and found not to "work." In such episodes it is possible that the way the world is participates in belief selection, even though there are vast resources to negotiate why it did not work. More borderline cases are those in which an idea is rejected because of reasons why it will not work, or because of rumors that a trusted researcher is known to have tried it and failed. If those reasons and rumors themselves have been coselected by the way the world is, then (still more indirectly) coselection by referent may have been involved. Latour's (1987) chapter "Laboratories" also provides several examples of beliefs being abandoned by the resistance encountered in laboratory practice.

Reflexively, we should use interest theory (Barnes, 1977) to critique such a data set. The scientists being reported on shared an ideology probably leading them to exaggerate the role of the referent in belief selection, even in their apparently unguarded gossip and shoptalk. However, the SSK authors of these works might have had an opposite bias, in favor of the dramatic and more publishable message of "social construction out of whole cloth." It is conceivable to me that they did *not* start out with this bias, that it was for them a discovery in the research process. Interview testimony on this point would abet our coherence-based discretionary judgment. However, because they were all pressed by the need to write brief, vivid, and publishable books and articles, in which only one tenth of their notes could be used, a postresearch selection bias in the "constructed out of whole cloth" direction is possible. Access to the full field notes might provide a less biased set of episodes. On the other hand, we might expect an opposite bias on the part of science ethnographers such as Hull (1988b) and Galison (1987). Although I have not done the systematic rereading and coding, I am sure that these properly venerated texts (Knorr-Cetina, 1981; Latour & Woolgar, 1979; Pickering, 1984) make it plausible that the purported referents of belief are participating to some extent in belief selection.

SUMMARY

The achievements of the sociologists of scientific knowledge are quite compatible with the skeptical analyses of our predicament as knowers retained within mainstream philosophy of science. Any case for "successful" sciences' having achieved improved competence of reference for their scientific beliefs must be based on presumptive coherence grounds, not proof. In particular, it must be made plausible that the purported referents of the beliefs could have participated in the historical selection of those beliefs.

PART IV

DESIGNS AND TECHNICAL ISSUES

In Part I, Campbell offered his vision of a future society in which social experimentation and hard-headed evaluation would be integral in the quest to achieve incremental but continuous progress toward societal improvement. Part II highlighted some of the problems associated with the methodologies now available to study social problems and social reforms—namely, the threats to internal and external validity. Part III offered Campbell's justification for the notion that we can increase the validity of our knowledge. He proffered arguments that the social processes of the scientific community can lead to increased validity. In his view, the scientific community accomplishes this by focusing on common problems while monitoring and critically evaluating the ongoing work of others in the field. Through this process, the threats to validity are enumerated, methods are developed to control for those threats, and studies with increased interpretability are offered.

Part IV is the result of the social process described in Part III. It is intended to provide the student of applied social research some tools to use in program evaluation. The designs offered in this part were developed through the cumulative experiences of the community of researchers involved with evaluation. One of the pitfalls in such research is the regression artifact that is often mistaken for a true effect. Chapter 11 discusses how this threat can be rendered implausible by using time-series data. Chapter 12 demonstrates how regression artifacts can occur when matching is used to adjust for pretreatment inequality. Posttest differences may be results of regression to the mean rather than to any treatment effect, and this regression artifact can extend over previous and subsequent years. The following chapter, Chapter 13, describes a design that can be used to assess the impact of an award when selection for the award is based on an eligibility score. The regression-discontinuity design was the product of a collaboration of scholars in several disciplines and at multiple institutions. Along with this useful design, the reader is given a view of the scientific community at work with its fits and starts, barriers and solutions. In the previous three chapters, the treatment groups were selected based on extreme scores on the pretest or by a cutoff point in an eligibility score. Chapter 14 offers a design that can be used when the treatment group is selected for other reasons, perhaps political reasons or merely convenience. The regression point displacement design is useful because a single experimental unit can be used with a limited number of control units. The next chapter, Chapter 15, offers a design that does not attempt to adjust for pretreatment differences but uses those differences to estimate the effect. This design is useful when there are repeated measures at regular intervals and those receiving the treatment are volunteers. Finally, Chapter 16 examines a retrospective design referred to as the case–control design, which estimates the relative risk of having been exposed to some risk factor. It is argued that including participants in the control group who are not likely to be exposed to the risk factor increases the plausibility of the threats to validity. This consideration outweighs any increase in precision from using such participants.

These chapters provide evidence that Campbell saw quasi-experimentation as a process—a quest for new or modified research designs that would increase the validity of the research results. The designs outlined in Campbell and Stanley (1966) and later in Cook and Campbell (1979) were not a definitive list, but merely a beginning. Until the end of his life, Campbell was involved in creating new designs and recombining features of established designs in ways that increased their ability to rule out the threats to validity. He devoted a significant amount of his time to critiquing the elements of quasi-experimental designs and emphasized that only the vigilance of the research community could ensure that the designs produced valid research results.

OVERVIEW OF CHAPTER 11

Suppose you are asked by the head of a health maintenance organization (HMO) to study whether offering psychotherapy to patients who have frequent medical visits might reduce their demand for medical services. You look at the number of visits in the previous month and suggest that psychotherapy be offered to the group that has the highest number of visits. In the next 2 months, you observe that the number of medical visits has indeed decreased significantly for this group. You confidently report back to the HMO director that psychotherapy does reduce the need for frequent medical visits. Campbell suggests that you may have led the director astray. You may have fallen into a methodological booby trap known as a regression artifact.

Past studies suggest that when mental health services are provided to patients, the result is a reduction of approximately 20% in their overall medical service demand. Those patients who received the psychotherapy, however, were individuals who were exceptionally high users of medical services. One criticism of these studies suggested that the high usage was temporary, and that the medical visits would have dropped even without psychotherapy. This phenomenon is known as *regression to the mean,* and evaluators would be remiss if they were to attribute the drop in medical services demand to the psychotherapy. It is merely a pseudo-effect referred to as a *regression artifact*.

Campbell demonstrated how this rival hypothesis of regression artifact can be ruled out with time-series data that plot usage prior to and after psychotherapy. Indeed, plots for two such studies showed that regression artifacts played some part in the decrease of medical visits for those receiving psychotherapy. By examining those plots carefully, however, one might be able to justify a small genuine effect. Campbell also suggested the use of control groups to estimate the regression to the mean and to rule out the rival hypothesis that the general population had a similar overall decrease in the specified period of time (history). Finally, using data from HMO records, already available and computerized, would provide the clearest possible picture of the effect of psychotherapy.

ELEVEN

REGRESSION ARTIFACTS IN TIME SERIES

My examples from a mental health service quasi-experiment will illustrate a recurrent methodological booby trap, rather than successful practice, yet they demonstrate the value of extended time-series data displayed in temporal detail. The Connecticut crackdown on speeding discussed in Chapter 3, Figure 3.2, provides a classic illustration of a *regression artifact* as a result of selecting a time-series peak to administer a treatment.

Jones and Vischi (1979) have reviewed studies in which mental health services were added at a specific point in time, after which the demand for medical service was reduced for those electing psychotherapy. The short, solid line of Figure 11.1 illustrates the most common type of data presented and the 20% decline typically found. Olbrisch (1977, 1980) has criticized such evidence

Campbell, D. T. (1984d). Hospital and landsting as continuously monitoring social polygrams: Advocacy and warning. In B. Cronholm & L. Von Knorring (Eds.), *Evaluation of mental health services programs* (pp. 13–39). Stockholm, Sweden: Forskningsraadet Medicinska.

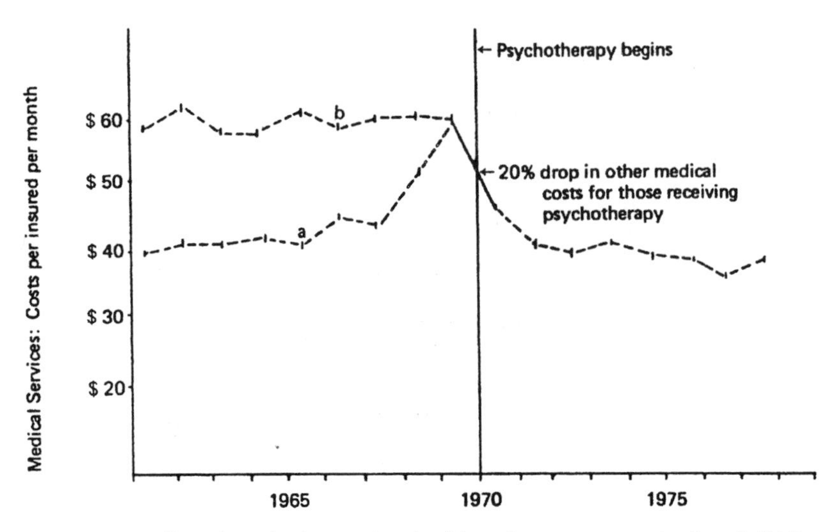

Figure 11.1. Effect of psychotherapy in a health maintenance organization. Solid line corresponds to effects noted in several actual studies (fictional illustration).

on the grounds that transiently high users of other medical services were referred to psychotherapy, or may have sought it out in a transient period of high disturbance that was accompanied by a need for medical services. Kogan, Thompson, Brown, and Newman (1975) have interpreted their own data as showing regression artifacts. From the records of the HMOs, it is possible to provide time-series data testing this criticism, even though such series are rarely examined. All studies agree that the psychotherapy participants had, on average, been exceptionally high users immediately prior to therapy being provided. The issue relevant to Olbrisch's argument is whether or not this high usage was a recent flare up, in which case a return to the mean would be expected without therapy. Figure 11.1 illustrates such in line *a*. On the other hand, if those receiving therapy had a sustained record of high usage, followed by the 20% drop, as illustrated in by line *b* of Figure 11.1, then the regression artifact interpretation seems to me ruled out. In this instance, the time-series becomes impressive evidence of genuine impact until someone comes along with a more plausible rival hypothesis.

Among the hundred or so data sets presented by Jones and Vischi (1979) there are two that present multiple time periods prior to psychotherapy. These

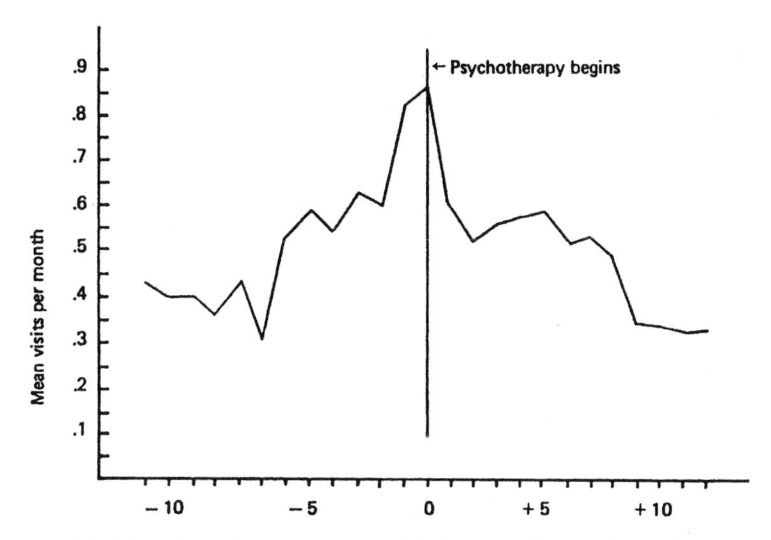

Figure 11.2. Mean physician visits per month before and after psychiatric intervention (*N* = 426; from Table 18, p. 54, in Jones & Vischi, 1979, based on Patterson & Bise, 1978, supplemented by personal communication from Bise).

are graphed in Figures 11.2 and 11.3. Unfortunately, both strongly support regression-artifact interpretation. Most of the gains shown are thus to be interpreted as pseudo-gains. Perhaps in addition there is some evidence of genuine effect. In Figure 11.2, the final plateau for months +9 through +12 averages around .34, whereas the plateau for –12 through –9 averages .40, which is 15% higher. If this was confirmed by much larger samples and longer time periods, it might be statistically significant. Providing this drop in physician visits was independent of a general population downward trend in usage, the difference might be interpreted as a result of psychotherapy. Thus, although making it plausible that the bulk of any before–after drop is a result of a regression artifact, there is also some possibility that a genuine benefit is also present.

A control group such as those in Figure 11.3 would help in this regard, if they confirmed that the general trend was slightly upward rather than downward. A different type of control group would also be useful. This would explore the nature of the time dependence or autocorrelation of medical service usage, and the shape of the peak that would be expected on the basis of

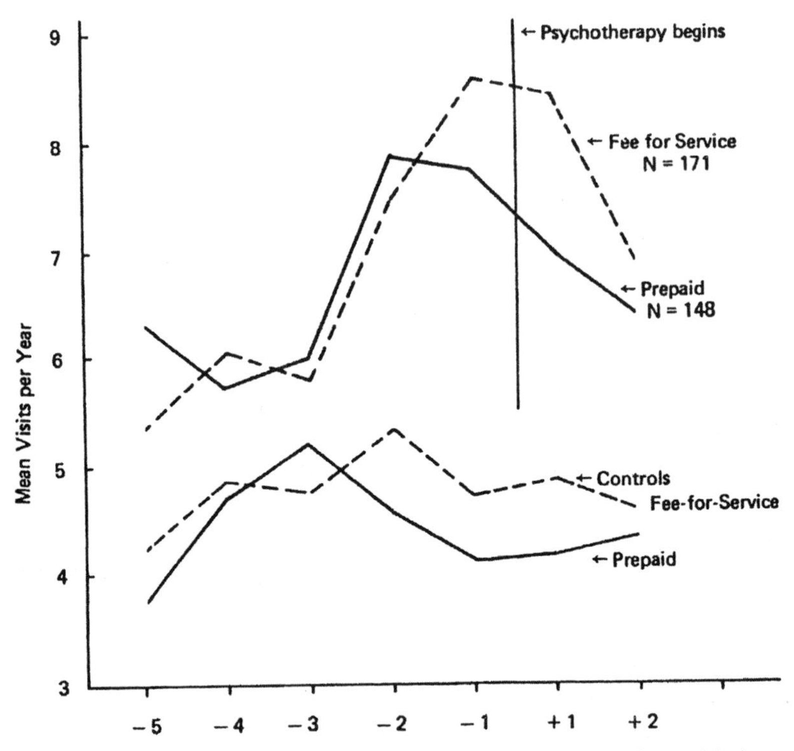

Figure 11.3. Mean outpatient visits by year, before and after initiation of psychotherapy, psychotherapy visits included. Plotted from data in Jones & Vischi (1979), Table 18, p. 54, based on Kogan et al. (1975).

a pure regression-artifact interpretation. For example, one might search the nontherapy cases in the program for those who had a month of use above .75 or .80 and plot the 12-month span surrounding each of these cases. Then by averaging the months prior and subsequent to the peak month for 500 such cases, one might have a portrait of a pure regression artifact as a result of selecting an extreme point to initiate therapy. But perhaps it is not the absolute level, but the high point relative to the previous level that motivates patients and advisors to try psychotherapy. Another comparison group selected by starting from the highest month of usage in a 5-year period would also be useful. What I would look for in such controls would be an asymmetry of the ascending and descending slopes. Do these show the sharper descent that may seem to characterize Figure 11.2? Or are they fully symmetrical, in which case a steeper posttherapy decline can be attributed to the therapy.

In terms of the economic arguments about psychotherapy coverage in health insurance, it is the costs per month or year, not the number of visits, that is important. Because these are already computed in most of the archival records, or computable from the treatment record, monetary costs should also be used in such analyses. The importance of the problem requires it. One wonders if the number of visits have been presented just because they show larger effects than costs in the simple before-and-after comparisons. But because the type of visits that provoke referral to psychotherapy are probably those low-costs visits in which no medically diagnosable ailment is found, the regression-artifact pattern might be relatively much less sharp for a time series of costs. We also need such cost figures in several forms: for example, including psychotherapy for the overall cost–benefit decisions, and excluding psychotherapy (and perhaps also diagnostic visits resulting in no medical treatment) for scientific interest in psychosomatic causation. We also need studies of total group plan costs from health maintenance organizations (HMOs) that have at various time periods had comparable employer or union groups that included or excluded psychotherapy in their coverage. That this rich resource of computerized records has only been used (e.g., Cummings & Follette, 1968; Follette & Cummings, 1967) to produce misleading results such as indicated by the solid line in Figure 11.1 is scandalous. All in all, here is an area in which extensive time-series data are available, making possible a rich exploration of causal hypotheses and avoiding the self-serving misanalyses that occur with simple before and after comparisons.

OVERVIEW OF CHAPTER 12

In the previous chapter, we saw that regression artifacts, not psychotherapy as claimed, explained a decrease in medical services demand when psychotherapy was administered to the group that had the highest number of visits in the month prior to the treatment. This chapter shows how regression artifacts can mistakenly make a treatment look ineffective, or perhaps even harmful, in some studies of compensatory education.

Suppose you administered a pretest to a group of children and plotted the distribution of their scores. Then you computed a mean for the children whose scores were in the highest 10% of the whole group and did the same for the lowest 10%. After a period of time, you administered a posttest to these children. When you plotted the posttest scores for the same group of children who made up the highest 10% on the pretest, you would note that the distribution seemed to flatten out, and the mean was lower than the pretest mean—that is, it was closer to the mean of the whole group. When you plotted the posttest score for the group who were in the lowest 10% of the pretest, the distribution here also seemed to flatten out, with the mean moving higher, also closer to the mean of the whole group. This is not a process working in time, but purely a methodological artifact that occurs whenever two variables are imperfectly correlated. Campbell's discussion of two measures of IQ administered a year apart to children in an orphanage offers an excellent illustration of this statistical phenomenon.

This gets more complicated when two matched groups are considered. Using a simulation with a "no treatment effect" built in, Campbell showed how a pure regression artifact might be mistaken for a true effect. When the distributions of two groups of differing ability are plotted, there is an overlap in which the individuals have similar scores. If these groups (representing the upper extreme of the disadvantaged group and the lower extreme of the advantaged group) are matched for the analysis, the pretest means will be approximately equal. However, the posttest mean of each of these extremes groups will regress toward their respective group mean. Because the means of these once comparable groups now appear to be more distant, this might mistakenly be interpreted as a treatment effect. If the treatment had been administered to the disadvantaged group, it would have appeared harmful; if it had been administered to the advantaged group, it would have looked wildly successful. This pure regression artifact has often been mistaken as a true effect. With an imperfect correlation, matching in this way underadjusts for pretest differences and may result in such pseudo-effects.

When long-term effects are examined by using repeated measures, regression artifacts must be considered. This is because true scores correlated in adjacent years will have a higher correlation than true scores of longer intervals. This is referred to as *proximally autocorrelated true scores,* in which there is a general pattern of decreasing correlations as time passes. This pattern is described as a first-order Markov process.

Campbell illustrated the regression artifact extending over previous and subsequent years when research participants were matched in a single year. Many evaluations of compensatory education exhibit these pseudo-trends of decreasing or increasing effects over the long term.

Another example of this autoregressive effect is given using data from Ashenfelter's (1978) study of the effect on earnings of job training. Based on regression-adjusted values derived from the earlier years, Ashenfelter concluded that an initial positive impact declined over the years until no differences were seen between the trained and untrained groups. Although Campbell believed that there probably was some diminution of the impact of job training over the years, he demonstrated that it is likely that some of the decline noted by Ashenfelter was a result of the diminishing correlations between the target year and later years. In fact, a simulation showed that in a "no effect" situation, Ashenfelter's method would have suggested that the training program was harmful.

Close attention to the examples and graphics in this chapter will provide you with a deeper understanding of this common methodological mistake. The question now becomes, "How might a study be designed to measure true effects and avoid the problem of pseudo-effects as a result of regression artifacts?"

TWELVE

REGRESSION ARTIFACTS IN REPEATED CROSS-SECTIONAL MEASURES

SYSTEMATIC UNDERADJUSTMENT OF PREEXISTING DIFFERENCES: MATCHING AND REGRESSION ARTIFACTS

In a paper by Campbell and Erlebacher (1970) on how regression artifacts in quasi-experimental designs can mistakenly make compensatory education look harmful, there was assembled a literature that has been persistent in the

Campbell, D. T., & Boruch R. F. (1975). Making the case for randomized assignment to treatments by considering the alternatives: Six ways in which quasi-experimental evaluations in compensatory education tend to underestimate effects. In C. A. Bennett & A. A. Lumsdaine (Eds.), *Evaluation and experimentation: Some critical issues in assessing social programs*. New York: Academic Press.

Campbell, D. T. (1991, May 6–7). Quasi-experimental research designs in compensatory education. In E. M. Scott (Ed.), *Evaluating intervention strategies for children and youth at risk: Proceedings of the O.E.C.D. Conference in Washington, DC.* Washington, DC: U.S. Department of Education, Office of Planning, Budget and Evaluation and the U.S. Government Printing Office psychology and education tests-and-measurement tradition (e.g., Lord, 1960; McNemar, 1940; Thorndike, 1942). This literature points out how, when the experimental group and the comparison group come from different populations, efforts to correct for pretreatment differences lead to regression artifacts. The problem is so recurrent and so little understood that it seems worth a brief review.

First, let us consider the simplest case of "regression to the mean" in a single group measured twice, as discussed in the classic paper by McNemar (1940). If we concentrate on extreme samples selected on the basis of the pretest, then, as shown in Figure 12.1, we will find that the subgroups that were so extreme and compact on the pretest will have spread out on the posttest, and that each subgroup mean will have regressed toward the total group mean. Presented as in Figure 12.1 this may seem like a rather mysterious process. The same facts presented in Figure 12.2 as a scatter diagram of the pretest–posttest relationship make it more understandable. The apparent regression is nothing more than a by-product of a biased form of entry into a symmetrically imperfect relationship. The bias comes from constructing groups on the basis of extreme test scores on *only one* of the variables—for example, the pretest. Inevitably, if the correlation between the two tests is less than perfect, the extremes on the pretest will be less well defined on the posttest. If, on the other hand, the extreme groups in Figure 12.2 had been defined by the posttest scores (e.g., horizontal slices with all those with scores of 20 and above on the posttest as one extreme, all those with scores of 8 and below as the other extreme), these groups would have been better defined and more extreme on the posttest than on the pretest. There would then have been "regression toward the mean" working backward in time. "Regression to the mean" is thus not a true process working in time, but a methodological artifact, a tautological restatement of imperfect correlation. To use McNemar's (1940) example, suppose the IQ of all the children in an orphanage is measured on two occasions, one year apart. If one looks at the children with initially high scores, they will have regressed down toward the mean on a second test, and will appear to have gotten worse. Those initially lowest will have improved. One may mistakenly conclude that the orphanage is homogenizing the population, reducing the intelligence of the brightest, increasing the intelligence of the less bright. However, if one were to look at the extremes on the posttest and trace them back to the pretest, one

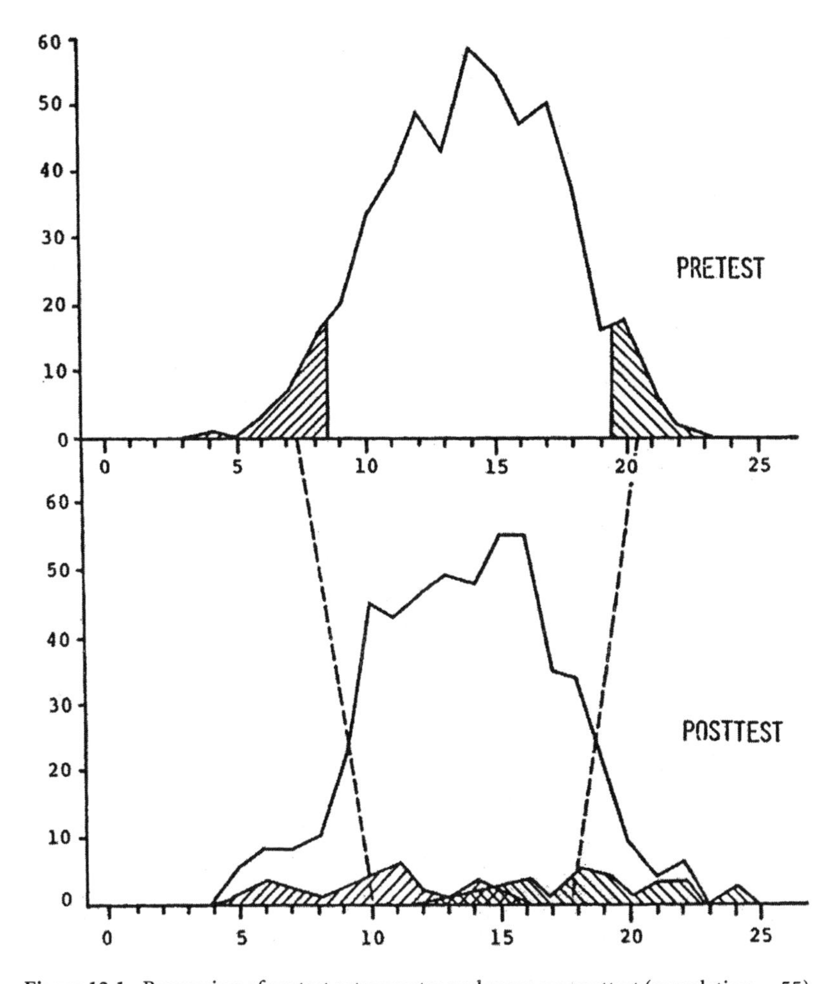

Figure 12.1. Regression of pretest extremes toward mean on posttest (correlation = .55).

would find that they had been nearer the mean on the pretest, thus implying the opposite conclusion. If the whole group mean and standard deviation have been the same on both testings (as in Figures 12.1 and 12.2), these dramatic findings truly imply only that the test–retest correlation is less than perfect.

Now let us shift to the more complicated two-group regression artifact case, such as that produced by matching. Figure 12.3 sets up a hypothetical case of two elementary schools with student populations that differ in ability, where

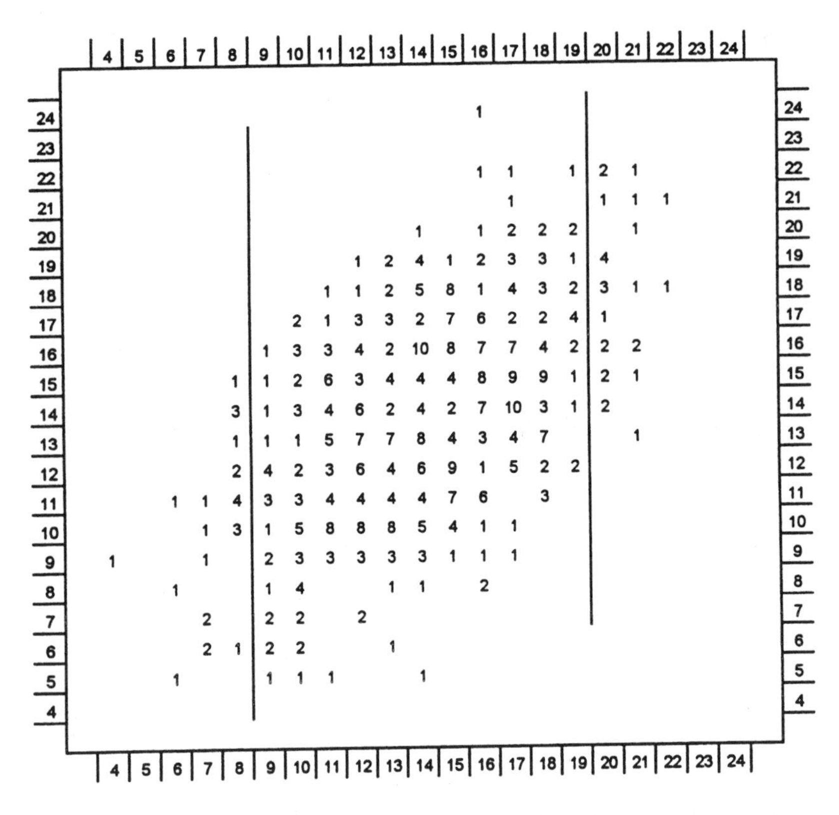

Figure 12.2. Scatter diagram with pretest extremes indicated. Same data as in Figure 12.1 ($r = .55$).

the treatment is being given to the less able student population, and the other group is being used as the control. As shown in Figure 12.3, the mean difference between the two groups and the variance within each group remain essentially the same on the pretest as on the posttest. Given only this evidence, the obvious conclusion would be that the hypothetical treatment given to the experimental group between the pretest and posttest had no effect whatsoever. The status of the two groups, relative and absolute, is the same on posttest and pretest. Although this conclusion involves some assumptions, it is the correct conclusion, in that it was built into our simulation.

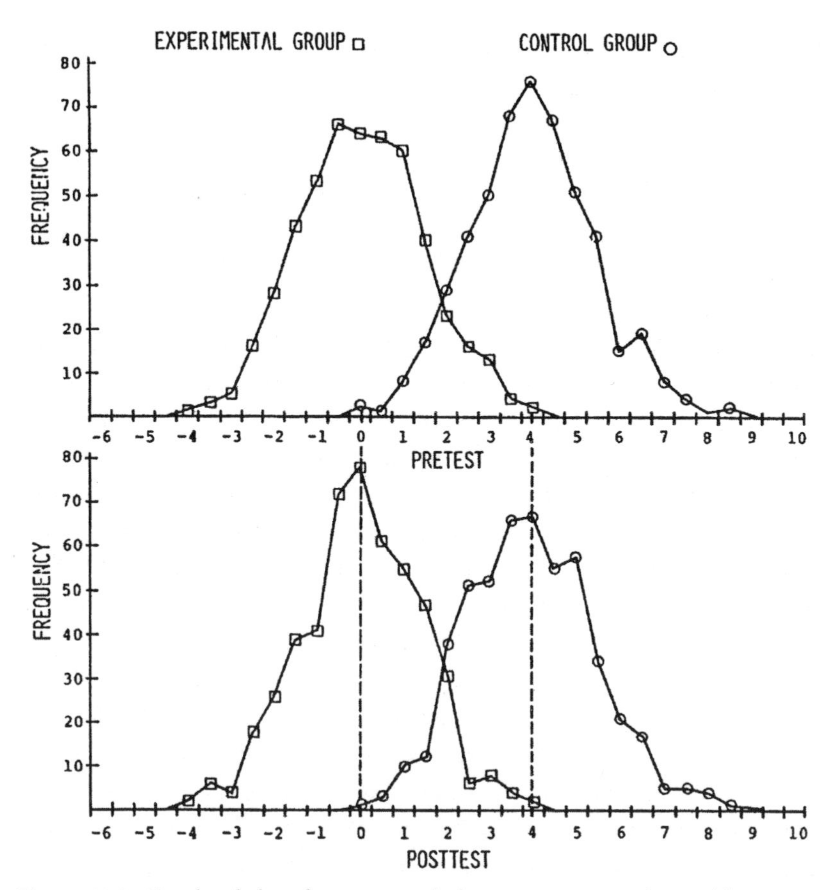

Figure 12.3. Simulated data for a nonequivalent group comparison with no true treatment effect and in which the means and standard deviations are the same on both pretest and posttest except for independent sampling of error on the two testings. The two dashed lines connecting pretest and posttest means are seen to be essentially parallel. The data have been generated using normal random numbers for both true score and error. For the experimental group, the formula was: Pretest = True + Pretest Error; Posttest = True + Posttest Error. For the advantaged control group, a constant of 4 has been added to the random true score. Five hundred cases have been generated for each group. For these data, the pretest means come out to be .045 and 3.986, the posttest means .011 and 3.924, where the universe values are 0 and 4. The four standard deviations are 1.436, 1.424, 1.411, and 1.469 for the data grouped as shown, where the universe value is 1.414. The pretest–posttest correlations are .535 for the experimental group and .485 for the control, where the universe value is .500.

In contrast to this conclusion, several seemingly more sophisticated approaches, such as matching and covariance, produce "significant effects"—actually pseudo-effects. We will attempt to illustrate this in intuitive detail in the case of matching.

Suppose one were bothered by the conspicuous pretest dissimilarity of our experimental and control groups, and felt that this dissimilarity made them essentially incomparable. It might be noted that in spite of this overall noncomparability, the two groups did overlap, and must therefore contain comparable cases. This would suggest basing the quasi-experimental comparison not on the whole group data, but only on subsets of cases matched on the pretest. Figure 12.4 shows what happens in this case. Each of the two purified, compact, matched subgroups on the pretest spreads out widely on the posttest, although each remains within the boundaries of its whole group distribution. Because the subgroups on the posttest are no longer as pure, extreme, and compact as they were on the pretest, their means have been distorted toward the mean of their respective whole-group distributions. The result is a separation in posttest means for the matched groups that has often been mistaken for a treatment effect. It is instead, once again, a product of a biased mode of entry into an imperfect relationship. The intuition to look for comparable subsets is not necessarily wrong, but the employment of the fallible pretest scores to define comparable subsets is fundamentally misleading. If one were to have used matched scores on the posttest and traced them back to the pretest, a pretest separation would have been noted. One would be unlikely to do this, however, because the temporal order of the testing tends to be associated with the temporal order of the hypothesized causal sequence.

Once again, approaching the problem through scatter diagrams may make the effect more comprehensible. Figure 12.5 shows the scatter diagram for the 500 observations relating pretest to posttest for the experimental group. Superimposed on this diagram is a line connecting the column means—in other words, the mean expected posttest value for each specific pretest value. (For convenience we will refer to this as the "unfitted regression line." The least-squares straight line fitting these data can be drawn in by connecting the two + points.) Note that this is only one of two such lines. The other would connect the row means—the mean *pretest* score corresponding to each specific posttest value. (The linear fit to this second regression line can be drawn by connecting

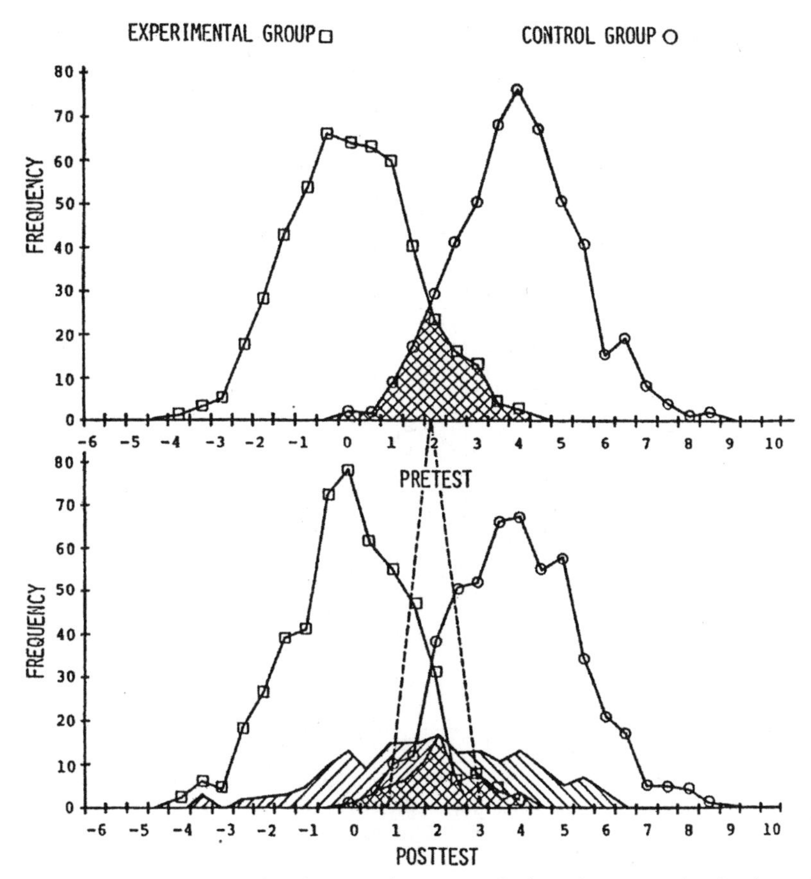

Figure 12.4. The posttest distribution of cases matched on the pretest for the data of Figure 12.3. To illustrate how the matching was done, consider score 0.0, for which there were only two control group cases. These two were matched with the two first-listed experimental group cases having that score, out of the 64 such available. Because all cases had been generated from independent random numbers, no bias was produced by this selection rule. There resulted a total of 86 matched pairs. For each of the two subgroups, the pretest mean was 2.104. The two subgroup posttest means are .878 and 3.145. The two subgroup pretest standard deviations are .805. The two subgroup posttest standard deviations are 1.331 and 1.323, reflecting the greater spread.

the two *'s.) Whereas the illustrated regression line has a slope of about 28°, the other regression line would have a slope of about 62°. Had the correlation been 1.00—that is, perfect, without scatter—both regression lines would have been the same, and at a 45° angle in this case. Figure 12.6 shows the pretest–

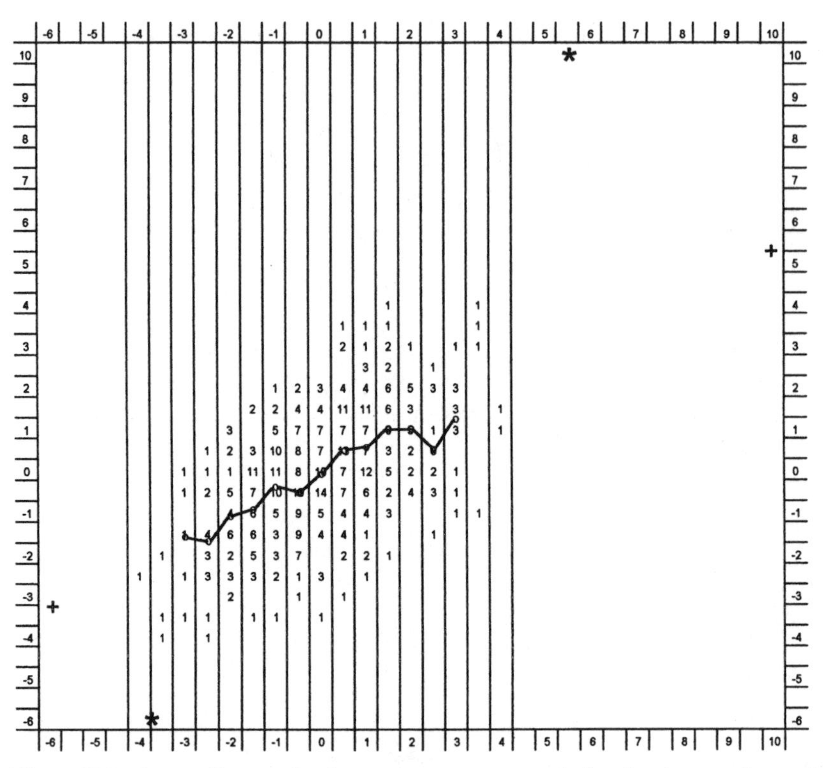

Figure 12.5. Scatter diagram for the pretest–posttest correlation for the experimental group of Figure 12. 3. The correlation coefficient is .535.

posttest scatter for the control group, and the regression line predicting posttest scores from specific pretest values. In Figure 12.7 these two scatters are superimposed. Although we have not found it convenient to tag the specific matched cases in this graph, the vertical distance between the two regression lines makes the point clearly. For example, for pretest scores of 2.0, right in the middle of the matching area, the 23 experimental cases have a mean posttest score of only .98, whereas the 29 control group cases have a mean posttest score of 3.02—two whole points higher. The expected values *in the absence* of any treatment effect thus show the inadequacies of matching on a fallible pretest. Note that as the correlation between pretest and posttest becomes higher, the underadjustment of the preexisting differences resulting from matching (or covariance adjust-

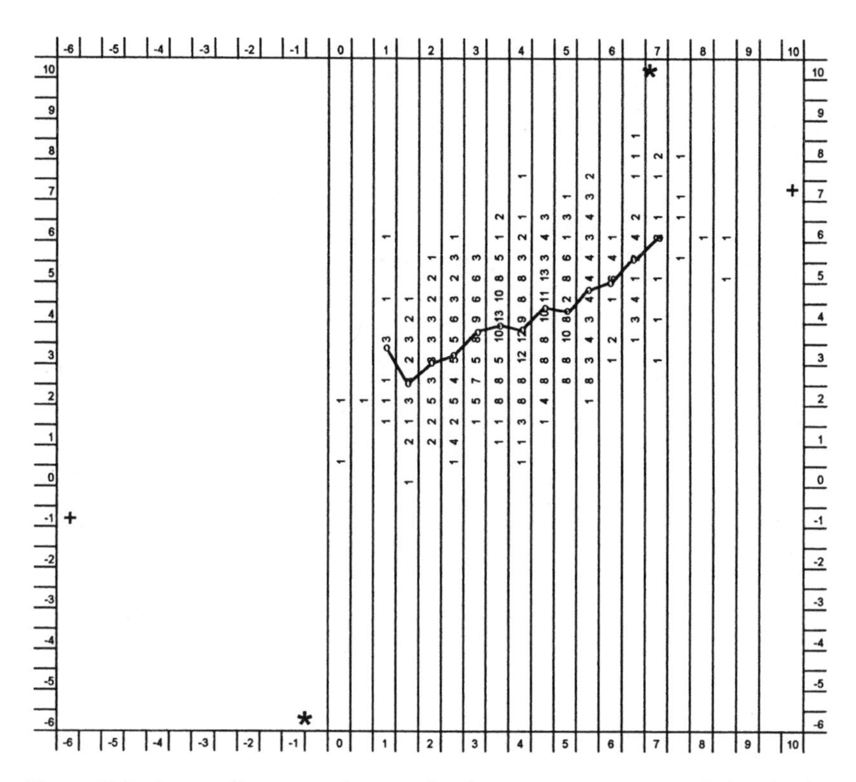

Figure 12.6. Scatter diagram, and so on, for the pretest–posttest correlation for the control group of Figure 12.3. The correlation coefficient is .485.

ment) becomes less. Before matching, the experimental and control groups differed by four points. After matching (or covariance adjustment) on the pretest, and when the within-group pretest—posttest correlations are .50, the group difference becomes two points. Had the pretest—posttest correlations been .75, only a 1-point difference would remain; had they been 1.00, the matching would have completely adjusted away the initial group difference. Referring back to Figure 12.7, note that as the correlation gets higher, the linear slope of the regression line gets closer to 45°. Were the correlations to reach 1.00, the two regression lines would connect, the expected posttest values for any given pretest value would be the same for both experimental and control groups, and matching would achieve its intended aim.

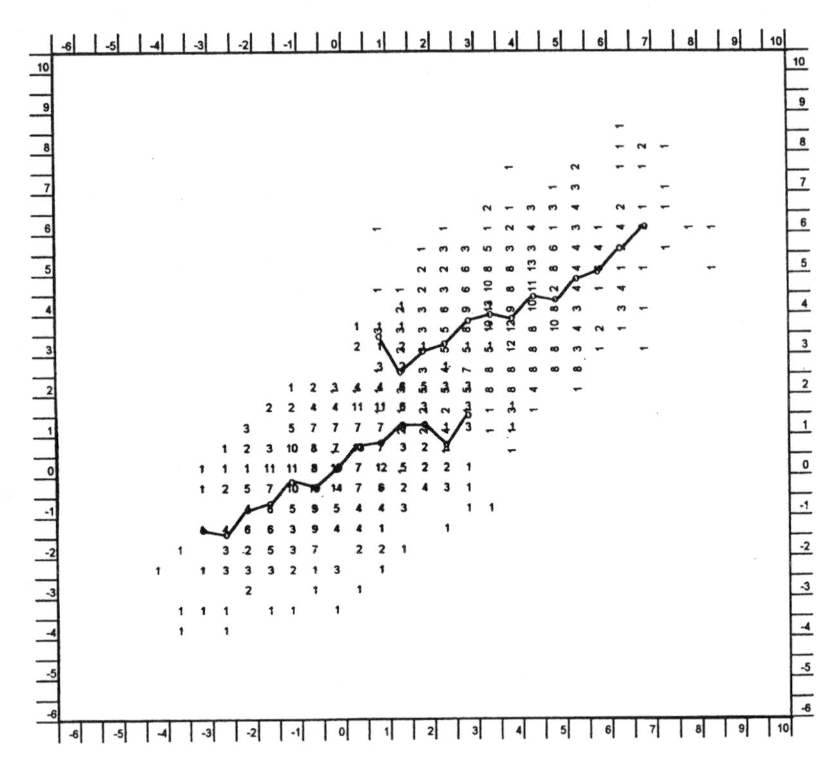

Figure 12.7. Superimposed scatter diagrams from Figures 12.5 and 12.6. The vertical differences between the regression lines indicate the magnitude of the "regression artifact" produced by matching on pretest scores.

PROXIMALLY AUTOCORRELATED TRUE SCORES

Repeated waves of measurements in longitudinal studies always show a proximally autocorrelated pattern. This means that adjacent years correlate more highly than nonadjacent ones. This pattern is sometimes called a *Guttman simplex.* Table 12.1 illustrates such a pattern. Such patterns *cannot* be explained as a result of an "error" superimposed on a constant "true score": It requires a *proximately autocorrelated true score.* Paying attention to this important feature makes an especially important difference in the mode of analysis recommended once one gets beyond two waves. What is important in Table 12.1 is the general pattern of decreasing correlation over longer elapsed time. Illus-

TABLE 12.1 Hypothetical Year-to-Year Correlations for a Vocabulary Achievement Test (Within a Disadvantaged Neighborhood or Within an Advantaged Neighborhood), as Used in Figure 12.8

Age	*2*	*3*	*4*	*5*	*6*	*7*	*8*	*9*
2	(.80)							
3	.64	(.80)						
4	.51	.64	(.80)					
5	.41	.51	.64	(.80)				
6	.33	.41	.51	.64	(.80)			
7	.26	.33	.41	.51	.64	(.80)		
8	.21	.26	.33	.41	.51	.64	(.80)	
9	.17	.21	.26	.33	.41	.51	.64	(.80)

trated here is a perhaps too-simple first-order Markov process, plus a time-specific error. The "instantaneous" reliabilities have been set at .80, a reasonable value for subpopulations restricted in range. The autoregressive coefficient has also been set at .8, thus the 1-year lag value is .8 × .8, the 2-year .8 × .8 × .8, and so on. If real-life examples show departures from this pattern, it is usually in the direction of decreasing decrements with longer lapses. That is, one might see higher correlations than those predicted by a first-order Markov process for the longest lags. Such a pattern would be produced if a second-order Markov process were involved, or if the measure was factorially complex, with the components decaying according to first-order Markov processes but with differing coefficients for different factors (Campbell & Reichardt, 1991).

Figure 12.8 is intended to expand the two-wave illustration of Figure 12.4, but it is turned on its side graphically, and it extends the regression artifact from matching on a single year to more years—earlier years as well as subsequent ones. In the top section are the group means on an annual vocabulary test standardized to remove growth and differential growth for two neighborhoods, *A* (advantaged) and *D* (disadvantaged). The two groups have the same overlap in each year to the degree shown by the normal distribution curves (turned on their sides in contrast to Figure 12.4). In the middle section, the expected values in subsequent years for subgroups matched on the Age-3 measure are portrayed. (The Age-3–Age-4 section corresponds to Figure 12.4, turned on its side, except that for Figure 12.4, the test–retest correlation for 1 year was .50 within each group. Here it is .64, as per Table 12.1.) The separation of the Age-3 matches grows larger as the number of years separating the two measures

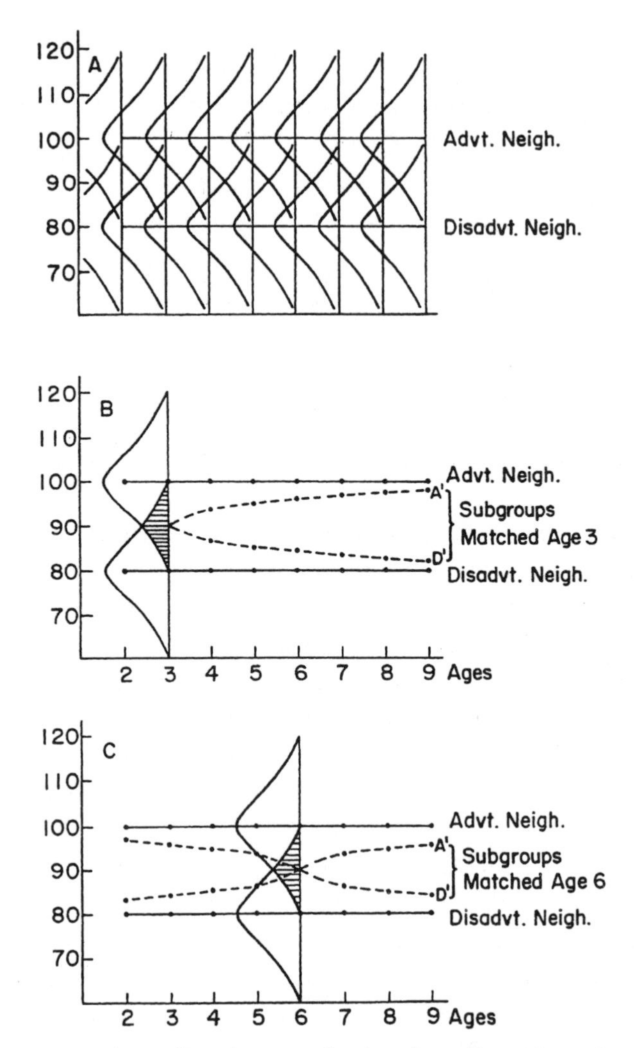

Figure 12.8. Regression artifacts (or regression-based no-effect estimates) as a function of years before and subsequent to matching or selecting a year as a covariate for covariance adjustment.

increases, in conformity to the lower correlation (as illustrated in Table 12.1). In the lower section of Figure 12.8, the effects of a hypothetical matching at Age 6 are shown. As can be seen, the proximately autocorrelated nature of the data generates not only pseudo-program effects, but also pseudo-trends of

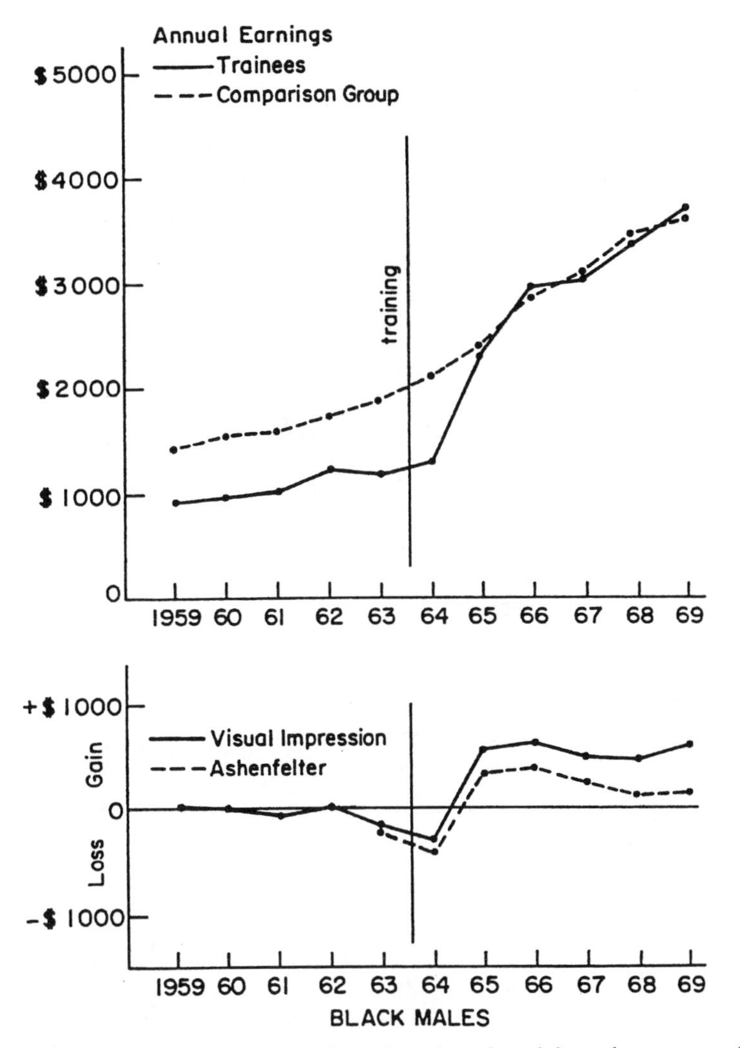

Figure 12.9. Effects of job training for Black males. Plotted from data presented in Ashenfelter (1978).

decreasing (or increasing) effects in long-term follow-up. Still, regression adjustments predominate in the quasi-experimental compensatory education literature.

Figures 12.9 and 12.10 come from a famous study of one job training program done by Orley Ashenfelter (1978). As graphed (not in Ashenfelter's

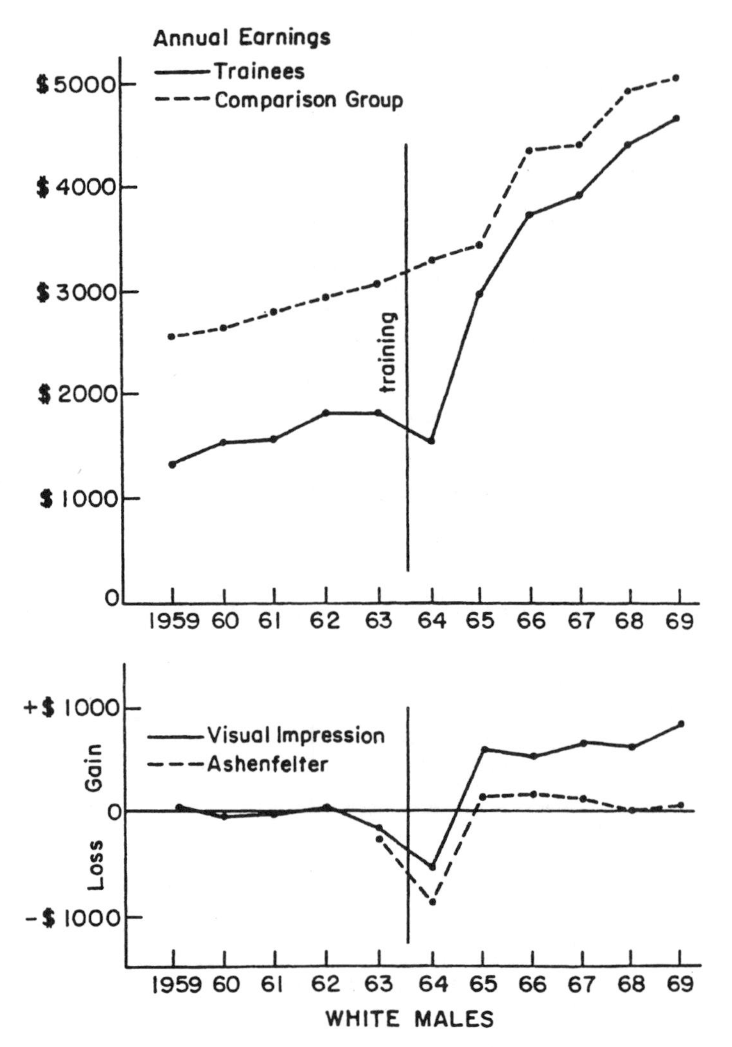

Figure 12.10. Effects of job training for White males. Plotted from data presented by Ashenfelter (1978).

papers, but by us [Campbell, 1979c; Cook & Campbell, 1979, fig. 5.9, p. 229]), this is the most effective intervention I know of. (Most job training studies show either no benefits or pseudo-harmful ones. See Director, 1979, for a review.) Ashenfelter used 10 years of records of earnings subject to withholding tax for individuals receiving job training in 1963–1964, and for age, gender, and race mates not receiving such training. From the graphs (at the top of Figures 12.9 and 12.10), there are sustained differences between the groups during 1959–1963, followed by dramatic recoveries in 1965–1969. It looks as though the effects are sustained in undiminished degree. Ashenfelter did not present these graphs, but instead regression-adjusted values based on years 1959–1962. Because they are based on four separate years, the correlation with a subsequent year is very high, and the adjustment initially almost entirely removes the group differences. But during the later years, 1965–1969, the multiple *R* between 1959–1962 and the target year undoubtedly becomes lower and lower because of the autoregressive nature of such data. The adjustment thus becomes less and less, showing in his statistics as steadily declining benefits. In the lower half of Figures 12.9 (for Black males) and 12.10 (for White males), I have graphed Ashenfelter's estimates of gains in contrast with the visual impression. For White males (Figure 12.10), Ashenfelter showed the effect entirely disappearing —quite in contrast with the visual impression shown at the top of the figure.

I am ambivalent in using this illustration, because I believe that in most cases the immediate impact of focal interventions will steadily diminish (Campbell, 1988b, pp. 308–312), in compatibility with Ashenfelter's reported results. But I believe equally strongly that Ashenfelter's method would have produced diminishing results even if they had, in truth, been undiminished. Figures 12.9 and 12.10 come from a presentation by Campbell and Reichardt (1983), and for that presentation, Reichardt did a simulation (shown in Figure 12.11) of what would have happened if in an Ashenfelter-like situation there had been no effect: *Ashenfelter's mode of analysis would have made the job training program look increasingly harmful!*

SUMMARY

This chapter discussed a common source of bias found in quasi-experimental designs for evaluating compensatory programs. Regression artifacts, seen in many major evaluations, produce pseudo-effects or harmful effects. Further, irrelevant reliable variance in variables (regression artifacts) increases over time. As the correlations get lower (proximally autocorrelated true scores), researchers must consider the serious problem of estimating persisting effects in longitudinal studies.

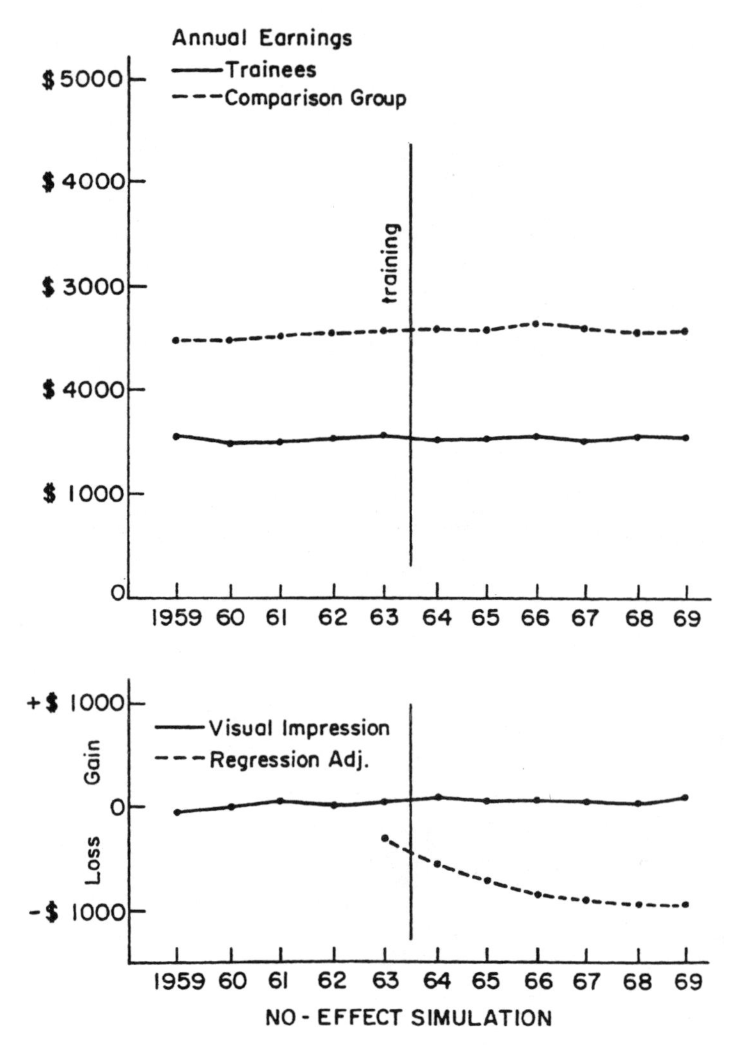

Figure 12.11. Simulation of a no-effect case using Ashenfelter's (1978) mode of adjustment. (Within each group, the data are structures as $Y_T = .7071\ Y_{T-1} + .7071\ E_T$. Where Y_T stands for earnings for year T and E_T is an uncorrelated residual. Sample size for each group is 500.)

OVERVIEW OF CHAPTER 13

In this Forward to William Trochim's (1984) book, *Research Design for Program Evaluation: The Regression–Discontinuity Approach*, Campbell gave a brief overview of the regression–discontinuity design. But the description of how it emerged is as important as the discussion of the design itself. Campbell describes an ideal situation in which scientists from different institutions, and with expertise in different fields, conferred to see how they might assess the impact of receiving an award such as the Merit Scholarship on minority students. They wished to develop a quasi-experimental design that could demonstrate the true effects of receiving such an award.

The initial and superior suggestion was a tie-breaking randomization design, whereby the individuals in a defined borderline group would be randomly assigned to the award or no-award condition. This was not feasible, and in the continuing discussions, the regression–discontinuity design was conceived. For this design, all applicants were to be ranked according to their eligibility score and a sharp cutting point was to be designated to determine who would receive the award and who would not. If outcome measures were available for the entire range, regression lines were to be fitted to awardees and nonawardees separately. If there were a true effect, the graphed data should show a discontinuity (or gap) between the regression lines at the cutting point. This difference is an extrapolation to the difference that would be detected on the tie-breaking random-assignment experiment. Campbell was convinced that a mere change in slope would not indicate an effect.

Campbell advocated using multiple designs, if possible, to provide better estimates of effects. He believed that using the tie-breaking randomization design along with the regression discontinuity design would increase confidence in the results. Unfortunately, the National Merit Scholarship Corporation could not implement either of these designs.

The researchers at Northwestern University continued to apply the regression–discontinuity design to other problems. One of the issues that came into question at this time was the assumption of linearity. If a linear regression line was fitted to data that were curvilinear, a pseudo-effect occurred. That is, when the quadratic or cubic regression line was fitted to the same data, the discontinuity at the cutting point disappeared. Under further scrutiny, however, it was discovered that when one extrapolated beyond the data with these higher order polynomials, the extensions were often wildly incredible. The scholars continued to work on this problem and offered several suggestions. One such suggestion involved beginning with higher order models and working down to a linear model. Others suggested estimating no specific curve, but rather, estimating something like a mean prediction of all possible curves within a large class. Later simulations with this model revealed flaws that required modifications.

Another cause for concern was that points far removed from the cutting point contributed to determining the regression line as much as points adjacent to the cutting point. It would be preferable for data closer to the cutting point to be weighted more heavily in determining the curves for the extrapolation. This is in keeping with Campbell's belief that nature is "sticky" or "viscous." This means adjacent regions in space, time, and attribute values are more similar than remote ones, and one can be more confident when generalizing to adjacent regions. In view of this phenomenon, models were explored in which the values nearer the cutting point were weighted more heavily than those at a distance for determining the curves.

When the regression–discontinuity design was applied in the Mesa Arizona School district, some of the expected problems were borne out. For example, the programs were not delivered uniformly, and those receiving the treatment were not always those below a particular cutoff pretest score. Also, curve-fitting was difficult because of ceiling and floor effects. Campbell then went on to describe the "fuzzy" alternative to the regression–discontinuity design. This can be used when the quantified measure was not the sole criterion for deciding who would be recipients and nonrecipients. Instead, unquantified criteria were used to alter the decisions made through purely quantitative rankings. The initial methods used to analyze these "fuzzy" regression–discontinuity data produced pseudo-effects that were actually a result of underadjustment for selection bias of the sort described earlier in this paragraph.

The scholars continued to explore different methods for analyzing the "fuzzy" case to see if they could distinguish between the "no effect" and "true effect" cases. One approach was to assume linearity and analyze the outcome scores based on a strict cutting point, even though some above the cutting point did not receive the award and some below it did. Simulations demonstrated problems with this approach, and it was abandoned. A second approach compared the variances of scores along the entire range. If the columns that included the "fuzzy" misclassified cases had a larger variance than the correctly classified cases, it was assumed that some of the greater variability was a result of the effect of the award. A third approach involved fitting an *S*-shaped curve by using the probability of being selected for the award at each pretest level. This latter design is discussed in greater detail in Trochim (1984).

Campbell listed many researchers and students who worked together and independently to develop the regression–discontinuity design. He described ideal conditions for a scientific community at Northwestern, where scholars from the mathematics and psychology departments shared students and pooled their expertise to uncover and solve the glitches. The result was a design that could be used in the evaluation of compensatory education programs. To use this design, there must be planning that ensures that there is a quantified pretest score, a defined decision-making process, and identification numbers for the participants for follow-up at a later time. Campbell suggested that in some instances this design might be applied retrospectively—for example, in reevaluating past programs designed for other purposes, but with the express goal of a fair distribution of scarce resources.

THIRTEEN

THE REGRESSION DISCONTINUITY DESIGN

It is a great pleasure to be allowed to introduce Bill Trochim's book. It is the first complete presentation of the regression–discontinuity quasi-experimental design. Bill and the Sage editors have invited me to use this occasion to present an anecdotal history of the development of the method in Evanston, Illinois. (No doubt it has been independently hit on several times. I will report later on the cases that we know about.)

Some background: In the period 1958–1968, the National Merit Scholarship Corporation, headquartered in Evanston, supported an impressive research commitment under the direction of John M. Stalnaker, with scholars such as Donald L. Thistlethwaite, Alexander Astin, Robert Nichols, and John L. Holland on the staff at various times. This group, and a similar group at the Association of American Medical Colleges (Helen H. Gee and Edwin B. Hutchins, among others), combined with the faculty of Northwestern Univer-

Campbell, D. T. (1984c). Foreword. W. M. K. Trochim, *Research design for program evaluation: The regression-discontinuity approach* (pp. 15–43). Beverly Hills, CA: Sage.

sity to produce a truly outstanding applied social research community with a special focus on quasi-experimental designs. For example, the cross-lagged panel correlation technique also grew out of this environment.

On a winter day in 1958, I met with Donald Thistlethwaite and others to discuss how to measure the career effects of receiving a special new Merit Scholarship designated for minority students. Whereas the regular Merit Scholarships went to students so promising and well supported that the Merit award could do little to augment their level of eventual achievement, these minority awards were expected to make profound career changes in many cases.

Our discussion quickly rejected any broad-spectrum use of random assignment, even for a small experimental sample. We spent most of the day trying to convince the program administrators to employ a "tie-breaking" randomization: That is, for those applicants whose scores or ranks were on the borderline between award and no award, one would define a class interval of measurement, within which all who fell would be designated as tied in eligibility, being sure that there were more such "tied" cases than there were awards to cover. Among these, one would break the ties by random assignment. For this narrow band of eligibility, one would have a random-assignment experiment.

But by the end of the afternoon, Don Thistlethwaite and I had become convinced that if awards were made entirely on the basis of a quantified eligibility score or a complete ranking of a substantial pool of applicants in a range that included borderline and less eligible cases, and if one had outcome scores for this full range, one should be able to extrapolate from above and below the cutting point to what a tie-breaking random-assignment experiment would have shown. This double extrapolation produces the *regression–discontinuity* design.

The hypothetical tie-breaking experiment has been both the expository and interpretative key in all of the presentations of the method in which I have participated (Campbell, 1969c, 1976; Campbell & Stanley, 1966; Cook & Campbell, 1976, 1979; Riecken et al., 1974; Thistlethwaite & Campbell, 1960), leading me to reject effects such as change of slope when not accompanied by a change of cutting-point intercept. If a tie-breaking randomized assignment would not have shown the effect, I am unwilling to credit a causal inference based on an other-than-intercept discontinuity.

The idea of randomization at the margin (tie-breaking) can be regarded as a special case of Boruch's (1975a) "Experiments Nested Within Quasi-Experiments"—that is, using multiple designs that are put together to achieve both greater statistical power and more methodologically independent cross-validation. The statistical properties of estimates that we get out of coupling designs can be better than those we get out of single-approach designs. To put this more concretely, even had a tie-breaking randomization been permitted, a supplementary regression–discontinuity analysis would also have been desirable. In addition, a tie-breaking randomization, no matter how few its cases, will always add inferential strength to a regression–discontinuity analysis.

It turned out that in its scholarship awards of any type, National Merit could not implement either the tie-breaking randomization or the regression–discontinuity design because of its commitment to the many funding groups it coordinated, each of which designated a panel to make the final decisions on the few scholarships it was funding. These panels looked over the quantitative and qualitative evidence and produced a three-category decision (award, alternates, and unconditional rejectees) with no metricizing within categories. Had they been asked to first rank-order the entire pool, and then to use these ranks to decide awards and priority order of alternates, then a subsequent regression–discontinuity analysis using ranks above and below the cutting point would have been possible, but the administrative staff judged this to be too demanding. Post hoc ranking within categories would have produced a discontinuity in the quality of measurement at each category boundary, undermining the fundamental assumption of measurement homogeneity within and across cutting points.

For the Certificate of Merit, a nonmonetary commendation, the requisites for the regression–discontinuity analysis were available. The award was based on a sharp cutting point on a quantitative measure of eligibility; there was a record of the eligibility scores of persons above and below the cutting point; and follow-up data existed for all. This information made it possible to examine the effects of the award. Donald Thistlethwaite had already used these data in a more traditional ex post facto impact study. He did regression–discontinuity reanalysis that we published (Thistlethwaite & Campbell, 1960), illustrating its clear superiority over the ex post facto method for causal inference.

My next Northwestern colleague in work on the method was Joyce A. Sween, whose 1971 PhD dissertation, "The Experimental Regression Design: An Inquiry Into the Feasibility of Non Random Treatment Allocation" (see also Sween, 1977), probed and expanded the method by extensive computer simulation, with special attention to appropriate tests of significance once one has abandoned the usually inappropriate assumption of linear regression. She declared that the regression–discontinuity design produced a *true* rather than *quasi*-experiment, in agreement with later publications by Goldberger (1972), Cronbach, Rogosa, Floden, and Price (1977), and Reichardt (1979), who also argue that the magic of randomization is that it enables one to model accurately the exposure to treatment, and that other explicit assignment rules can achieve this same result. (Although I admire the brilliance of this perspective, the choosing of an appropriate estimate of the regression lines involves such great difficulties and strong assumptions that I still would emphasize the quasi-ness.)

Sween explored the use of higher order polynomials to achieve curvilinear regression, and ended up recommending essentially the same procedures Trochim (1984) presented in his volume. In the normal curve-fitting strategy, one uses the lowest order curve not statistically significantly demonstrated to be too simple. This results in a parsimonious bias toward underfitting. For the double-extrapolation t-test analysis developed by Sween, underfitting (e.g., using a linear model where the true form is quadratic) can lead to pseudo-effects, whereas the reverse error, overfitting, or using too high a polynomial, should not. Thus Sween recommended starting with a cubic or higher order model and working down.

Two anecdotes illustrate the sensitivity of the analysis to slight degrees of curvilinearity. For a long time, the most extensive presentation of the design was in my "Reforms as Experiments," which first appeared in the *American Psychologist* in 1969. In that first presentation there was a profoundly mistaken footnote 8, deleted in all subsequent presentations (e.g., Struening & Brewer, 1983). Within a week or so of its appearance, both William Kruskal and Harry Roberts of the University of Chicago Department of Statistics had independently contacted me to tell me I was wrong. There ensued a series of meetings in which Kruskal, Roberts, James Landwehr, Myron Straf, and others from the University of Chicago met with Joyce Sween, Edward Kepka, Donald Morrison, and me from Northwestern, working to discover the source of the problem. In the end, this turned out to be a subtle mistake in our simulation. Because I never got around to publishing the planned errata in the *American Psychologist*, I will take space to do that now.

The exposition will be clearer if I use the very same illustrations employed in that article. I have always regretted the labels of the abscissa and ordinate, a regret that increased with each reprinting of that article. Using the terms "pretest" and "posttest," instead of "eligibility score" as I should have, encouraged the misunderstanding that the design was only appropriate where the same or highly comparable measures were used for both eligibility and outcome. Instead, although this is a limitation for many quasi-experimental designs, including the two most frequently usable, the interrupted time-series and the pretest–posttest nonrandomized comparison group design, it is not at all a limitation on the regression–discontinuity design. Indeed, in the original Thistlethwaite and Campbell (1960) paper, the eligibility measure was an achievement test score, and one outcome measure was the distance from the home of the university attended.

Figures 13.1 and 13.2 show simulated regression–discontinuity studies, with no true effect (Figure 13.1) and with a true effect of some two points (Figure 13.2). When Joyce Sween and I extended no-effect runs such as Figure 13.1 to 1000 or more cases, we regularly obtained significant pseudo-effects in the direction of a beneficial impact of a treatment given to those with the higher eligibility scores. Footnote 8 was a misguided effort to explain this apparent bias.

The correct explanation was as follows: The simulation involved 20 discrete whole-unit true scores (ranging from 0 to 20). To these true scores were added errors of a "continuous" nature—that is, normal random numbers with five significant figures. Separate independent errors were added to the true scores to achieve the eligibility and the outcome scores. Thus all of the obtained scores lower than the lowest true score of zero, and all scores higher than the highest true score of 20, were purely a result of error, and hence had a perfectly horizontal regression line, adding a slight curvilinearity, which when fitted with a straight line tilted the two regression lines apart at the cutting point. All of this can be seen if one looks carefully at Figure 13.1. In Sween's dissertation, done later, "continuous" normal true scores were employed, and this subtle source of curvilinearity was avoided in the intended linear cases.

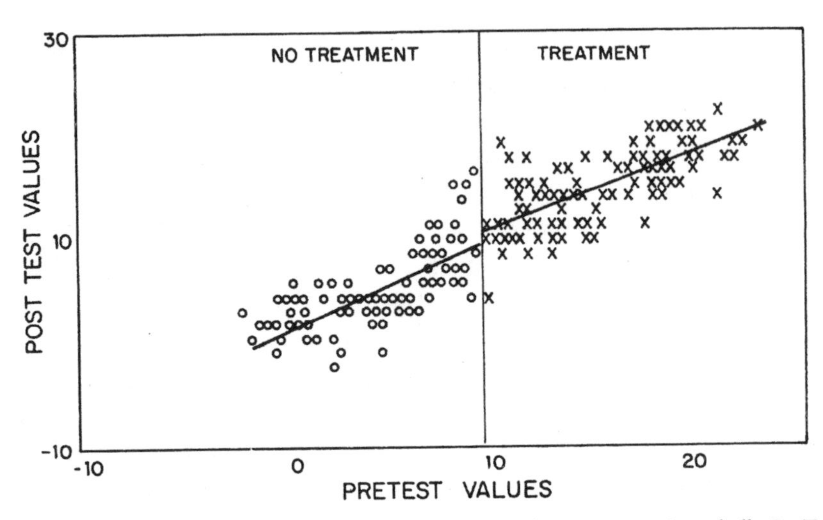

Figure 13.1. Regression-discontinuity design: No effect. Source: Campbell, D. T. (1969). Reforms as experiments. *American Psychologist, 24*(4), 409–429. Copyright 1969 by the American Psychological Association. Reprinted by permission of the publisher and author.

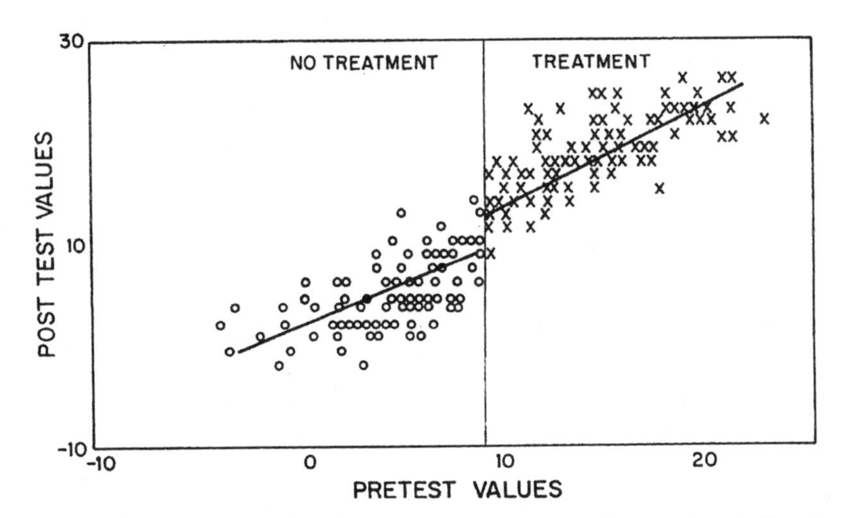

Figure 13.2. Regression-discontinuity design: Genuine effect. Source: Campbell, D. T. (1969). Reforms as experiments. *American Psychologist, 24*(4), 409–429. Copyright 1969 by the American Psychological Association. Reprinted by permission of the publisher and author.

Here is another example of the subtle effects of overlooking slight degrees of curvilinearity. In the Cook and Campbell (1976) chapter in Dunnette's *Handbook,* Seaver and Quarton's (1976) study of the impact of being on the Dean's List on subsequent grade-point average was used to illustrate the regression–discontinuity design. By the time this material was revised as a separate publication (Cook & Campbell, 1979), Seaver and Quarton's data had been reanalyzed by Sween allowing for a curvilinear fit, totally removing all indications of any effect.

Sween's and my work as reflected in "Reforms as Experiments" (Campbell, 1969c) and her dissertation (1971) were supported by the National Science Foundation continuation grant GS1309X, initiated in September 1966. For many years I had continuous NSF support for work on quasi-experimental methodology. For the second 5-year continuation grant, beginning in 1971, Robert F. Boruch was coprincipal investigator, and with his arrival at Northwestern University, new phases of exploration of the regression–discontinuity design began. Boruch and I were both involved in the Social Science Research Council committee that produced Riecken et al.'s *Social Experimentation: A Method for Planning and Evaluating Social Intervention* (1974). Although that volume focused on randomized assignment to treatment, its one chapter on quasi-experimental designs highlighted regression–discontinuity. The most dramatic graph in the whole book for illustrating results from a social program (Medicare made available only to the lowest income group) can be interpreted as a regression–discontinuity analysis, although we did not do so in the text.

The economist Arthur Goldberger consulted with the committee, and through this we became aware of his independent discovery of the method (1972; see also Cain, 1975), in papers focused on the problem of error in variables. His conclusion was that bias (in my language "regression artifacts" or "packaging underadjusted selection bias as treatment effects") occurred when assignment to treatment was based on a latent true score, but could be avoided by assignment on the basis of a known variable, even if fallible, as well as by random assignment to treatment. The illustrations were based on linear models and covariance analysis, and were the same as the linear regression–discontinuity analysis.

The independent rediscovery of the central idea of the regression–discontinuity design also occurred in the "special regression" model for compensatory education evaluation presented by Tallmadge and Horst (1976) and Tallmadge and Wood (1977). Their design varied in several ways, such as estimating the treatment effect at the treatment group pretest mean rather than cutoff, and the absence of any attempt to address curvilinear relationships. Although I was critical of these variations, the frequent use of their version in Title I compensatory education evaluations has provided us with the richest database of applications yet produced, much of which Trochim has reanalyzed according to our tradition.

The addition of Boruch at the faculty level and Charles Reichardt as a graduate student greatly augmented our Psychology Department group's formal statistical training. Meyer Dwass and Jerome Sacks, statisticians in Northwestern's Mathematics Department, had already been giving us occasional advice, but the presence of Boruch and Reichardt greatly increased the level of this interaction. Jerome Sacks decided that the problem of appropriate estimates of effects and tests of significance had a fundamental enough challenge to merit his professional attention. There emerged a 7-year period of close collaboration, most of it devoted to this problem.

The ultrastatistics they brought to the task are discontinuous with the statistical concepts social science methodologists are exposed to, and certainly beyond my mastery, as the least trained of our team. But I will make an effort to indicate some of the issues that motivated our high-morale collective search. First, curve fitting by composite higher order polynomials is a very unsatisfactory procedure. Bill Trochim, who joined our team for the last several years, introduced a procedure I recommend to others as a routine aid to inference. He plotted extrapolations of the resultant curve beyond the data it was designed to fit. These extensions were almost always wildly incredible. Because regression–discontinuity analysis, more than other uses of curve fitting, depends on extrapolation, this is at least a significant conceptual weakness. (Sween emphasized the minimal extrapolation involved and doubts that any practical liability ensues.)

Second, data points far removed from the cutting point contribute to the determination of the curve fully as much as do data points adjacent to the cutting point, and hence to the extrapolation generating the results of a hypothetical tie-breaking experiment. This weakness is shown in the "Footnote 8" episode. On many grounds, one would prefer to have the data nearer the cutting point weighted more heavily than remote values in the determination of the curves and extrapolations. In the somewhat analogous problem in the interrupted time-series quasi-experimental design, the methods of Box and Tiao (1965, 1975) weighted periods adjacent to the time of impact much more heavily than data from remote periods in generating the predicted values about what would have been observed in the absence of the impact. Although I have never understood how their differential-weighting parameters were derived, they are consistent with assuming a positively proximally autocorrelated true score, where time points closer in time are more similar. I cannot now remember whether or not we ever tried out the Box–Tiao transfer function approach as though it were appropriate to regression–discontinuity data. In vision research, "spatial" autocorrelation concepts are now being used, and perhaps they could be rationalized for the regression–discontinuity setting on the basis of assuming a proximally autocorrelated structure in *attribute* space. This would be consistent with a presupposition of induction that I believe most of us share: In place of the assumption that nature is orderly, we assume that nature is "sticky" or "viscous" and that more adjacent regions in space, time, and attribute values are more similar than are remote ones, and can be generalized to with greater confidence (Campbell, 1966b; Raser, Campbell, & Chadwick, 1970, pp. 197–199).

I interrupt the discussion to provide more details on the working context. This introductory essay is intended not only as a history of the ideas, not only as a partial agenda of unfinished problems, but also as evidence for a sociology and psychology of interdisciplinary collaboration. The period of collaboration that I am about to describe I regard as ideal, even though it did not promptly produce the practical new methods we hoped for. Myer Dwass's organization in 1972 of a Center for Statistics and Probability at Northwestern University created a faculty community in which Bob Boruch and I were invited to make a series of presentations on unsolved problems in tests of significance for quasi-experimental designs. In the summer of 1973 and for two following summers, Clifford Spiegelman, a PhD candidate in mathematical statistics under Sacks and Dwass, was employed by our NSF Grant. He spent most of his first summer going over Joyce Sween's PhD dissertation, arguing vigorously with her and the rest of us about it. Although in the end he approved of her procedures, it was a very time-consuming process, primarily because of the differences in statistical traditions employed. This experience greatly facilitated future communication. Out of this grew an ad hoc seminar on the problem, led by Jerome Sacks and another math department faculty member, Rose Ray. Such ad hoc seminars characterized our summertime interaction for a number of years. Sacks and Ray devoted great effort to the regression–discontinuity problem, their efforts supported by their own NSF grants. As a general strategy, I do strongly recommend such summer cross-disciplinary employment. A one-summer commitment permits cross-disciplinary collaboration on a tentative basis. These "shared" graduate students greatly facilitate faculty communication.

Sacks, with Rose Ray for the first year or so, took a general line of approach that I can only crudely characterize. In continuity with Sween's and my approach, the focus was on the sharp cutting-point model and the double-extrapolation technique—that is, extrapolating an estimated value at the cutting point from points below the cutting point and comparing this with a similar independent estimate based on the observations above the cutting point. All of the several approaches Sacks and Ray explored had the feature that observations nearer the cutting point (nearer the to-be-predicted value) were weighted more heavily than more remote observations. They also had the characteristic that no specific curve had to be estimated or assumed. (It helped me feel that I had a glimmer of their approaches to note that in extrapolating just one unit beyond the observed data, one will probably get very similar predicted values no matter what curve is employed.) They began first with linear and higher order splines, and then moved to still more complex techniques. Sacks's second approach (Sacks & Ylvisaker, 1978) was to estimate something like a mean prediction from all possible curves within a large class, without ever specifying any one of the curves. Although the method is classified as "nonparametric," it produces a predicted value in the metric of the original measures. Although costly in degrees of freedom or power, in many situations of application, there would be sufficient numbers of observations so that this cost could be met. George Knafl worked for two summers developing a computer program based on this analysis. Again and again, computer simulations revealed flaws requiring fundamental changes in the model. Knafl's "Implementing Approximately Linear Models" (1978) reported on this stage.

These approaches are only half of the story of this interdisciplinary collaboration. Clifford Spiegelman explored two different approaches, devoting his PhD dissertation to one of them, and continuing these developments subsequently (Spiegelman, 1976, 1977, 1979a, 1979b). Almost immediately Spiegelman abandoned dependence on the fact of assignment by known rule or measure, characterizing the "sharp" or "true" regression–discontinuity analysis, and his methods became a general procedure for assignment by latent unmeasured variables, purporting to avoid selection bias (regression artifacts) in the pretest–posttest nonequivalent control group design (Campbell & Stanley, 1966; Cook & Campbell, 1979), as well as in the "fuzzy" regression–discontinuity setting.

Our Northwestern Psychology Department group was simultaneously active in work on the method all through this period. Bob Boruch, in addition to being our major participant in the recurrent series of ad hoc meetings with Dwass, Sacks, and their students, was active in both methodological developments and applications (Boruch, 1973, 1974, 1975b, 1975c, 1978b). In 1973, Boruch and James S. DeGracie, a statistician and Director of Evaluation for the Mesa Arizona School District, began collaboration or field tests of the design in eleven schools. The context was Title I reading programs in which children were supposed to be assigned to programs on the basis of pretest scores. Their findings on first-, third-, and fifth-grade students anticipated some of the problems in application of the design during the late 1970s: Programs are not delivered uniformly (some children spend only a few weeks in a "nominally" yearlong program); children are assigned to special services despite high scores on pretests; and there is difficulty in fitting curves to the *R* data as a result of floor and ceiling effects (Boruch & DeGracie, 1975, 1977; DeGracie & Boruch, 1977).

Especially important was Boruch's 1973 unpublished paper "Regression–discontinuity Design Revisited," which recommended the comparison of a single overall regression model with the two separate regressions above and below the cutting point. This paper became the "in-house" reference on statistical analysis of the regression–discontinuity design.

I believe it will be helpful to future investigators to give a brief report on our explorations of the "fuzzy" case, joined at an early stage by Trochim. To make these intelligible, I will start back again with the earlier presentations.

The concept of a "fuzzy" alternative to the regression discontinuity design was born on the same day in 1958 as the "sharp" or "true" design. The National Merit decision panels were supplied with quantified examination data and quantified school grades, integrated into a single composite eligibility score. Had the panels been lazy, they could have merely ratified the division into winners, alternates, and also-rans that these scores dictated. Instead, although they made decisions highly related to these scores, they added unquantified evidence such as letters of recommendation and interviews by some panel member with each of the finalists. This evidence was used to change the rankings provided by the quantitative measures in some cases. Were one to have plotted the data organized by the quantified decision scores and the outcome measures, the outcome would be more or less as shown in either Figure 13.3 or 13.4. (Figure 13.3 was based on a no-effect simulation.)

The presence in some cases of the award has made no effect on the relationship between the eligibility score and the outcome measure. This can be seen by comparing the envelope of the scatter diagram with that of Figure 13.1. Yet in Figure 13.3, the regression line for the award recipients lies significantly higher than that for the nonrecipients. This significant effect is a result of underadjusted selection biases and is a "regression artifact," a pseudo-effect if misinterpreted as an effect of the award. My major emphasis then, and now, is that almost all modes of analyses of "fuzzy" regression–discontinuity data (with the possible exception of those of Spiegelman, Trochim, Barnow, Cain, and Goldberger) will produce this artifact, this "packaging of underadjusted selection differences as though they were treatment effects." This effect follows if some of the rank reversals made by the decision panel are "valid," either through picking up some current symptom of promise reflected in the later outcome but not present in the eligibility score, or more generally and less easily comprehended, by tapping the same factor or factors as did the quantitative decision score, but through a partially different route, partially independent of the error component in the eligibility score. This need not at all imply that the award panel's decisions were more valid than the quantified eligibility score that the staff prepared. They could have been substantially less valid, considered as one single metric compared with another, and this regression artifact pseudo-impact would still occur if the large error component in the fallible award decision did not include all of the error of the eligibility score with which they

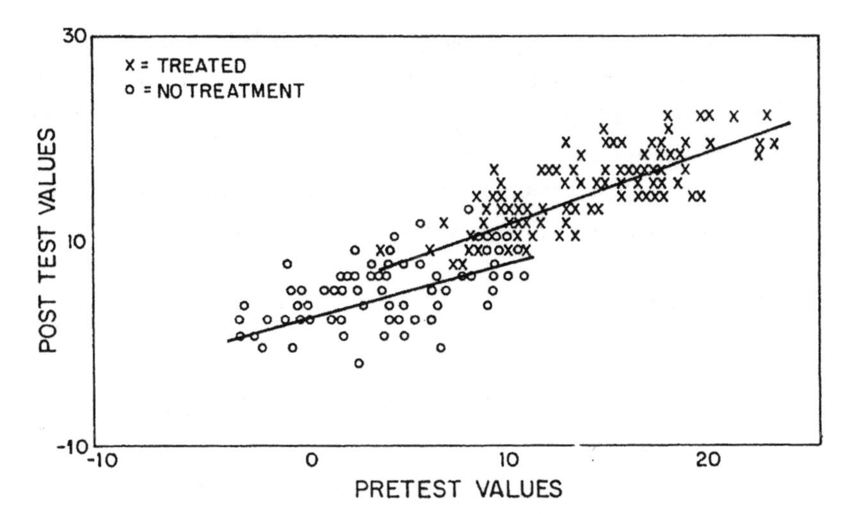

Figure 13.3. Regression-discontinuity design: Fuzzy cutting point, pseudo-treatment effect only. Source: Campbell, D. T. (1969). Reforms as experiments. *American Psychologist, 24*(4), 409–429. Copyright 1969 by the American Psychological Association. Reprinted by permission of the publisher and author.

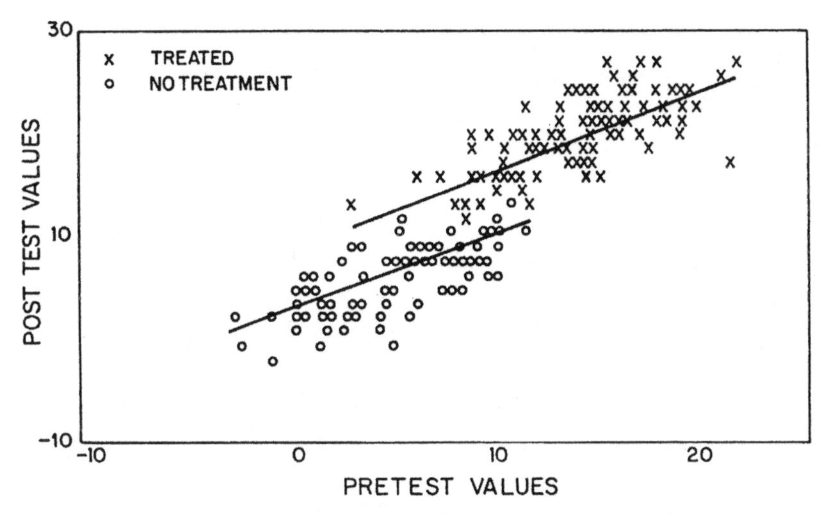

Figure 13.4. Regression-discontinuity design: Fuzzy cutting point, with real treatment plus pseudo-treatment effects. Source: Campbell, D. T. (1969). Reforms as experiments. *American Psychologist, 24*(4), 409–429. Copyright 1969 by the American Psychological Association. Reprinted by permission of the publisher and author.

had been provided. This being so, then for a given eligibility score, the award winners with that score would average higher than the nonwinners on the latent true score, and hence higher on the outcome score for that reason, even in the absence of any genuine award effect.

It will make these conditions and assumptions clearer to specify two alternative models for simulation of the fuzzy regression–discontinuity design.

Let E = Eligibility score
O = *Outcome measure*
A = Award, 1 if award, 0 if no award
I = Impact of award (zero in Figures 13.1 and 13.3, 2 in Figures 13.2 and 13.4)
e = Error component (in Figures 13.1 to 13.4, a normal random number)
T = Latent ability "True Score" (in Figures 13.1 to 13.4, a whole number between 0 and 20, selected at random, in later simulations, a normal random number multiplied by a constant.)

In these simulations, E and O are measures of parallel structure, sharing for each individual the same true score, but with independent error.

$$E = T + e_E$$
$$O = T + e_O + I$$

For the Sharp case, as in Figure 13.1,

$$A = 1 \text{ if } E < 10, 0 \text{ if } E > 10$$

For the fuzzy case of Figures 13.3 and 13.4, A^* is a third variable constructed, is paralleled with E and O, then dichotomized,

$$A^* = T + e_A$$

$$A = 1 \text{ if } A^* < 10, 0 \text{ if } A^* > 10$$

For Figures 13.2 and 13.4, the I, the impact of the award, = 2.

Although I believe that the case of Figure 13.3 is the ubiquitous one, and the one appropriate for the National Merit award process, I feel the need to make explicit another possibility: It could have been that, merely to demonstrate their authority, the award panel deviated from the quantitative scores available in a purely haphazard, irrelevant way, adding error to the eligibility score by an essentially random process, using no independent channels to the true score. In such a case, the formula for the award would have been

$$A^* = E + e_A.$$

In other words,

$$A^* = T + e_E + e_A$$

$$A = 1 \text{ if } A^* < 10, 0 \text{ if } A^* > 10.$$

In this event, for the null case of Figure 13.3, the regression lines for the awardees would have been the same as for the nonawardees. The whole scatter diagram envelope relating E to O would not have changed had this been the case, for Figures 13.3 and 13.4, but the regression lines would have. If one could be sure that the awards were based on a $T + e_E + e_A$ basis rather than a $T + e_A$ basis, then an ordinary covariance analysis would be appropriate in the "fuzzy" case too. (Most of the complications one would like to add to make the assignment processes more like reality, and outcome measures factorially complex in ways not exactly paralleled in the eligibility measure, will have implications making no-impact outcomes look like Figure 13.3.)

Reichardt, Trochim, and I initiated explorations of possible analyses of the fuzzy case seeking to find a way of distinguishing between null cases such as Figure 13.3 and true-effect cases such as Figure 13.4, within the tradition of higher order polynomial curve fitting. These explorations are reported in a 1979 report of 35 pages, actually written by Trochim (Campbell, Reichardt, & Trochim, 1979). I would like to convey to future explorers in these areas the nature of our attack and its problems.

Our first attack was in imitation of the old experimental statistician's rule of thumb when not all of the experimental group receive the treatment, and some of the controls get it on their own: "Analyze 'em as you randomize 'em." A conservative test, because the impact estimate is diluted by the untreated and the overtreated, but one unlikely to produce pseudo-effects. Our version of this we called a *pseudo-sharp* analysis. With Figures 13.3 and 13.4 in front of one, assume a pseudo-sharp cutoff point in the middle of the fuzzy transition region (the value 10 will do) and analyze as though all cases to the right received awards, all to the left did not, assuming linearity. Figure 13.3 will produce a no-effect outcome as did Figure 13.1. Figure 13.4, we thought, would show some effect, underestimated because of the misclassified cases. This degree of underestimation might be estimated and corrected for, if one assumed that the award had a constant effect regardless of eligibility-score level, because one knows how many have been misclassified for each eligibility score. Two considerations lead to the rejection of this approach. First, curvilinear fitting eliminates the effect in Figure 13.4, and we have many reasons to insist on rejecting a linear fit in most settings. (Floor effects and ceiling effects, differentially present in eligibility and outcome, are among them.) Second, in a simulation involving larger error (or unique) components, the overlap of award and nonaward cases would not be restricted to a central area, but would extend the entire range. In such a case, even a linear pseudo-sharp fit to a Figure 13.4 would show no effect. Evidence incidental to the Certificate of Merit analysis in Thistlethwaite and Campbell (1960, Table I) shows that the distribution of scholarship awards may have had this characteristic. Another hope was that the linearity or curvilinearity might be established by analysis of the awardees and nonawardees separately, and then the pseudo-sharp analysis limited to this level of polynomial complexity, excluding higher order ones. Again, if the fuzzy area were restricted to the central region, this might work, but not if it extended throughout the range of observations. In any event, the composite higher order polynomials approach came to seem to us so unpredictably fickle as to recommend avoiding extreme dependence on them. (Divgi's [1979] demonstration of the extreme costs in power of using higher order polynomials augmented our dissatisfaction.) In contrast, for the "pure" genuinely "sharp" regression–discontinuity design, the graphic presentation of scatter and column means provides nondeceptive visual evidence reducing greatly the blind dependence on curve-fitting statistics.

We considered one other approach to the fuzzy case that remains unexplored. A visual comparison of Figures 13.3 and 13.4 will help communicate the basic insight. One of the clues that indicate that in Figure 13.4 a substantial effect of the award has been built in is that for the columns containing a mixture of Xs and Os, the column variance is larger than in the pure-X or pure-O columns, a feature that is absent in Figure 13.3. If one assumes homoscedasticity, then deviations from uniform column variance that correspond to the mixture of treated and untreated cases could be solvable for the treatment effect. By further assuming that the award would have a constant level of effect regardless of the true score or eligibility score, the impact-augmented variance would be greatest in those columns in which half of the cases were treated. In general, homoscedasticity assumptions seem obviously untenable for both dimensions in a curvilinear plot. Perhaps for some such data sets, a limited homoscedasticity might be plausibly assumed, in the form of equal column variances in the absence of treatment effect. Having made this one assumption, no other assumption of curve form nor any estimate of it would be required in this analysis. Probably the lower the correlation between eligibility and outcome (that is, the larger the error components), the more reasonable are both assumptions of linearity and homoscedasticity, although this would not hold for the curvilinearity induced by floors or ceilings in the measures.

With my move to Syracuse in 1979, I became less directly involved in the continuing pursuits of an acceptable analysis for the fuzzy case, but received regular reports from Trochim on his own work and his collaboration with Spiegelman. Trochim began with Goldberger's (1972) observation that in the fuzzy case, the true regression lines in each group will be nonlinear. When joined, the overall regression line in the null case will approximate an *S*-shaped curve. Trochim concluded that a fuzzy case analysis would only work if one modeled the *S*-curve adequately. Trochim's basic insight was that a plot of the probability of assignment to treatment given the pretest score would, in the fuzzy case, typically yield an *S*-shaped curve of the type described by Goldberger. He tested this idea out in simulations that involved dividing the pretest scores into columns and, for each column, computing the percentage of cases assigned to treatments. This "assignment percentage" is a rough estimate of probability of assignment to treatment and is used in place of the dummy-coded assignment variable in the analysis. These simulations convinced him that the approach had promise and, through Bob Boruch, he contacted Cliff Spiegelman to obtain some statistical consultation on the feasibility of this analysis. It turned out that Spiegelman (1976, 1977, 1979c) had recommended an approach that was mathematically related but more exact, although his recommendation had largely gone unnoticed. Spiegelman's version involved computing a moving average of the dummy-coded assignment variable across the range of the pretest and substituting this estimate of probability of assignment to treatment in the analysis. Trochim and Spiegelman (1980) then collaborated on a paper that presented the statistical argument and computer simulations (Trochim, 1984). This "probability of assignment to treatment" approach to the fuzzy case (or, in Trochim's terms, the "relative assignment" approach) was independently suggested by Barnow, Cain, and Goldberger (1980) who recommended the fitting of a probit function to the relationship between the dummy-coded assignment variable and the pretest. The central difference between their approach and the Trochim and Spiegelman version is their selection of a probit function to model the *S*-shaped curvilinearity (Trochim and Spiegelman used essentially distribution-free strategy). Although further work is needed to explore the appropriateness of these fuzzy regression–discontinuity analysis strategies, they appear to have great promise for offering a potential solution to one of the most critical problems besetting the design.

As with randomized experiments, the design is best used prospectively, with quantified decision processes, recorded eligibility scores, and individual identification records permitting follow-up on outcomes at a later time. Usually its implementation will require much effort and change of customary admission processes as would a randomized assignment to treatment. But with both methods, we should also be alert for retrospective applications as a result of administrative arrangements designed for other purposes, such as fairness in the distribution of scarce resources.

The regression–discontinuity design has received its greatest use to date in the evaluation of compensatory education programs. Trochim has investigated more than 200 such regression–discontinuity analyses, has illustrated extensively the difficulties involved in implementing the design and the implications of these problems for estimates of treatment effect. His work is particularly valuable for its description of the interaction of social and political issues with the use and validity of the regression–discontinuity design. But this history is best told by Trochim himself in his book, *Research Design for Program Evaluation* (1984).

OVERVIEW OF CHAPTER 14

Suppose you are an administrator of a state-run program in your county that provides counseling and social services to juvenile offenders. You and your staff develop a special approach that you believe will reduce recidivism, and you are given permission to implement this approach in your county. If it is successful, the program director would like to disseminate your approach to the other counties in the state. How will you demonstrate that your program is successful?

This chapter, coauthored by Trochim and Campbell, may provide an inexpensive yet valid way to determine whether your approach had an impact. The authors discuss the Regression Point Displacement Design (RPDD), which can deal with large units such as schools, cities, or counties using aggregated instead of individual data. A single unit receiving an ameliorative treatment can be compared to a number of control units. Pretests and posttests are required, although it is not necessary that the same measure be used. The RPDD regresses the posttest on the pretest and fits a regression line for the control units only. A *t*-test determines if the distance (or displacement) of the treated unit is significantly different from its expected value on the regression line. If there was no treatment effect, one would not expect the treatment group to differ at greater than chance levels from the regression line. In the situation described previously, this design could be applied rather effortlessly, because it is likely that recidivism rates at regular intervals are already available for every county in the state and can provide the pretest and posttest measures.

The RPDD is characterized by four features: A single treatment unit is often used, aggregated data are generally used rather than individual data, the groups do not have to be equal prior to treatment, and the observed regression line is used for the analysis rather than as a statistical adjustment for pretreatment differences.

The usefulness of the design is demonstrated in a discussion of the Medicaid Regression Point Displacement Design and the Schizophrenic Reaction Time Study. In both of these applications, the RPDD showed that the treatment groups differ significantly from the regression line. The authors warn, however, that because this is a quasi-experimental design, rival explanations for the noted effect must be ruled out. In the Schizophrenic Reaction Time Study, for example, the effect is not apparent when the data from the individuals (rather than the grouped data) are analyzed. The authors argue that the RPDD can never have greater power than such a micro-level analysis, and that the significant effect noted using the RPDD is a result of the small error term that occurred when the regression line was fitted to only three points.

The basic requirements for the RPDD are (a) multiple control groups, and (b) pretest and posttest measurements. There are many variations of this design, however, based on five dimensions: how the treatment group is chosen, the entity that will be measured, the number of treatment groups, same or different measures for pre- and posttest, and the number of covariates.

The authors devote the final section of the chapter to the threats to validity that are most problematic when using this design. Both selection bias and regression artifacts can be ruled out if the treatment groups are chosen randomly or selected based on a sharp cutoff point. Doing so makes the RPDD akin to the Randomized Experiment or the Regression Discontinuity Design, respectively. Bias can be introduced, however, if there is measurement error in the pretest. When this occurs, the regression line is rotated slightly, and the bias becomes more extreme as the distance increases between the treatment group pretest mean and the overall pretest mean. Instrumentation can be a problem if the measurement process has changed in the treatment group between the pretest and the posttest, and local history should be examined to determine if differences exist in the treatment group setting that might affect posttest performance. The power of the RPDD is diminished somewhat because of the few points typically used; however, this is offset by a gain in power resulting from more reliable aggregated data. The authors feel that this issue needs further investigation. The *t*-test used in the statistical analysis of this design assumes that the control groups are a random sample of the population. Because this assumption can never be met, control groups should be selected based on their variability.

Unfortunately, the external validity of this design is low. What does this mean, then, if you found that your approach to reduce recidivism for juvenile offenders was effective? You might suggest that the approach be tried first in counties with similar pretest recidivism rates, because you can generalize with more confidence to similar groups. Unfortunately, you cannot be sure that your approach would be as effective in markedly different counties.

FOURTEEN

DESIGN FOR COMMUNITY-BASED DEMONSTRATION PROJECTS

This chapter describes an old but neglected quasi-experimental research design. Figure 14.1 reprints its most widely distributed exemplar, used by Riecken et al. (1974, p. 115), and Cook and Campbell (1979, pp. 143–146), neither source presenting any statistical analysis. They cite Fleiss and Tanur (1972) from whom we borrow Figure 14.2, in which is claimed an effect significant at the $p < .05$ level. They in turn cite H. F. Smith (1957) and Ehrenberg (1968) as at least partial predecessors. In the examples of Figure 14.1 and 14.2, the measures employed on the *x*- and *y*-axes are quite different. This option is shared with the regression–discontinuity (RD) design (Trochim, 1984), which is related to the design described in this chapter. Our Figure 14.3 illustrates this application, drawn from the same data set as Figure 14.1. Although generally the interpretability of the design will be

Trochim, W. M. K., & Campbell, D. T. (1996). *The regression point displacement design for evaluating community-based pilot programs and demonstration projects.* Unpublished manuscript.

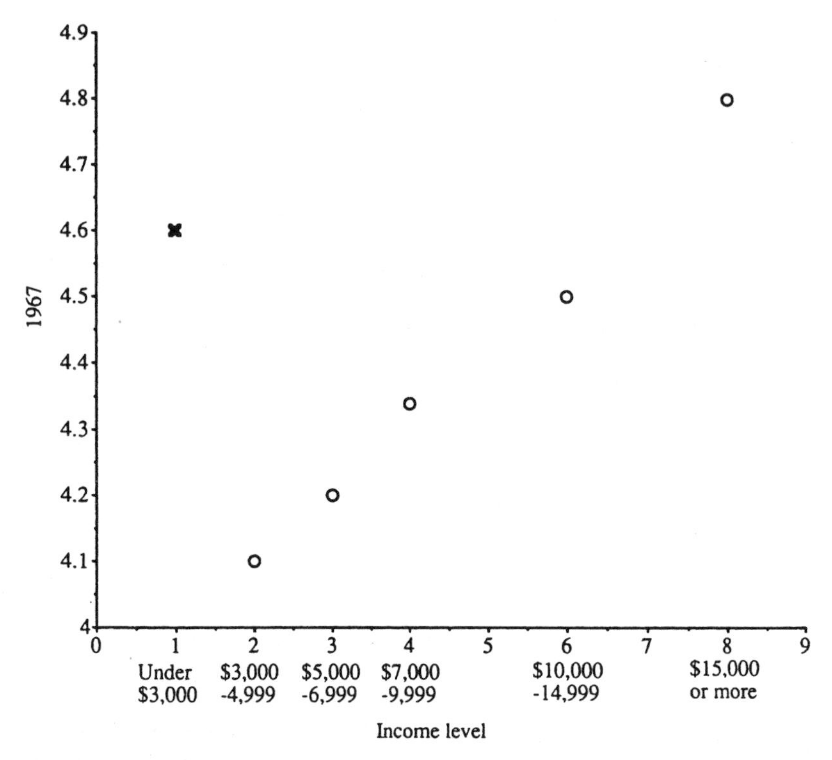

Figure 14.1. Medicaid example modified from Riecken et al. (1974). Used by permission of Academic Press.

substantially greater when the same measure is used before and after, we will not argue that this is so for Figure 14.3.

We envisage a typical application as employing repeated (e.g., annual) rate measures, for cities or some larger or smaller reporting units, with one (or several) units receiving an intensive ameliorative effort not given in the others. Because of this anticipated usage, we shall refer to the x-axis as the *pretest* and the y-axis as the *posttest* in what follows. Some treatment or program, very often likely to be a community-based pilot program or demonstration project, is administered to the treated unit. Although in Figure 14.1 the treated unit was the most extreme on the pretest, the method is applicable no matter where in the distribution of pretest values the "experimental" unit falls. Indeed, the statistical power and interpretability is likely to be better for mid-distribution

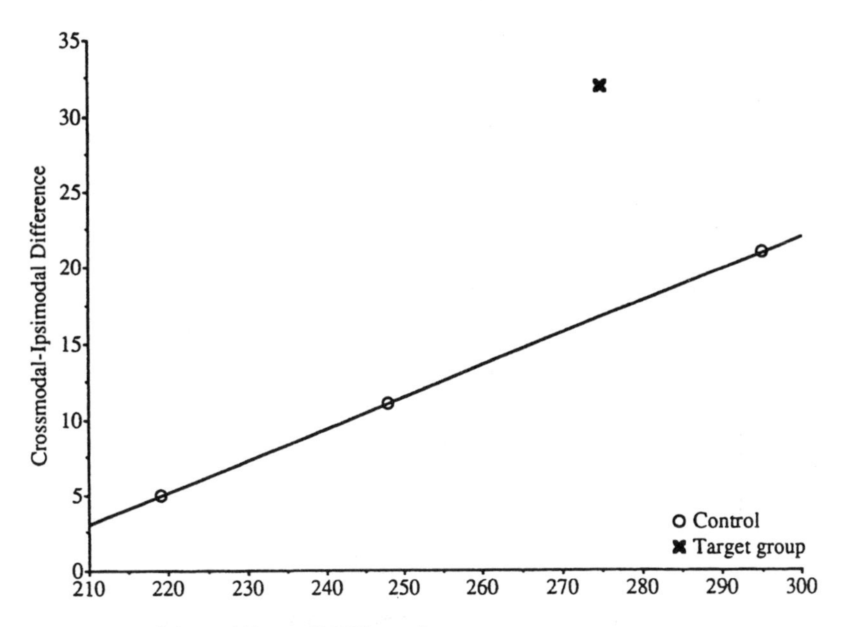

Figure 14.2. Fleiss and Tanur (1972) graph.

demonstration sites. The untreated units we will identify as *control units* or *control groups.* The analysis fits a regression line to the control units and tests the significance of the departure of the experimental unit from that regression line. The name suggested for this design is the Regression Point Displacement Design (RPDD).

Demonstration programs and pilot projects usually receive only very weak and methodologically suspect evaluation. Often there is only a comparison of rates for that one unit for a time period before the special effort and a time period afterward. Or a single comparison unit (e.g., another city similar to the demonstration one) is employed, inevitably differing in many ways. The potential power of the RPDD comes from using numerous untreated units as "controls," and from not requiring pretreatment equality between any one of them and the experimental unit.

The RPDD remains very much a quasi-experimental design, for which many of the common threats to validity must be examined on the basis of contextual information not included in the statistical analysis. If a statistically

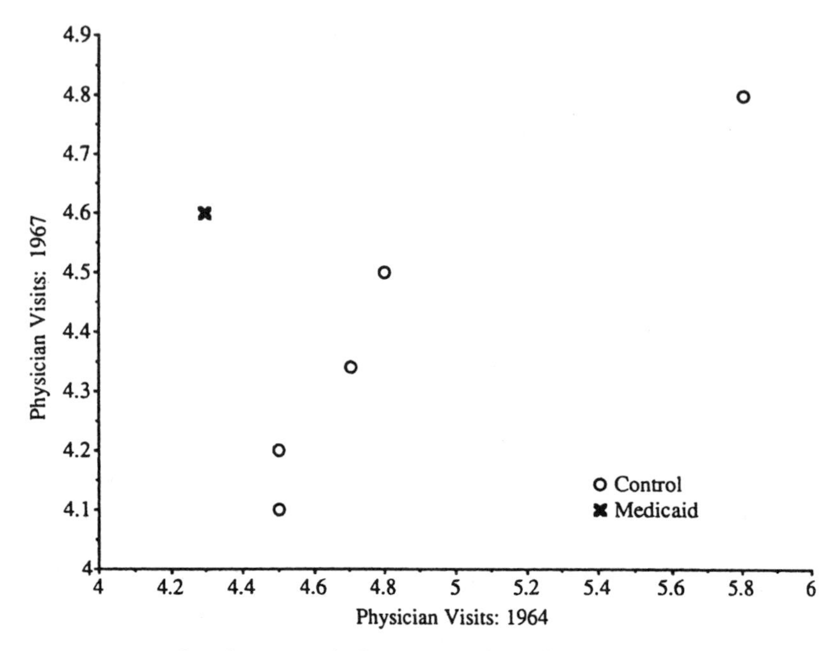

Figure 14.3. Medicaid RPDD with physician visit for both pre- and posttreatment from Riecken et al. (1974). Used by permission of Academic Press.

significant displacement is shown, there are many other possible causes that need to be considered, over and above the demonstration program.

The RPDD can be illustrated with a simple hypothetical example. Consider a single site at which a treatment is administered. Furthermore, assume that there are arbitrarily ten other sites that will not get the treatment (control sites) but will be measured. All available persons at both the treatment and control sites are measured before the treatment and at some specific time after the treatment. Note that as a result of normal turnover rates and absenteeism at each site, the persons measured at the pretest may not be the same as those measured at the posttest. The resulting data might look like the simulated values depicted in Figure 14.4.

The figure shows the linear regression line of the ten control group pretest–posttest pairs of means. The vertical line indicates the posttest displacement or "shift" of the treatment group from the regression-line predicted value.

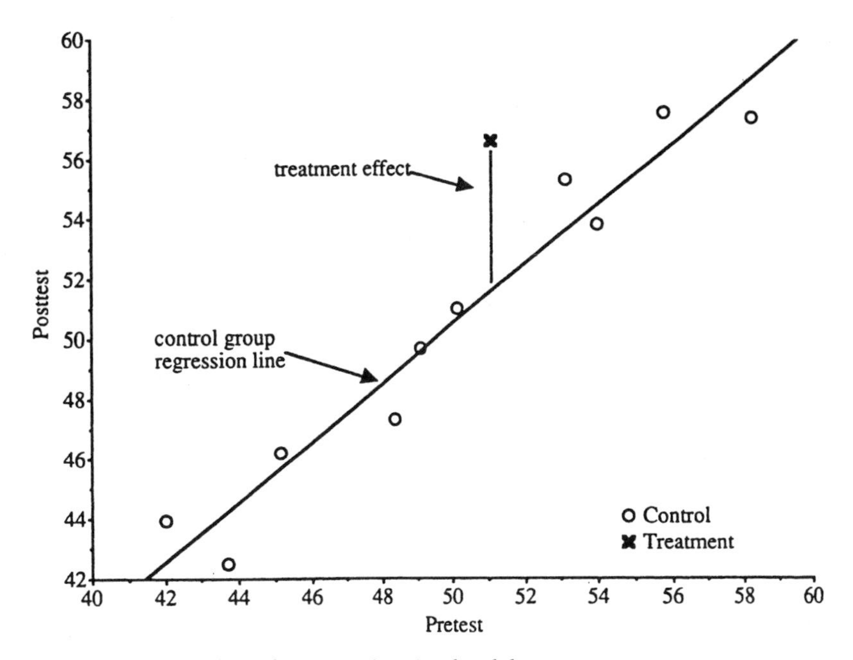

Figure 14.4. Hypothetical RPDD using simulated data.

In this case, it is visually clear that the displacement probably exceeds the normal variability one might expect around the regression line and indicates a likely treatment effect.

The central idea of the design is that in the null case, one would not expect the treatment group to differ at greater than chance levels from the regression line of the population. There is evidence for a treatment effect when there is a posttest (vertical) displacement of the treated group point from the control group regression line. This led to the name *regression point displacement.* Of course, this evidence does not imply that the treatment of interest is what "caused" this vertical regression shift. One must always assess the plausibility of other potential causes for such a shift.

The RPDD is characterized by four major features: (a) the use of a single treated unit instead of many; (b) the use of aggregate-level data instead of individual-level; (c) the absence of any need to ensure (or attempt to achieve by statistical adjustment) pretreatment equality between treated and control groups; and (d) the avoidance of "regression artifacts" or underadjustment as a result of "errors in variables" by employing the observed regression line, rather than using it as a means of adjustment. Feature d, and to a lesser extent c, are shared with the regression–discontinuity (RD) design. The first two are not absolutely necessary. We would still probably classify a study with two or even three demonstration sites as an RPDD. The key distinction is that in alternative designs there are enough points to allow one to fit the same model to both groups, whereas the RPDD typically does not. For instance, in the RD design, one usually has enough points in both groups to be able to estimate a within-group slope, whereas in the RPDD, that is not possible or justifiable, given the few available treatment group points. The higher aggregation level (e.g., cities) is also a typical RPDD characteristic, although it is by no means required. One can envision an RPDD design involving only a single person who receives a treatment, with multiple control persons.

THE MEDICAID REGRESSION POINT DISPLACEMENT DESIGN

The first example comes from the Medicaid study discussed in Riecken et al. (1974, p. 115) and Cook and Campbell (1979, pp. 143–146), part of which was shown in Figures 14.1 and 14.3. The original data are shown in Figure 14.5 (taken from Lohr, 1972; Wilder, 1972, p. 5, Table B).

The figure shows the average number of physician visits per person per year in the United States for the years 1964, 1967, and 1969, broken out by family income ranges. The Medicaid program was introduced in 1964. The legislation mandated that only those families with an annual income of less than $3,000 were eligible to receive Medicaid. Overall, it appears that the annual average number of physician visits is declining for most income groups with two notable exceptions—the lowest income group shows an increase over both time intervals and the second lowest group increases between 1967 and 1969.

The central question is whether the introduction of Medicaid is associated with a significant increase in the average number of physician visits per year. Several RPDDs can be constructed from these data. The first is identical to that shown in Riecken et al. (1974, p. 115) and Cook and Campbell (1979, pp. 143–

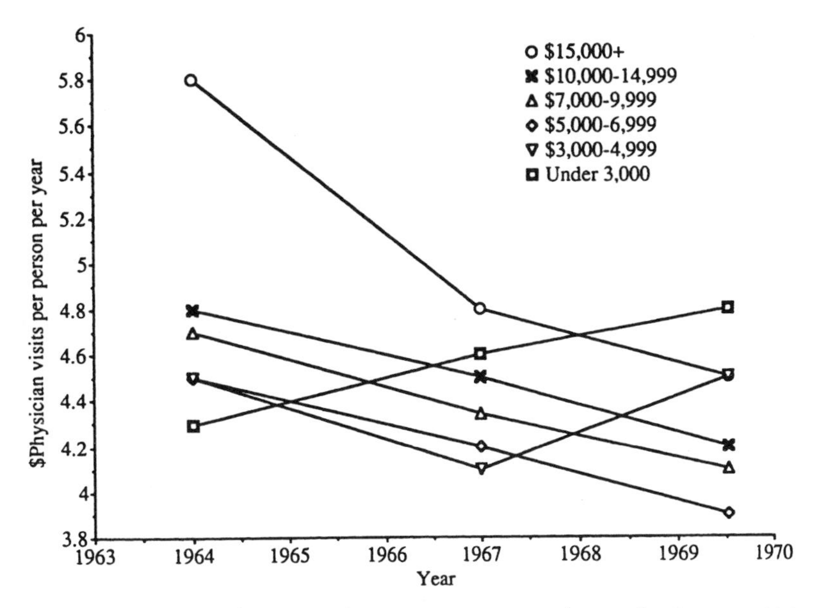

Figure 14.5. Average number of physicians per person per year for the years 1964, 1967, and 1969, by family income ranges (taken from Lohr, 1972; Wilder, 1972, p. 5, Table B).

146) and displays income group along the horizontal axis and physician visits on the vertical as shown earlier in Figure 14.1. One problem in analyzing these data concerns the metric for the pretest. We know that income distributions tend to be nonnormal and as a consequence we may need to transform the pretest variable before conducting the analysis. We also see that the highest income group has no upper-income limit given. To analyze these data, we decided to use the logarithm of the upper and lower limit for each pretest income interval (with an upper limit for the high pretest group set arbitrarily at $50,000 and the lower limit for the low income group set to $1000) and then use the midpoint between these logs as the pretest value for each group. The transformed data are graphed in Figure 14.6. The ANCOVA estimate of effect (in log pretest units) is β_2 =.824 (t = 21.03, p = .0002).

A second RPDD can be constructed from these same data by graphing posttest (1967) physician visits against pretest (1964) ones for each income group as shown earlier in Figure 14.3. The lowest income group is by definition

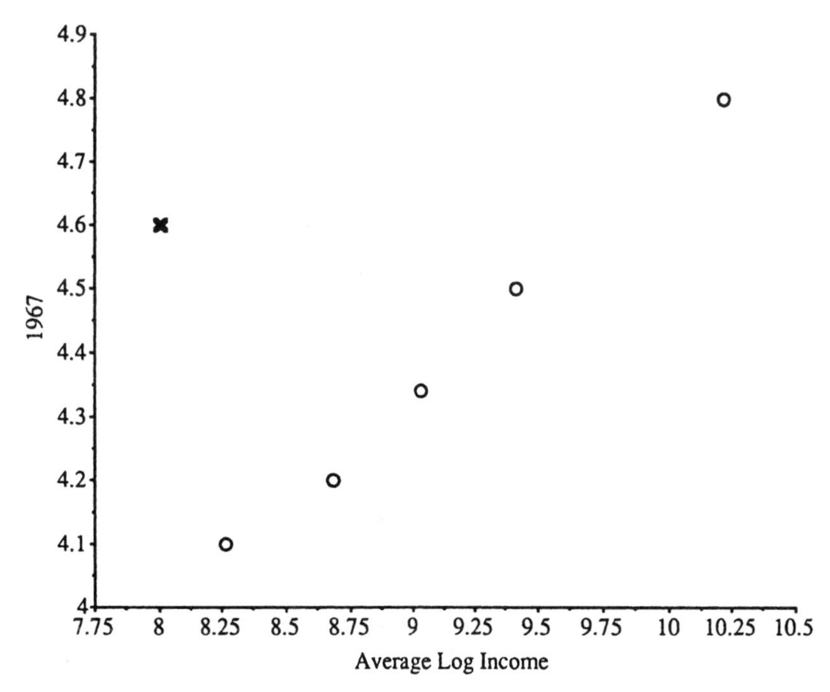

Figure 14.6. RPDD for the study of the effects of Medicaid on physician visit rates with logs of income as pretest.

the Medicaid treatment group indicated by an x on the graph. The other income groups are shown with an o and can be considered comparison or control groups. The Medicaid group had the lowest pretest average number of physician visits. The question is whether their posttest level is significantly higher than would be predicted given the control group pre–post levels. The ANCOVA estimate of effect is $\beta_2 = .479$ ($t = 3.63$, $p = .036$).

Medicaid appears to be associated with a significant rise in annual physician visits, but one still cannot conclude that Medicaid is what caused this rise. In order to reach this conclusion, one has to rule out any plausible alternative causal explanations for the observed effect (Cook & Campbell, 1979). Several possibilities suggest themselves. First, it could be that the regression line that is fitted to the data does not accurately reflect the true regression for the population in question. The question is whether the apparent significant effect results from specification of the wrong regression model. This could arise in some contexts because the control groups do not represent the population of interest, an unlikely event in this case because the control group means include the entire U.S. population income ranges (although the use of a single group to indicate the annual physician visits for all persons with incomes of more than $15,000 may very well distort the shape of the graph). The more plausible problem is that the control group pre–post relationship is not linear in the population, but is instead quadratic or some other functional form. Following Darlington (1990, p. 295), the polynomial regression (including both the linear quadratic terms in the regression model) is fitted to the data. The resulting equation is

$$Y_i = -13.56 + 6.6X_i + -.59Y_i^2$$

Neither of the *X*-coefficients is statistically significant (at $p < .05$), a result that is probably attributable to the small number of points and the consequent low statistical power associated with each estimate (note, however, that the linear term alone is significant in the original ANCOVA model). This linear plus quadratic regression line is shown in Figure 14.7.

The polynomial regression fits the control group points better than the linear one did. However, it is impossible to know whether it is the better model for the population with so few points available for prediction (remember, one could fit the observed points perfectly with a fourth-order polynomial model). In judging plausibility one must examine the assumption that the straight-line fit is misleading in that a curvilinear plot would reduce the departure from expectancy. In this example, even if the linear model is not the best, it is clear from visual inspection that no reasonable model would rule out the sensibility of concluding that there is a significant effect for the Medicaid group. The linear fit reduces the $Y_0 - \hat{Y}_0$ discrepancy from that which any reasonable curvilinear fit would produce. With the polynomial model, the observed treatment group posttest will be even further from the predicted value. We feel that in this case, the linear fit is conservative, rather than misleading. A curvilinear fit should of course be the privileged fit in cases in which it reduces the apparent effect. Note that these problems are minimized in cases in which the experimental unit falls in the middle of the controls in as much as the $Y_0 - \hat{Y}_0$ discrepancy will vary little as a function of the curvilinear plot chosen.

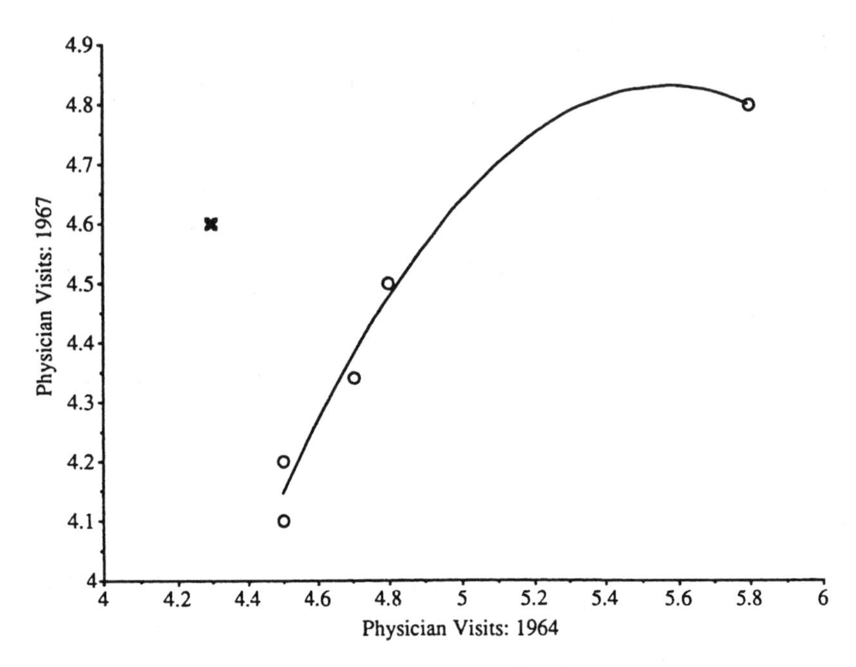

Figure 14.7. Second-order polynomial regression line for the RPDD data from the study of the effects of Medicaid on physician visit rates.

A second threat to the causal inference is an issue of internal validity. There may have been some other factor affecting only the lowest income group that increased their annual physician visit rate. For instance, if there is another low income subsidy program, such as the WIC nutritional supplement program, it may be that the rise in physician visits is attributable to the increased gynecological care subsidized by that program rather than to the Medicaid subsidies. There is no way to rule out that threat with these data, although one could examine it by comparing physician visit rates for WIC versus non-WIC recipients if such data were available. The problem in this context is that there are likely to be many federal or state programs that are targeted to the lowest income group (i.e., those who fall below the official poverty line). Although we would usually argue that an RPDD that gives the treatment to an extreme case is preferable to uncontrolled treatment assignment, it is not preferred when the implicit cutoff is a well-publicized, frequently used value such as the federal poverty level, which is used as the criterion for assignment in many national programs that constitute alternative potential causes for any observed treatment effect. Thus, in this case, it is impossible to be confident that the observed effect is a result of Medicaid alone. Nor does the apparent jump in physician visits for the next highest income group from 1967 to 1970 (evident in Figure 14.5) solve the problem because it too may very well result from raises in the Medicaid eligibility cutoff values, raises in the official poverty income level (and the consequent eligibility for other federal programs), or both.

SCHIZOPHRENIC REACTION TIME STUDY

Fleiss and Tanur (1972) described a version of a Regression Point Displacement analysis that explored reaction time in schizophrenics. The purpose of this analysis was to examine whether clear-cut schizophrenics differ from other groups in their cross-modal–ipsimodal reaction time difference. The data are shown in Figure 14.2. The ANCOVA estimate of effect is $\beta_2 = 16.11$ ($t = 109.59$, $p = .0058$).

Many readers will join us in being surprised that such power can be obtained from an application with just three control group points. There are several variables affecting such a p-value. One is the degrees of freedom (in this case, $N - 2 = 1$ df). A second is the dispersal of values in the control groups from the fitted line. A third is the magnitude of the departure of the treated group from the center of the fitted line (i.e., the error term is larger the greater the distance of the treatment group pretest mean from the mean of the control group average).

Let us consider the second component. In this case, the linear fit is essentially perfect, producing a very small error term, and hence a very large t-value. But with only three points, are not such perfect fits going to happen by chance very frequently? Or, to put it another way, in repeated samplings from the same universe, is not the error term going to fluctuate widely from replication to replication? Can we be sure that the small-sample values for the t-test are still appropriate for this application, for $df = 1$? Note that Fleiss and Tanur failed to find a significant effect when using within-group variance, in contrast to the Figure 14.2 analysis using only group means. We would argue that the relative power of an RPDD *can never be greater* than the power of a more microlevel analysis (e.g., using individual data points instead of group means) on which it is based, even though we may serendipitously find extremely significant estimates in a given RPDD, as in this example. Thus, although our presentation has been inspired in part by Fleiss and Tanur's (1972) seminal paper, we regard the specific application with caution just because the *perfect* linear fit is so out of line with ordinary experience. Someone interested in applying their finding would be well advised to explore the relationship between the ipsimodal versus cross-modal–ipsimodal differences over a larger number of diagnostic groupings in an effort to get a more plausible error term.

REQUIREMENTS FOR THE RPDD

To qualify as an RPDD, there must be multiple comparison or control groups and pre–post measurement. But given this restriction, there are many alternative versions of the design that are possible. Many of the variations can be described in terms of five major dimensions, where a different RPDD can be constructed for different combinations of these dimensions. The dimensions follow.

1. Method of Assignment of Treated Unit

For almost any two-group pre–post design it is possible to construct a RPDD analogue. If the single experimental unit is randomly assigned (from a pool of potential candidate units), the RPDD is analogous to a Randomized Experimental (RE) design. When the experimental unit is chosen because it has the most extreme value (high or low) on the pretest, this is essentially equivalent to assignment by a cutoff (the cutoff in this case is usually implicit and consists of the pretest values that distinguish this extreme case from the others), making the RPDD analogous to the regression–discontinuity (RD) design. Finally, when the treatment group is chosen arbitrarily, for political reasons or personal favoritism, or for any other unspecified reason, we can consider the assignment to be by an unknown or unspecifiable rule, and the RPDD is most analogous to a Nonequivalent Group Design (NEGD). For most of the experimental or quasi-experimental designs, it is possible to construct RPDD analogues, and useful to do so because consideration of analogous designs and the literatures that have grown around them will help raise validity issues that ought to be considered in the RPDD analogue.

2. The Unit of Measurement

The unit of measurement refers to the entity represented in each pre–post point in a RPDD. Usually, these will be broad units—states, cities, communities, socioeconomic groups, diagnostic groups—not individual persons. However, an RPDD design can be constructed using either aggregated individual data or group data. For instance, many readily available databases consist of already aggregated frequencies, rates, proportions, or averages across geographically or demographically defined groups. In the Medicaid example of Figures 14.1 and 14.3, the average number of physician visits per year per person for six different income groups is used. The RPDD analysis does not require physician visit rates by individual (nor changes in such rates)—it operates in this case on the group averages. Restricting the data analyzed to repeated measures from the same individuals adds power to some statistical analyses. If the group means in each case are based on the same persons in each year, we might expect a smaller error term (see Cook & Campbell, 1979, pp. 115–117). Yet in most instances, if a survey is conducted in a number of communities before and after an educational intervention in one of the communities, the people measured on the pretreatment survey are not likely to be the same as those measured afterward (unmatched). The RPDD is perhaps the strongest design available for studying community-level interventions in which different persons are sampled on each occasion. It is also possible to use the RPDD when the pretest and posttest scores are based on the same person. For example, the same (matched) patients could be measured before and after, but for the RPDD, all such patients at a given site (or clinic, hospital) would constitute a unit and their average scores used. This pre–post same-individual RPDD might arise in the case in which the same students within a school are measured before and after some treatment is implemented in a single classroom. Classroom average scores could be used in conducting a RPDD analysis, or individual scores (grouping all control group cases together) could be used in a more traditional nonequivalent group analysis. One must be careful in using data from separate pre- and posttest samples because of the potential bias that can arise. For instance, if pretest and posttest average scores for a group are used, it is likely that those nondropouts present on the posttest are unrepresentative of the original pretest group. Contrast this with two different but repeated random samplings over time from the same community. Although even this may be biased in the sense that the basic demographic structure of the community may have evolved between measurements, it is much less plausible in this case, especially when the time span is relatively short. Random, rather than opportunistic, sampling on each occasion can help to ensure some degree of equivalence when repeated measures are not obtained for the same group of people.

3. The Number of Treatment Groups

In the simplest case, the RPDD involves only a single treated group or site. The treatment would be administered in one community or classroom. When the design involves enough treated points that it is reasonable to estimate the same functional form as for the controls, the RPDD essentially transforms into one of the other pre–post designs—randomized experiment, nonequivalent group design, or regression-discontinuity—depending on the method used to assign units to treatment or control condition.

4. The Same Versus Different Pre–Post Measures

In general, the same variable is measured before and after the treatment (matched measures), but different measures can be used. For instance, if one is looking at the effect of an educational program, it might not make sense to measure content-related performance on the pretest because students would not be expected to know any of the content (and may not even understand the questions). One might use a general measure of prior intelligence or academic achievement (GPA, standardized achievement test scores) as the premeasure, with a treatment content-specific outcome measure. Thus, pretest and posttest in this case are different, or unmatched, measures. In Figure 14.1, the pre- and posttest measures are still less similar. Whether matched or not, statistical power is likely to be greater when the premeasure has a strong linear or monotonic relation with the outcome variable.

5. The Number of Covariates

In the simplest case, the RPDD uses a single pretreatment variable. But it is also possible to use multiple pretreatment variables that can simultaneously be entered as covariates in the model. The major problem with multiple pretreatment covariates is that, because each covariate costs one degree of freedom, using multiple covariates requires more control groups. However, the use of multiple covariates when many control groups exist will be an important mechanism for improving the statistical power and efficiency of the treatment effect estimate.

THREATS TO VALIDITY

Selection Bias

Probably the most important threat to validity in the RPDD is the potential for selection bias that stems from initial between-group differences that affect the posttest and are unrelated to the treatment. The plausibility of such a threat rests on the method used for assigning (or selecting) the treated group in the RPDD. The method of assignment determines which traditional multiple unit pre–post designs the specific RPDD is most like. If the RPDD units are randomly assigned, the design is most analogous to a RE. In the RPDD case, however, random assignment is not used to ensure probabilistic pretest equivalence as much as to minimize the chance that a unit might be opportunistically chosen because it is well-suited, politically favored, likely to be successful, or any other number of factors that could bring about an apparent effect even if the program is never administered. Random assignment helps to guard against the many pretreatment correlates that might bias the outcome, wittingly or unwittingly.

If the RPDD experimental unit is assigned solely on the basis of its extremity on the pretreatment measure, this is analogous to a RD design because the assignment is by means of an implicit *cutoff rule*. This is the case with the Medicaid study as described in Figure 14.1. There, Congress allocated Medicaid explicitly using an income cutoff rule. Note, however, that this is not the case for other RPDDs constructed from the Medicaid data (shown in Figure 14.3) where, for instance, 1964 average physician visits constitute the pretest and 1965 values the posttest. Even though in this case the experimental group also turns out to be the lowest in pretest average physician visits, *physician visits were not the basis for the allocation of the Medicaid treatment.*

Where the RPDD is structured like the RE or the RD designs, the selection bias problem is largely mitigated by the fact that we know perfectly the rule that determines the assignment to treatment (probabilistic in the RE case and cutoff-based in RD). Just as in those designs, only a factor that correlates perfectly with the known assignment rule poses a legitimate selectivity threat. Of course, as the Medicaid study shows, there can be many such factors because the same implicit cutoff (i.e., the poverty rate) is used to allocate multiple programs.

This should be contrasted with the third assignment strategy—uncontrolled assignment—that yields an RPDD most analogous to the NEGD. In this case, the rule for assignment (i.e., selection) of the experimental unit is not explicit or able to be controlled for perfectly in the statistical model. As a consequence, one is less sure that the observed treatment effect is attributable to the treatment as opposed to any of the many possible selection factors that might also affect the posttest. For instance, assume a study in which there are ten possible treatment sites for some presumably beneficial treatment. Further assume that the selection of the experimental site is intensely political with each site lobbying to be the first to receive the experimental. The city that is ultimately selected is likely to differ in many ways from those that were not selected. It may be more highly motivated, have greater resources, have more political clout, and so on. If these (and other) factors can affect posttest scores, it will not be possible to say with great confidence whether any observed treatment effect is a result of the treatment or of these inherent differences between this city and the controls. Measuring all cities, including the experimental, on a pretest is likely to improve our inferential ability because posttest differences can be adjusted for pretest ones, but our experience with such adjustments for selection bias warns us that we should be cautious about attaching too much credibility to treatment-effect inferences in this case.

In an ideal situation, the demonstration site for a pilot program in a RPDD would be chosen purely at random, perhaps in a public lottery. Although with an *N* of 1 in the experimental group one would not be getting the benefit of plausible pretreatment equalization, one would be reducing the plausibility that a systematic difference on other variables not only determined the choice of the pilot site, but also determined the exceptional departure from expectancy on the outcome variable.

The discussion so far has assumed that the experimental unit was *not* selected *after* its eccentricity on the outcome variable was known. Nonetheless, that possibility needs discussion. Consider a case in which ten cities have measures on HIV-positive rates over successive years, and one notices that one of them is exceptionally far below expectancy for the second year. The interpretive problem is that each city has had AIDS prevention programs, all slightly different, so that there is an "experimental program" to be credited with the effect, no matter which city is exceptional, even if that exceptionality is a result of chance.

Although such interpretive opportunism is to be discouraged (especially if it is disguised from the reader), the strategy of locating a "truly exceptional" city (or site) first on the basis of posttest scores, and then speculating on what "caused it" should not be prohibited entirely. But in this case, the ordinary p-value for a given t-value cannot be used. Instead, a correction on the order of that for "error-rate experimentwise" is needed. The simplest approach would be to use a Bonferoni correction of the p-values (Darlington, 1990, pp. 250–257; Dunn, 1961; Ryan, 1959, 1960). If for a specified in-advance site, for a given df (e.g., $df = 8$ for the ten cities assuming a linear fit) a t-value of 2.306 is required for $p < .05$, when we want a comparable p-value (i.e., 1/20 for testing the exceptionality of any one of the ten points from the regression line determined by the nine others (not specifying which one in advance), we need a t-ratio corresponding to $1/(20 \times 10)$ or 1/200, or $p < .005$.

The RPDD, unlike other quasi-experiments, does not require pretest equivalence between the treated group and the controls. The treated group theoretically could come from anywhere along the pretest continuum. The design rests on the assumption that the treated group posttest mean does not differ significantly from the regression line prediction. As a consequence, the traditional concerns about selection bias take on a slightly different form in this context. Here, the key issues are whether the control groups yield an unbiased estimate of the true population regression line and whether the treatment unit is a member of the control group population. This could be assured by randomly sampling control groups from the population, a circumstance that will not be feasible in many situations. If the sample of control groups is not representative of the theoretical population or the regression line is incorrectly estimated, the estimate of the treatment effect will be biased. There is no solution to this problem, although it might best be minimized by selecting many control groups with wide pretest variability. For instance, in their study of schizophrenics, Fleiss and Tanur (1972) gave this advice for selecting control groups:

> A more efficient approach would call for the identification of many of the factors that distinguish schizophrenics from normals: having a mental disorder, being hospitalized, having been treated with drugs some time in the past, and so on. Samples of subjects from groups defined in terms of various combinations of such factors would be drawn and studied. These samples would have one feature in common: they would all consist of subjects who are not schizophrenic. (p. 525)

Measurement Error and Regression Artifacts

The general problem of regression artifacts (or error in independent variables) is taken care of in the RE or RD analogues of the RPDD, because such regression is displayed and accounted for by the inclusion of *X* in the regression analysis (Cappelleri, Trochim, Stanley, & Reichardt, 1991; Trochim, 1984; Trochim, Cappelleri, & Reichardt, 1991). Nonetheless, when choosing the experimental unit involves unknown but systematic variables on which the experimental group differs from the control group in ways that would affect the posttest differentially from the pretest, one might mistakenly conclude that the treatment was effective when, in fact, the apparent effect should be attributed to measurement error and the resulting regression to the mean.

This is similar to the misleading interpretations that can occur in relation to the "fuzzy" RD design (see chapter 13, this volume). If in fact, the choice of the experimental unit had been based on a latent decision variable related to the pretest by the addition of a pretest random error component, and to the posttest by the addition of a posttest random error component, and if the award of the experiment is based on extremity on the latent "true score," then a mistaken inference comparable to that graphed in Figure 13.3 is possible.

The problem is a manifestation of the familiar effect of pretest measurement error in NEGDs as described by Reichardt in Cook and Campbell (1979, Fig. 4.4, p. 161). Reichardt showed that pretest measurement error attenuates the within-group slope. In the RPDD, however, because there is only a single experimental group point, it is impossible to estimate a treatment group slope (and thus, the "slope" cannot be attenuated because of pretest measurement error). However, the true control group regression line would appear to be rotated clockwise slightly (assuming a positive relationship). The further the experimental group is away from the control group pretest mean, the greater will be the deleterious effects of such measurement error—bias will be greater.

The issue of how the treated unit is chosen is so central to the interpretability of the RPDD that it warrants some belaboring. The central concern is this: Can the treated unit be considered a member of the control unit population prior to treatment, or does it come from some different distribution? If we select the treated unit randomly or because of its extremity (implicit cutoff), it is reasonable to infer that the unit is sampled from the control group population. Here, just as in the RE or RD designs, estimates of treatment effect will be unbiased by the measurement error. The regression of the posttest onto pretest accurately describes the amount of regression to the mean expected for all units, treated and control. But when selection of the treated group is not controlled, it is plausible that the treated unit does not come from the control unit population, but rather from some population that differs systematically from the controls. In this case, we must assume that the two populations may differ in their overall pretest averages and, as a consequence, measurement error would affect the populations differently and there would be regression to different population means, just as in the NEGD.

In any regression analysis, random measurement error on the pretest will attenuate pre–post regression line slopes. This is not likely to be a serious problem in the RPDD, because presumably the group means are less influenced by random error than individual data. Nevertheless, this needs further investigating. It is likely, for instance, that such an investigation would lead to the conclusion that random-measurement error introduces greater bias in treatment effect estimates when the treatment group pretest mean is located further away from the overall pretest mean, where the attenuation most affects point predictions. Traditional adjustments for random measurement error have to be modified for the RPDD and may be problematic, especially when there are relatively few control points for estimating reliability.

However, we can state *unequivocally* that the deleterious effects of measurement error will be less manifest in the RPDD than in an individual-level NEGD analysis of the same data because the average values used in the RPDD must (by definition) have less variability or error than the individual data on which they are based. We expect that the power and efficiency of the NEGD will be the upper limit for a comparable RPDD (because the loss of degrees of freedom will outweigh the gains in reliability), but that measurement error will be reduced and the bias in estimates because of it will be correspondingly less in the RPDD.

Another regression artifact comes where the choice of the experimental unit is triggered by the error component in the pretest. During 1956, Connecticut endured an extreme crackdown on speeding and subsequently claimed a dramatic reduction in fatalities. But we know that the 1954–1955 increase in Connecticut's traffic fatalities was the largest in its history and that the 1954–1955 increase caused Governor Ribicoff to initiate the crackdown. Campbell and Ross (1968) concluded that the purported effects were merely a return to trend, a regression artifact. They also present the effect in the context of other nearby states, as in Figure 14.8.

Were one to use the 1955 and 1956 data as an RPDD, one potentially could get a significant pseudo-effect, as plotted in Figure 14.9. In this case, there are too few control states to produce significance, but the danger is illustrated (the actual t-value is -1.50, $p = .26$).

Instrumentation

As for any quasi-experimental design, the range of rival hypotheses should be examined in case a significant effect is found. For pilot studies and demonstration sites, one of the most frequently troubling will be that the program effort has changed uniquely the measurement process between the pretest and posttest for the experimental unit. The pressure on the law enforcement system to show a good effect may lead, for example, to the downgrading of felonies to misdemeanors (e.g., Seidman & Couzens, 1974). Equally frequently, program attention to a problem such as child abuse may lead to increased thoroughness

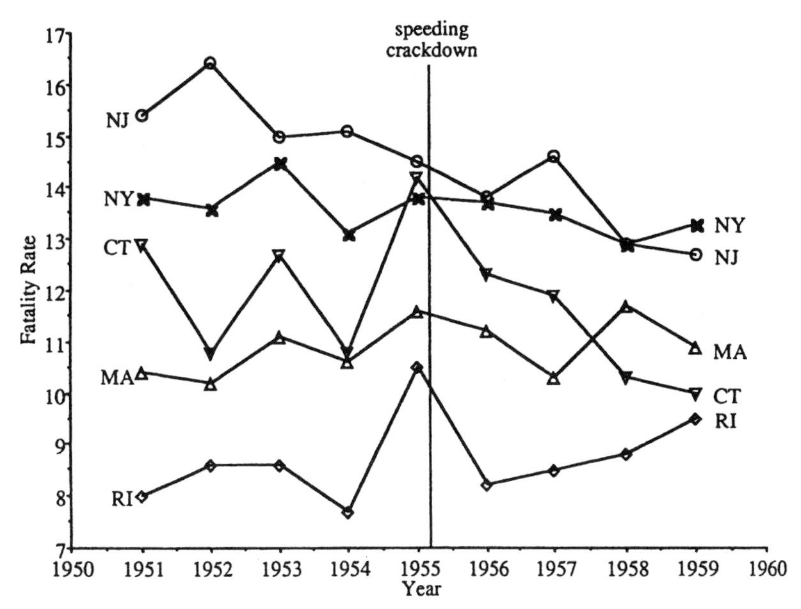

Figure 14.8. Traffic fatalities for five Northeastern states: 1951–1959.

of reporting, and a pseudo-increase (a pseudo-harmful effect) specific to the experimental unit.

Statistical Power

Although at first glance it appears that the RPDD design suffers from low statistical power because of the relatively few pre–post points that are typically used, group means are generally more stable and precise than within-group data. Fleiss and Tanur (1972) compared a traditional pre–post ANCOVA with the RPDD analysis using the same data and found that the ANCOVA results were not significant, whereas the RPDD results were. They commented,

> The difference between the analysis of covariance performed at the beginning of this chapter, where significance was not found, and the regression analysis just performed, where significance was found, is that predictability in the former was determined by covariation within groups, whereas predictability in the latter was determined by covariation between groups. (p. 525)

Two major issues related to statistical power need to be investigated. First, what is the power of the RPDD design as it stands? This should be relatively simple to determine and would make it possible to estimate the needed number of control points given some initial estimates of probable treatment effect size and desired level. An important factor in statistical power is where the treatment group scores on the pretest continuum. Statistical power will decline as the treated group pretest occurs further from the overall pretest mean. Second, an analysis needs to be done of the power of the RPDD relative to within-group ANCOVA alternatives. This should reveal whether one would ever gain statistical power in the trade-off between within-group variability in the ANCOVA framework and the presumed lower variability in the between-group-oriented RPDD.

Violating Assumptions of Statistical Tests

The RPDD may also be subject to violations of the assumptions of the *t*-test that is used. Fleiss and Tanur (1972) pointed out that the analysis is technically valid only when the control groups used to estimate the regression are a random sample from a population of such groups. The population in this case would be hypothetical (there are an infinity of potential groups that could be entered as controls) and as a consequence, this assumption can technically never be met. Instead, as Fleiss and Tanur pointed out, "One must be sure to select groups defined by the presence or absence of enough factors to assure that the variability of their mean responses is high" (1972, p. 525).

One benefit that accrues in the RPDD derives from the usually higher aggregate values used for the data. For instance, when group means are used (as opposed to individual-level values), we can more reasonably expect that the statistical assumption of normally distributed variables is likely to be met. This is because of the well-known statistical property of the central limit theorem, which holds that with sufficient sample sizes, sampling distributions are normally distributed. The advantage, of course, is that one needs to worry less about this distributional assumption, which is critical to many statistical tests.

Local History

A key threat to internal validity in this design is "local history" (Cook & Campbell, 1979). Whenever the treatment group consists of persons who are

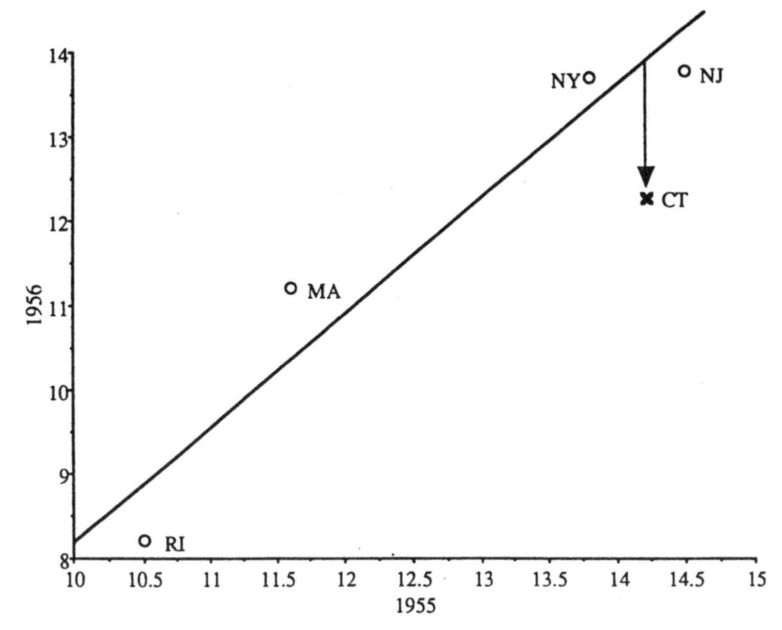

Figure 14.9. Connecticut Speeding Crackdown, RPDD.

treated together (such as at the same site) and distinct from control group persons, any factor in the setting that affects posttest performance can lead to a pseudo-effect in the data. If treated persons receive multiple treatments, or experience a markedly different setting from controls, or have a change in instrumentation (e.g., clinicians change their implicit judgement standards at the treatment site between pre- and posttest, but not at control sites), a pseudo-effect can result. These threats are not as serious an issue for the control groups because, with lots of such groups, setting variability is increased and the potential for systematic bias declines.

External Validity

Finally, the RPDD is not strong in external validity or generalizability. Although generalizing to other potential treatment groups that have the same pretest level may be reasonable, it is impossible to know whether any observed treatment effects would hold for groups with other pretest levels. Put another way, it is impossible with this design to study treatment interaction effects. If the treatment effect changes for different pretest levels, it will not be possible to know from the RPDD. Nevertheless, if the treatment group is typical of the potential target treatment group of interest (especially if it is a unit randomly sampled from the population of interest), it will be reasonable to generalize to other similar target groups. One might not be interested in whether the treatment might work for groups having markedly different pretreatment levels (see the Fleiss and Tanur example).

CONCLUSION

The RPDD has a very specific potential range of applicability. It is limited to contexts in which one has pre- and postmeasurement and multiple control groups. It is strongest when applied to routinely collected multiwave administrative data, where it is either too costly to match individual cases or may not be possible. Because such data are widely available and cost constraints a constant factor in our society, it is likely that the RPDD would be widely applicable. In fact, there are probably many instances in which the requirements of the design have already been met and for which a post hoc analysis could be simply constructed.

The design has several important weaknesses that need to be anticipated. Where few control groups are available, one is likely to have low statistical power. It is recommended that a power analysis be routinely reported with any analysis of the RPDD that fails to show significant treatment effects. In terms of internal validity, there are several possible threats that could lead to pseudoeffects in various situations. Care needs to be taken in selecting a heterogeneous set of control groups. Whenever possible, the treated unit should be randomly selected from the population. This will tend to minimize any deliberate selection factors that might threaten internal validity and is also likely to be the variation that has the greatest statistical power. Failing that, selection of the most extreme case is preferred over arbitrary or convenience-based selection, especially when the same measure is used for before and after measurement.

The RPDD has great potential for enhancing our ability to conduct research in natural social contexts. It is relatively inexpensive to apply where appropriate administrative data exist. It is based on well-known statistical models that can be estimated with almost any statistical computing package. It extends our ability to evaluate the effects of community-level programs where other designs are often not readily available. Although much work is yet needed to explore the implications and variations of the RPDD, it is clearly a useful addition to the methodological tool kit of the researchers of the experimenting society.

NOTE: This is an abridged version of an original article, "The Regression Point Displacement Design for Evaluating Community-Based Pilot Programs and Demonstration Projects." The complete version can be accessed through the World Wide Web at the following address: http://trochim.human.cornell.edu/research/rpd/rpd.htm

OVERVIEW OF CHAPTER 15

In the previous four chapters, we discussed designs to evaluate whether or not an impact occurred after some treatment had been administered. There was some discussion regarding how the treatment group would be chosen, whether through random assignment, being at the upper or lower end of a distribution, through a sharp cutoff point, or chosen for convenience, politics, or some other reason. To this point, there was no guidance on how to evaluate a situation in which the research participants volunteer. We can be sure that those who volunteer probably differ in some ways from those who do not. That difference would likely affect their posttest scores, and perhaps even the way they react to the treatment. This chapter discusses a design that could be used to assess the impact when volunteers are involved, providing repeated measures at regular intervals are available before and after the treatment.

In the past, researchers dealt with pretreatment differences by adjusting them away through matching or statistical methods. As we have seen, these methods tend to underadjust and can actually make ameliorative programs look harmful. Campbell suggested that we live with the pretreatment differences and use them in our design to estimate treatment effects. The pretreatment difference is measured and a t-test applied, and then the posttest difference between the two groups is determined and tested. The treatment is judged to have an effect not if the posttest difference is greater than zero, but rather if there is a significant difference in the pretest and posttest *t*-values.

When it is likely that the outcome is correlated with the levels of the treatment variable, Campbell suggests a correlation to determine the impact of the treatment. When the treatment variable consists of only two levels, for example, taking or not taking the treatment, a biserial correlation coefficient is used. This test would be interpreted like the *t*-test described previously. For example, with a zero pretest-treatment correlation, we would perceive an effect if the correlation between the posttest and the treatment was significantly greater than zero. If the pretest and treatment levels were significantly correlated, however, the posttest-treatment correlation must be significantly different from the pretest-treatment correlation to establish an effect.

Campbell points out that the treatment is not the only factor that affects the posttest-treatment correlation. All relationships weaken over time; this is referred to as temporal erosion. Two assumptions are made regarding temporal erosion: (a) all relationships fade in time, and (b) the erosion rate is constant. Each variable in a correlation has an erosion rate associated with it, and this is to be distinguished from the dissipation rate for the true effect of the treatment.

To demonstrate how the researcher can deal with this issue of temporal erosion, Campbell discusses an example of how English vocabulary scores might be affected by taking ninth-grade Latin. This illustration was used because English vocabulary scores were available at yearly intervals before and after taking Latin. To characterize the no-effect case, the erosion rate over time was estimated by graphing the year-to-year correlation between taking Latin and English vocabulary scores. There is a point in this graph at which this correlation was the highest, and each successive and previous year decreased. This peak was associated with the point in time when the decision was made to take Latin. This graph provided the basis for comparing the slope of the actual posttest-treatment correlations to determine if taking Latin had an impact. It is pointed out that determining the peak is a crucial decision, because it will affect the interpretability of the results, especially when the no-effect slope and the posttreatment slope are not widely divergent. Matrixes that depict test–retest correlations of a variable over time can help in estimating peaks and erosion rates.

This design can also be used to estimate the effects of remedial programs. However, the interpretation of such studies are not as clear-cut because the treatment effect and the erosion rate are in the same direction. Campbell warned of the pitfalls in using a single pretest–posttest design for studying ameliorative programs, because it is easy for the unwary researcher to mistake the erosion rate for a treatment effect.

As with all the quasi-experimental designs that have been covered in these chapters, threats to validity must be considered for this design as well. One such threat is a treatment artifact that may be a result of mismatches of the persons taking the pretest and the posttest. Still, this design can be a useful one that avoids the problems associated with attempts to establish pretreatment equivalence.

FIFTEEN

TREATMENT-EFFECT CORRELATIONS

This chapter has two general goals. The first is to present some quasi-experimental designs particularly appropriate to the utilization of educational records and data from longitudinal or multiwave panel studies. The second, and perhaps more important in the long run, is to search for experimental designs appropriate for situations in which people volunteer for experimental treatments. At the present time there are no designs available that will adequately distinguish between treatment effects and cosymptoms of the selection differences that volunteering produces. Yet The Experimenting Society of the future (Chapter 1) must also be a voluntaristic one, avoiding the coercive control implied in randomized assignment to treatments (Janousek, 1970). We are each of us convinced, in terms of

Campbell, D. T. (1971c). Temporal changes in treatment-effect correlations: A quasi-experimental model for institutional records and longitudinal studies. In G. V. Glass (Ed.), *Proceedings of the 1970 Invitational Conference on Testing Problems.* Princeton, NJ: Educational Testing Service.

our own experience, that treatments we have volunteered for—the jobs, wives, curriculums, psychotherapies, and so on, that we have chosen—have changed us. Although part of this may be a causal–perceptual illusion akin to the statistical regression artifact, surely not all of it is. Eventually the ponderous processes of science should also be able to see what is thus visible to the naked eye.

Consider a study in which attributes of children (such as vocabulary, mathematical skills, problem-solving ingenuity, and so on) are repeatedly measured on the same children over a substantial number of years, and in which specific experiences not uniformly shared (such as courses in new math, Head Start, Follow Through) are recorded. Although in a true experiment these experiences, these potential change agents, can be assigned at random to a subsample and withheld from an equivalent group, in our situation this has not been possible. Instead, selection and treatment are confounded; those getting the treatment differ systematically even before the treatment.

The usual approach to such initial differences is to attempt to adjust them away. Not only have such adjustments proven inadequate; they have, as a by-product of the chronic underadjustment, produced results with systematic biases. For that class of treatments given to those who need them least (such as accelerated tracks, honors courses, and university education), these may often seem benign errors, merely exaggerating the efficacy of treatments we know in our hearts to be good. But for a treatment we give to those who need it most (such as remedial reading or Head Start), the bias is in the direction of making the treatment look harmful, and thus of underestimating or swamping any true effects. It seems to me certain that the Westinghouse–Ohio University evaluation of Head Start (Cicirelli et al., 1969) contained such a bias, a tragic error when one considers that this study was used to justify the destruction of the Head Start program, and was probably the most politically influential statistical evaluation ever done up to that time (Campbell & Erlebacher, 1970). Not only do "matching," ex post facto analysis, and "control" by partial correlation produce such regression artifacts (for example, Brewer, Crano, & Campbell, 1970; Meehl, 1970), but so does analysis of covariance (Campbell & Erlebacher, 1970; Cronbach & Furby, 1970; Lord, 1967; Porter, 1967).

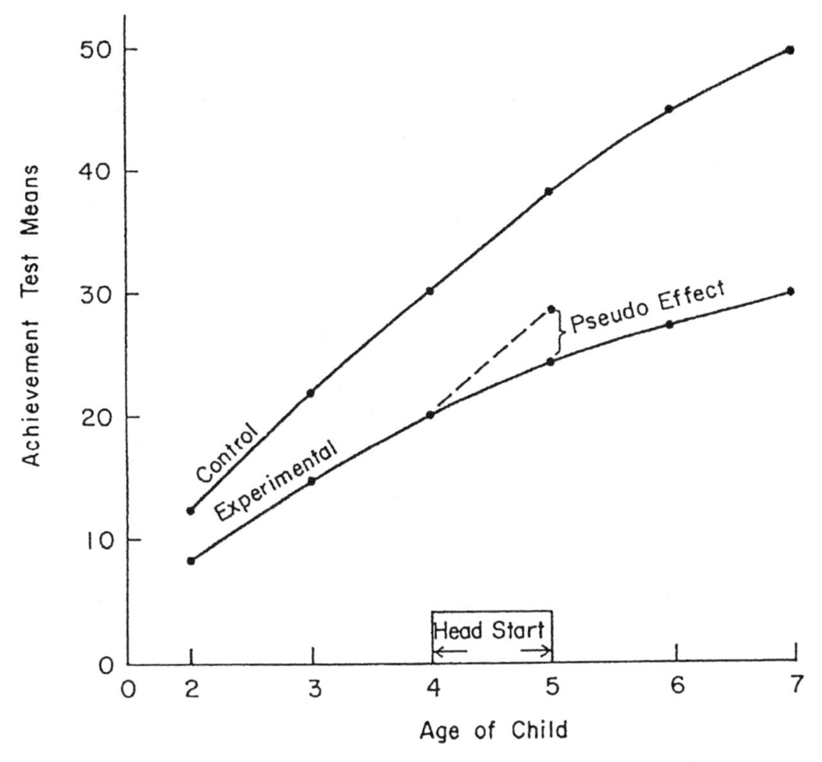

Figure 15.1. Illustration of the pseudo-effects possible if the differential growth rates associated with initial mean differences are disregarded.

LIVING WITH PRETREATMENT DIFFERENCES RATHER THAN ADJUSTING THEM AWAY

One basic recommendation in this chapter is that we give up trying to adjust away pretreatment differences. Rather, we should live with them, use them as a baseline, and demand that an effective treatment significantly modify that difference.

There are numerous statistical symptoms of an experimental treatment effect (Campbell & Clayton, 1964). The common ones of mean differences or differences in change scores must be ruled out for growth data on children because pretreatment differences almost certainly imply preexisting differences in growth rates as well, as illustrated in Figure 15.1. Such divergent growth rates no doubt occur within groups as well as between groups, the increased separation of means being accompanied by increased variability of groups, in what we can call the *fan-spread hypothesis* (Campbell, 1967a). Indices such as *t* or *F*, which express mean differences relative to variability, avoid the difficulty. Thus, the recommendation becomes that of computing the pretreatment *t* between experimental and control groups, and comparing the posttreatment *t* with it, an experimental effect being shown as a significant difference in *t*s, rather than a posttest *t* significantly different from zero.

In what follows, instead of *t* or *F*, an *r* between the treatment taken as a dichotomous variable (i.e., treatment, nontreatment) and the dependent variable will be used. This *r* is also an expression of mean differences relative to variability. For example, a biserial *r* is computed from the same ingredients as are found in a *t*. The preference for *r* over *t* or *F* is arbitrary, but it has the advantage of being descriptive of the strength of relationship independently of the number of observations employed. More important, *r* makes conceptual contact with the correlation–causation problem as explored in the lagging of time-series correlations (Hooker, 1901; Schmookler, 1966) and in the cross-lagged panel correlation (Campbell, 1963; Kenny, 1970a, 1970b; Pelz & Andrews, 1964; Rickard, 1972; Rozelle & Campbell, 1969).

In case *r* seems an unusual measure of an experimental effect, Figure 15.2 is provided. The top scatter diagrams illustrate pretest and posttest distributions for an experiment involving four degrees of the treatment variable, plus a control condition. For the pretest, because of random assignment (from sets of five matched pretest scores in this case), all groups have the same mean and standard deviation. The correlation between treatment levels (0 = control, 1, 2, 3, 4) and the pretest scores is thus zero. For the posttest, *r* has acquired a high positive value. If the effects had been nonordinal, one would need to use a curvilinear or nonordinal measure of relationship, such as eta, or a contingency coefficient. The effect, of course, might be negative rather than positive. In any case, if a true experiment was done in a case in which there was a treatment effect, the correlation would start at zero for the pretest, and the correlation on the posttest would go to some value positive or negative, significantly different from zero. In the lower half of Figure 15.2 is portrayed the more usual situation in which there is only one experimental group and one control group. Here,

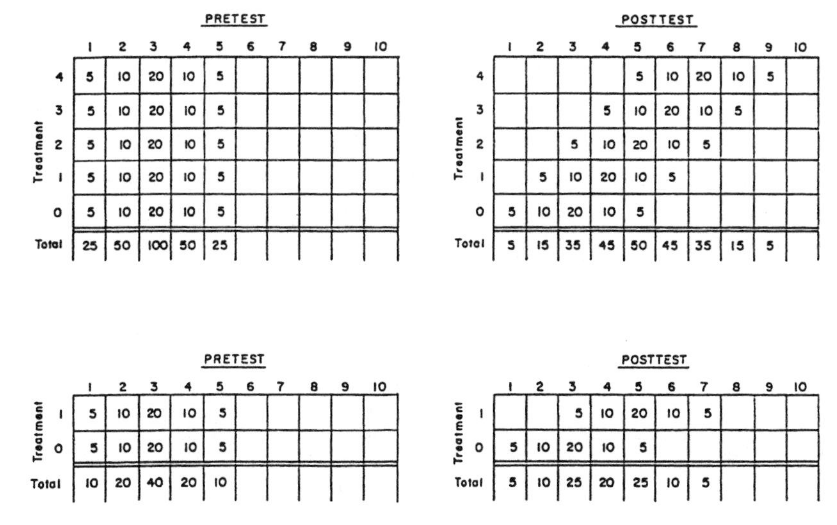

PRETEST

Treatment	1	2	3	4	5	6	7	8	9	10
4	5	10	20	10	5					
3	5	10	20	10	5					
2	5	10	20	10	5					
1	5	10	20	10	5					
0	5	10	20	10	5					
Total	25	50	100	50	25					

POSTTEST

Treatment	1	2	3	4	5	6	7	8	9	10
4					5	10	20	10	5	
3				5	10	20	10	5		
2			5	10	20	10	5			
1		5	10	20	10	5				
0	5	10	20	10	5					
Total	5	15	35	45	50	45	35	15	5	

PRETEST

Treatment	1	2	3	4	5	6	7	8	9	10
1	5	10	20	10	5					
0	5	10	20	10	5					
Total	10	20	40	20	10					

POSTTEST

Treatment	1	2	3	4	5	6	7	8	9	10
1			5	10	20	10	5			
0	5	10	20	10	5					
Total	5	10	25	20	25	10	5			

Figure 15.2. Experimental effects as changes in the treatment-effect correlation. In these hypothetical "true" experiments, the correlation for the pretest is zero.

too, one can use the correlation concept. The biserial r (and the t) start at zero for the pretest, and move to a substantial positive value for the posttest.

For quasi-experiments in which the correlation does not start at zero, it is here proposed that we give up as misleading all statistical efforts to adjust it back to zero (by matching or covariance, and so on) and instead demand that a treatment effect show itself as a significant change in the treatment-effect correlation, a significant increase or decrease.

TEMPORAL EROSION

But experimental treatments are not the only processes that change treatment-effect correlations. All relationships tend to weaken with time, a process we have previously designated as *temporal attenuation* (Rozelle & Campbell, 1969), but to avoid confusion with ordinary reliability processes we now call *temporal erosion* (Kenny, 1970a, 1970b).

Let us first consider a series of repeated measures in the middle of which a treatment has been given. Annual September English vocabulary scores and

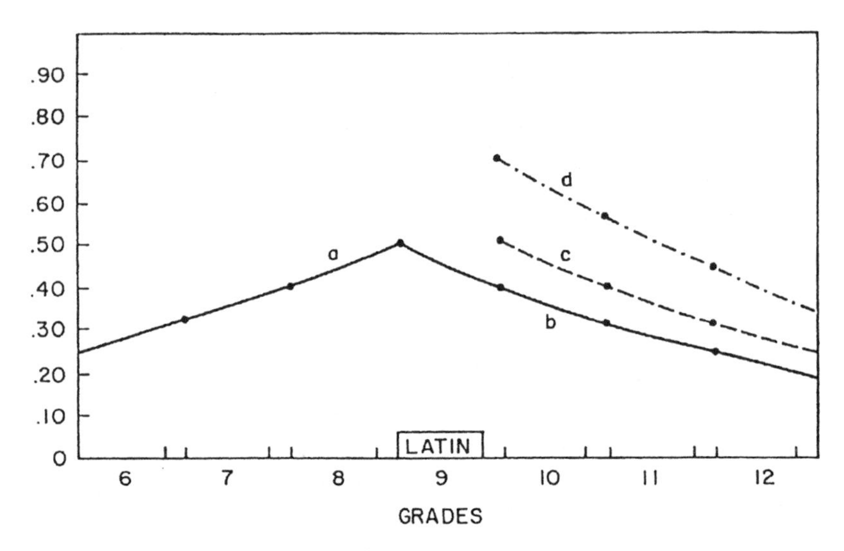

Figure 15.3. Biserial correlation of annual September vocabulary tests with taking Latin. Line *b* is a clear-cut case of no-effect, line *d* is a clear-cut case of incremental effect.

a ninth-grade course in Latin can be used for illustration. The biserial correlation of vocabulary with the presence or absence of Latin is computed. In Figure 15.3, a no-effect outcome and an incremental effect of Latin are plotted.

Figure 15.3 presumes that (a) all relationships erode in time, and (b) that erosion rate is constant over equal time periods. In the graphed values, the erosion rate is .80. (The no-effect values are .50, .40, .32, .256, and .2048. The effect values of line *d* are .70, .56, .448, and .3584.) The assumption of constant erosion rate means the slopes would appear linear when plotted in logarithms.

The erosion rate for a correlation is presumably a product of erosion characteristics of both variables. Because the "measure" called Taking Latin occurs only once, we have no additional grounds for estimating its rate. (The erosion rate for Taking Latin as a measure is also to be distinguished from the dissipation rate for the real "effects of Taking Latin," if any. Figure 15.3 assumes that the composite of Taking Latin as a measure, Latin effect, and English vocabulary as a measure attenuates at .80.) The correlations among the vocabulary measures provide bases for evaluating its rate and the validity of assumptions (a) and (b). The matrix of such relationships should be "proximally autocorrelated" (Rozelle & Campbell, 1969) or of a "superdiagonal" type (Lubin, 1969) or a quasi-simplex (Guttman, 1955; Humphreys, 1960) in form. That is, the correlations between adjacent time periods should be higher than those spanning two periods, and these higher than those spanning three periods, and so on. This corresponds to a uniform rate of erosion, a uniform rate of degrading the relationship by substitution of error or mismatching persons. (If there has been an effect of Latin, this might affect the intercorrelations of the vocabulary tests. We should accumulate experience from true experiments on this. Is the test–retest correlation higher in the experimental or control group?) If there are grounds for ascertaining erosion rates separably for each variable, the erosion rate for the correlation might be assumed to be the geometric mean of the two, in analogue to the correction for attenuation in reliability, and on the assumption of homogeneous erosion of all components within a given variable.

LOCATING THE "CORRELATION PEAK" FOR THE NO-EFFECT CONDITION

With this background, we can begin to consider the problems of any specific instance. The pretreatment correlations with Latin are a result of the fact that taking Latin is a symptom of common determinants that also produce high English vocabulary scores. The peak in this correlation comes at the point of simultaneous "measurement." If "intention to take Latin at the first opportunity, that is, in the ninth grade" were measured in the sixth grade, the correlation of vocabulary and this "Sixth Grade Intention" would peak in the sixth grade. Note in Figure 15.3 that we have peaked the no-effect curve at the beginning rather than at the middle of the year of Latin. It should be peaked at the point at which the decision was made, at registration if Latin is optional. What if Latin is an obligatory part of a track system and all pupils on one track receive it? Then presumably the peak is at the last point of actual or potential revision of track membership prior to Latin. Note in this case that it is of help to have the several pretest measures. If the tracks were fixed when pupils entered junior high in the seventh grade, then the correlations should peak at 7 tapering off through 8, 9, and by extrapolation, 10 and 11.

The judgment about when the decision or determination was made will be important in interpreting weak effects, such as outcomes lying between *c* and *b* in Figure 15.3. The coarse grain of the measurement series (the wide spacing of measurements) will increase the ambiguity. Almost certainly, the decision, and hence the peak, will occur prior to the treatment with how much prior being the question. Thus, an outcome such as *d* will stand as an unequivocal effect whatever decision point and whatever temporal erosion rate one assumes. An outcome such as *c*, even though the correlation after is the same as before, is usually also symptomatic of a positive effect for reasonable fixings of decision point and temporal erosion rate (but see later discussion).

Types of decision processes vary in their temporal location and sharpness of focus. In Figure 15.3, we have assumed a voluntary choice of courses made at the beginning of the term, maximally symptomatic of the pupils at that moment. At another extreme, the assignments would be decided by the high school staff at the beginning of the term, based on the pupils' grades of the prior year. In this case, the decision point, and the correlation peak under the no-effect case, lies sometime in that prior year, depending on the weighting given to various semesters and the intercorrelation of grades from semester to semester. Not only is the peak earlier, but it is also less focused, more spread out. Intermediate and more characteristic conditions would include setting prior-performance prerequisites for Latin or heavy influence of teacher's advice, the latter being based on prior performance, and so on. All of these move back and spread out the time in the pupil's career maximally symptomatized by the decision to take Latin.

In a situation in which pupils can freely drop or transfer out of Latin, and in which considerable numbers do, staying in Latin becomes a selective diagnostic of ability and interest, and so on, which has its time of maximal symptomicity toward the end of the Latin course. If the situation were completely fluid, with each day of Latin requiring a new commitment made without cost in either direction, then the symptom of attending the last day of Latin would have its peak at the end of the treatment. Probably all reasonable analyses would show that decisions in or out are greatly increased in difficulty and rarity once the term has begun, and that later-term drops are a result of the symptom load of early-term performance; hence, no reasonable model would put the correlation peak later than the middle of the Latin treatment because a middle

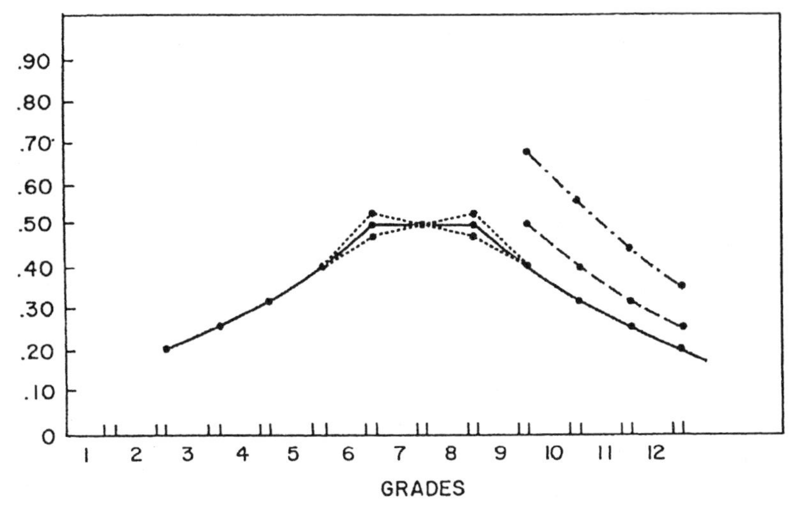

Figure 15.4. Figure 15.3 modified for revisable seventh-grade tracking into Latin.

placement jeopardizes the interpretation of outcome *c* in Figure 15.3, but not outcome *d*.

More likely than complete fluidity of decision, or homogeneity of redecision in time, is a stepwise process of major decisions and reluctant revisions. These would create erosion patterns with plateaus in them. Figure 15.4 illustrates a case in which all those in the top junior high track take Latin, the tracking decision being made at entry to seventh grade, but with minor revisions and transfers made each year.

Getting into a track at the beginning of the seventh grade is much easier than changing in or out in the eighth or ninth grade. There results some kind of correlation plateau in the seventh-to-ninth region. Whether this tilts up toward ninth or up toward seventh depends on the relative strengths of the selective factors. A procedure that let no more in, but continually purified by elimination the group selected at seventh, might correlate higher at the end of the process, at ninth grade.

A sharp-focused peak will result from assignment to Latin on the basis of a test, given on a specific date, which correlates with the English vocabulary test. The date of that test will be the peak. The sharpest peaking would result from using the English vocabulary test itself as the basis of assignment to Latin.

TABLE 15.1 Cross-Temporal Correlations of Equal Erosion Rate (.80) and Intercept (.90)

	Grades						
	6	*7*	*8*	*9*	*10*	*11*	*12*
6	(.90)						
7	.72	(.90)					
8	.58	.72	(.90)				
9	.46	.58	.72	(.90)			
10	.37	.46	.58	.72	(.90)		
11	.30	.37	.46	.58	.72	(.90)	
12	.24	.30	.37	.46	.58	.72	(.90)

This would produce a peak at the level of 1.00, making it impossible to achieve an unequivocal evidence of effect—that is, a posttreatment *r* higher than the peak. The lower the pretest-treatment correlations and the lower the presumed peak, the clearer the experimental inference. A decision base that correlates zero with English vocabulary would be as good as randomization, with no peak, all pretest values, and erosion slopes flat at zero.

Hidden peaks are a threat to this analysis. Because the sharp-peaked decisions will occur before the onset of the treatment, an immediate pretest such as assumed in Figure 15.3 will protect against a hidden peak masquerading as a treatment effect. But if the nearest pretest were in June of the previous year, and if the decision were made on a September language aptitude test at the beginning of the ninth grade, then the failure to ascertain this peak might lead to an underestimation of the no-effect level for posttest values.

Estimating Erosion Rates and Intercepts

It should perhaps be announced that problems of both peak location and erosion rates are probably exaggerated in the .80 rate used in Table 15.1. Analyses of the data from the big ETS STEP-SCAT longitudinal study, covering seventh through eleventh grades show biannual erosion rates of .95 to be typical, .90 to be minimal (T. L. Hilton, personal communication, 1969; Kenny, 1970a). Such high rates mean that the peaks are only slightly higher than the other values, that equivocalities in the location of the peaks, or in estimating

TABLE 15.2 Cross-Temporal Correlations With Constant Erosion Rate (.80) and Increasing Intercepts

	Grades						
	6	*7*	*8*	*9*	*10*	*11*	*12*
6	(.65)						
7	.54	(.70)					
8	.56	.58	(.75)				
9	.58	.60	.62	(.80)			
10	.59	.62	.64	.66	(.85)		
11	.61	.64	.66	.68	.70	(.90)	
12	.63	.65	.68	.70	.72	.74	(.95)

TABLE 15.3 Cross-Temporal Correlations With Constant Erosion Rate (.80) and Increasing Rates (Indistinguishable Without Information on Reliability from Table 15.2)

	Grades						
	6	*7*	*8*	*9*	*10*	*11*	*12*
6	(.80)						
7	.54	(.80)					
8	.56	.58	(.80)				
9	.58	.60	.62	(.80)			
10	.59	.62	.64	.66	(.80)		
11	.61	.64	.66	.68	.70	(.80)	
12	.63	.65	.68	.70	.72	.74	(.80)
Rate	(.65)	(.70)	(.75)	(.80)	(.85)	(.90)	(.95)

rates, create only narrow ranges of equivocality in estimating the no-effect expectation for posttreatment values.

One approach to estimating peaks and erosion rates is through the pattern of correlation across time within a single variable. In Tables 15.1, 15.2, and 15.3 are some patterns that might be looked for in the intercorrelations among repeated measures of a variable such as English vocabulary. Table 15.1 shows the simplest condition, in which there is uniformity in time for both the erosion rate and the intercept. (The *intercept* is something like a reliability, a repeated measure correlation with no lag in time, and no enhancement as a result of memory for specific items or temporally correlated error. It is a value extrapo-

TABLE 15.4 Cross-Temporal Correlations With Junior High–High Break Between 9th and 10th Grades

	Grade						
	6	*7*	*8*	*9*	*10*	*11*	*12*
6	(.90)						
7	.72	(.90)					
8	.58	.72	(.90)				
9	.46	.58	.72	(.90)			
10	.30	.37	.46	.58	(.90)		
11	.24	.30	.37	.46	.72	(.90)	
12	.19	.24	.30	.37	.58	.72	(.90)

lated from the lagged values to the point of no temporal erosion at all.) In a limited way, these assumptions can be checked within the data.

If there are systematic trends toward higher and higher 1-year test–retest correlations, as there may be in some longitudinal studies, this may be interpreted as either a case of increasing intercepts with constant rate, which we currently favor (Kenny, 1970a), shown in Table 15.2, or as a constant origin with increasing rate, shown in Table 15.3, containing identical values as Table 15.2 except for the diagonal. Insofar as the intercept conceptually corresponds to a synchronous test–retest correlation without memory for specific items, and is therefore like an internal consistency reliability, such reliabilities if computed on the same *Ss* would be relevant to choosing a model. (In the STEP-SCAT longitudinal data, no incremental pattern seems indicated; Table 15.1 could be assumed, with some unevenness of reliabilities and intercepts for the yearly testings but of no orderly pattern.)

Shifts in schools, as between junior high and high school, may create greater erosion than the normal 1-year erosion rate. Such outcomes as Table 15.4 should be looked for.

Remedial or Compensatory Programs

In the previous illustration, the selection bias and the treatment effect operated in the same direction, and the erosion effect operated in the opposite direction. In many remedial or compensatory cases the reverse is the case, and

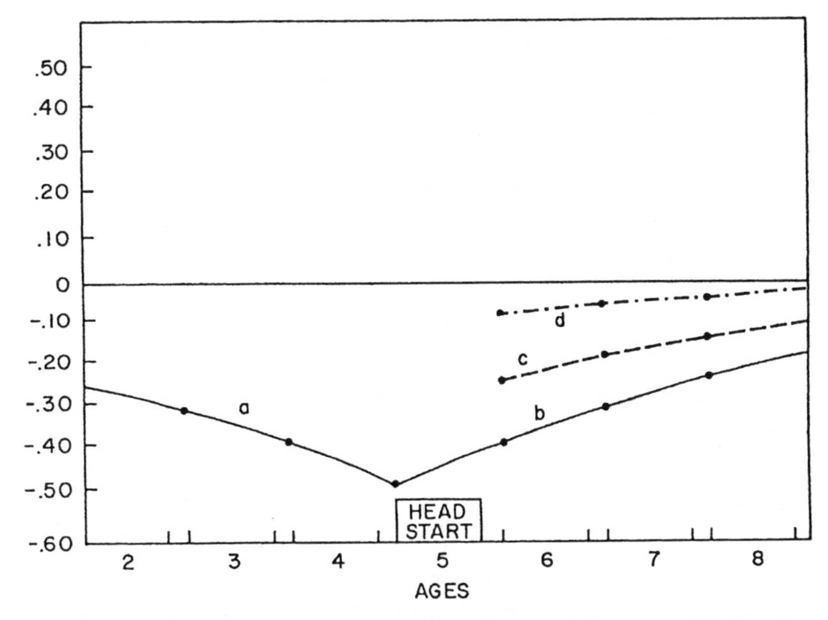

Figure 15.5. Biserial correlation of annual September vocabulary tests with Head Start experience. Line *b* is a clear-cut case of no-effect, line *d* is a clear-cut case of effect.

the effects of treatment and temporal erosion may be in the same direction. This probably means that unequivocal evidence of effects is rarer, but the analysis should still prove relevant.

One such case comes from the current ETS preschool longitudinal study in which some children receive Head Start. If Head Start is given to those who need it most, as a compensatory program should be, then the pretest correlations with Head Start exposure are negative. A successful treatment makes this correlation less negative. Temporal erosion also makes it less negative. Figure 15.5 plots such a situation. The values for lines *a* and *b* are those of Figure 15.3, except negative (.50, .40, .32, .256, .2048). Line *d* starts with a .30 difference from line *b*, and this treatment effect dissipates at .80 (producing the reduced increments of .24, .192, .064, .0512). The net effect is for lines *d* and *b* to come closer together while both approach zero. If the treatment effect were to dissipate more rapidly, the net effect could actually be an increase in the negative magnitude of the correlation.

Figure 15.5 has been plotted with as sharp a peak as Figure 15.3. No doubt this presents an exaggerated view of the erosion and peak location problem. Probably the sharpest peak will come from selection decisions based on individual pupil attributes. If the decision is based on neighborhood or school attributes, the cross-temporal neighborhood correlations will be higher than person correlations and will show less erosion. The decisions are not apt to be time-specific as far as individual children are concerned. Longitudinal data give us the power to ascertain these facts.

PROBLEMS WITH DATA LIMITED TO ONE PRETEST AND ONE POSTTEST

Imagine in Figures 15.3 and 15.5 that only one pretest measure and one posttest measure are available. In Figure 15.3, with outcome *b* one would not be tempted to claim a positive effect, whereas in Figure 15.5, outcome *b,* one might be. The fact that treatment counters attenuation has made an outcome like *d* unequivocally an effect in Figure 15.3, but interpretable as rapid erosion in Figure 15.5.

Thus, for those instances in which the initial correlation and the treatment effect are in the same direction, treatment and attenuation have opposite effects, and a simple one pretest, one posttest analysis (Campbell, 1967a; Campbell & Clayton, 1964; Seaver, 1970) is interpretable, albeit with excess conservatism. In the other instances, the one pretest, one posttest design is extremely vulnerable to mistaking erosion as a treatment effect, and the need for longitudinal data is extremely great.

ARTIFACTUAL SOURCES FOR AN INCREMENT IN TREATMENT

The "plausible rival hypotheses" approach to quasi-experimental design demands that we look for likely sources of a correlation increment, as in Figure 15.3, other than a treatment. In Campbell and Clayton (1964), it was argued that the co-occurrence on the same interview of the posttest and the ascertainment of exposure would create a higher posttest exposure correlation than pretest exposure correlation, whether or not the treatment had an effect. In that case, the treatment group was seeing an anti-antisemitism movie and the dependent variable was an antisemitism scale. In all panel studies, some persons are misidentified, with different persons providing the pretest data than the posttest data. This lowers the exposure–pretest correlation but not the exposure–posttest correlation in which exposure is retrospectively ascertained in the posttest interview. Furthermore, forgetting one has seen the movie, or erroneously reporting that one has, are attitude symptoms, and attitude measures occurring in the same instrument and testing situation always correlate higher than when the same measures are separated in time.

The same problem could occur in causal analysis in longitudinal studies. Consider another ETS interest, the impact of the *Sesame Street* children's educational television serial. Here the longitudinal data of the Head Start study could be used, ascertaining which children have seen the series. The occasion of ascertainment should be kept separate from the testing program.

CORRELATIONAL ANALYSIS WHERE THE TREATMENT OCCURS IN DEGREES

In the *Sesame Street* and Head Start examples, and many others, one will have wide ranges in degree of treatment, number of days attended, or programs seen. There is no reason why the correlational analysis here described should not be employed, using the treatment as a continuous variable (with a mode, unfortunately, at zero). But for this analysis, the presumed correlation peak in the no-cause condition should be conservatively placed in the middle of the treatment period, as indicated in the earlier discussion of decision times.

PARTIAL AND MULTIPLE CORRELATION, MATCHING, AND COVARIANCE ANALYSES

These techniques represent pathetic efforts to artificially reconstruct a zero pretest–treatment correlation by "controlling for," "covarying," or "partialling out" the pretest correlation from the posttest. As many have demonstrated (Lord, 1960, 1967; Thorndike, 1942), and the others have reviewed (Brewer et al., 1970; Campbell & Erlebacher, 1970; Meehl, 1970) these statistical procedures are inappropriate to the task. If the reader doubts this, let him apply his favorite analysis to the no-cause conditions illustrated in Figure 15.3 and Table 15.1. Erosion is not at issue here. Even if there was no cross-temporal erosion, there would still result nonzero pseudo-effects in our no-cause conditions, a positive increment or positive partial correlation in the Figure 15.3, Table 15.1 case.

OVERVIEW OF CHAPTER 16

This article is a collaboration between two researchers from different traditions, Louise Potvin from the epidemiological tradition, and Campbell from the quasi-experimental tradition. Although their backgrounds and approaches differ, they agree that the randomized trial is the design most likely to lead to valid interpretation of results. This chapter discusses the case-control design, a design used in epidemiological studies to determine if exposure to a certain factor increases the risk of some undesirable outcome. This design is applied retrospectively by sampling study participants from two populations. The "cases" are the individuals who have exhibited the outcome; the "controls" have not exhibited the outcome at the time of the study. By examining and comparing the odds of having been exposed to the factor under study in the two populations, and under certain assumptions, it is possible to estimate the relative risk associated with that factor.

A debate surrounding this design has arisen in the epidemiological literature, centered around whether research participants who would not be likely to be exposed to the risk factor should be entered into the study. To explore this issue, a hypothetical study was used in which the goal was to determine whether recent exposure to oral contraceptives increased a woman's risk of myocardial infarction. On one side of the debate, Schlesselman argued that the results would be biased if women who were known to be infertile were included in the study. Poole, on the other hand, argued that including infertile individuals in a case-control study would increase the precision in estimating the risk, much like increasing the number of controls in a randomized trial.

Potvin and Campbell agree with Schlesselman, but for different reasons. Although they agree that a simulation by Poole was technically correct in saying that there was an increase in precision when sterile women were included, they held that his simulation did not mirror a true case-control study, because the number of women in the case and control groups was different. If the data were reanalyzed with equal numbers in the groups, the loss of precision was not as great as Poole had demonstrated. The authors felt that the threat to validity that is introduced with sterile women far outweighs the small gain in precision. Three arguments are given as to why this is so.

First, if the experimental paradigm were applied to this problem, women would be randomly assigned to use oral contraceptives or not. If sterile women were included, two hypotheses are possible. One is that the sterile and fertile women will comply with their assignment; the other is that there will be differential compliance, fewer sterile than fertile women using the contraceptive. This latter hypothesis seems more likely, although both hypotheses are extreme because there may be some noncompliance among fertile women as well. A simulation of these hypotheses shows that if sterile women assigned to the contraceptives group do not comply, the estimated risk is biased downward.

Second, Poole's simulation assumes there is no confounding between fertility status and myocardial infarction. By using path analysis models and their associated correlations, Potvin and Campbell show that even when fertility is not related to myocardial infarction, a small correlation occurs between fertility and myocardial infarction. Poole would assume that this correlation was spurious. Further, by using Poole's simulated data, Potvin and Campbell demonstrate that one cannot rule out, a priori, that fertility status is not related to myocardial infarction. The ratio of exposure to oral contraceptives cannot be estimated because the denominator is zero (that is, there are no cases of known sterile women using oral contraceptives). Yet the odds of having a myocardial infarction is 1.5 times greater for fertile women than for sterile women. The authors assert that if different subgroups have a differential probability of being exposed, one cannot rule out the possibility that they might also have differential risk for the outcome condition.

Third, Potvin and Campbell argue that the experimental contrast is sharper if both groups are homogeneous. The added complication of fertility status in the nonexposed group makes the attribution of a significant risk ratio more difficult, because the sterile–fertile factor must also be considered.

The case-control design is a quasi-experimental design for which the threats to validity must be considered. Potvin and Campbell demonstrate that these threats are made more plausible by including individuals who are not likely to be exposed to the risk factor under study.

SIXTEEN

THE CASE-CONTROL METHOD AS A QUASI-EXPERIMENTAL DESIGN

The stimulating debate between Schlesselman (Schlesselman, 1982; Schlesselman & Stadel, 1987) and Poole (1986, 1987) has intersected with a larger agenda we are working on. One of us (L. P.) comes from the epidemiological tradition shared by Schlesselman and Poole; the other (D. T. C.) represents the quasi-experimental design tradition (Campbell & Stanley, 1966; Cook & Campbell, 1979). Our joint agenda is a limited merger of these two traditions, on the one hand adding the case-control study to the list of useful quasi-experimental designs (including borrowing some of the threats to internal validity that have been assembled around it) (Feinstein, 1985; Sackett, 1979); on the other hand, elaborating for the case-control study the errors in variables (pseudo-effects in the form of regression

Potvin, L., & Campbell, D. T. (1996). *Exposure opportunity, the experimental paradigm and the case-control study.* Unpublished manuscript.

artifacts remaining after matching) that are a major preoccupation of the quasi-experimental tradition. The fact that both traditions regard the randomized trial as the ideal to be approximated strongly favors the merger. In this agenda we have found the Schlesselman–Poole debate helpful in clarifying our own thinking. We hope that we can also help in resolving the debate.

THE NATURE OF THE DEBATE

Schlesselman (1982) states that in a case-control study estimating the relative risk associated with an exposure, if that exposure is caused by a factor unrelated to the disease under study, one should control for it by including only individuals who would normally have had the opportunity to be exposed. "Failure to do so would dilute the effect, biasing the odds ratio toward unity" (Schlesselman, 1982, p. 71). For example, in a hypothetical study of the effect of recent exposure to oral contraceptives on myocardial infarction, the inclusion of cases and controls known to be sterile would dilute the effect, as women who are aware of their infertility would normally not have been exposed to oral contraceptives.

Poole (1986) challenged this view and, by means of a simulation, showed that if there is no causal relationship between fertility and myocardial infarction (that is, if fertility is not a confounder), the inclusion of infertile cases and controls in a case-control study would not bias the estimate of the relative risk. He maintained that it would increase the precision, in a way analogous to enlarging the number of control participants in a randomized trial. His recommendation is to exclude from the eligibility criteria the opportunity of having been exposed to the factor under study, exactly the opposite of Schlesselman's previous recommendation. The subsequent discussion (Poole, 1987; Schlesselman & Stadel, 1987) does not, in our opinion, present any convincing new arguments. Poole's conclusion has been accepted by Rothman in a recent textbook (Rothman, 1986, p. 66).

In the end, we agree with Schlesselman, but for different reasons. Although we begin with a critique of Poole's simulation, the central aim of this chapter is to present new arguments favoring Schlesselman's policy of excluding from case-control studies those cases and controls who would normally not have

TABLE 16.1 Incidence of Myocardial Infarction (MI) in Relation to the Use of Oral Contraceptive According to Fertility Status in a Hypothetical Study Base

	Sterile		*Fertile*		*Total*	
	OC^+	OC^-	OC^+	OC^-	OC^+	OC^-
No. of MI cases	0	100	90	70	90	170
Person-years	0	500,000	150,000	350,000	150,000	850,00
Incidence*		2.0	6.0	2.0	6.0	2.0
Rate ratio			3.0		3.0	
90% confidence interval			(2.3-3.8)		(2.4-3.7)	

SOURCE: From Poole, C. (1986). Exposure opportunity in case-control studies. *American Journal of Epidemiology, 123*, 352-358.
NOTE: OC = oral contraceptive.
*Cases per 10,000 person-years

been exposed to the causal factor under study. We will provide three overlapping arguments: (a) loyalty to the experimental paradigm, (b) the inability to ascertain the independence of fertility and myocardial infarction rates, and (c) the problem of sharpening the experimental contrast.

POOLE'S SIMULATION

In Table 16.1 we reproduce Poole's simulated data. To simulate a case-control study in which sterile and fertile women are included, Poole took 260 cases of myocardial infarction from the *total* study base. He then sampled 260 cases that did not suffer myocardial infarction, who were to act as controls, with a sampling fraction that is equal to 260/1,000,000. The results are reproduced in the left part of Table 16.2; χ^2 is used as the statistical test with the 90% confidence interval. To mimic a case-control study excluding sterile cases and controls, he subtracted from the unexposed *(oc-)* subsample those cases and controls who would be sterile. These results are shown in the right on Table 16.2.

First, let us make a minor criticism of the strategy used by Poole. The right part of Table 16.2 is not what one would use in a case-control study, because there are fewer control participants (130) than there are case participants (160). One would rather select the 160 cases of myocardial infarction and would sample 160 controls, only within the subpopulation of fertile women, with a sampling fraction equal to 160/500,000. The resulting odds ratio shown in

TABLE 16.2 Poole's Results of a Case-Control Study Using (a) the Total and (b) the Fertile Portions of the Hypothetical Study Base

	Total			*Fertile*	
	MI$^+$	*MI*$^-$		*MI*$^+$	*MI*$^-$
OC$^+$	90	39	OC$^+$	90	39
OC$^-$	170	221	OC$^-$	70	91
N =	260	260	*N* =	160	130
OR =	3.0 (2.1–4.3)		OR =	3.0 (2.0–4.5)	
χ^2 =	25.77		χ^2 =	18.96	

NOTE: MI = myocardial infarction; OR = odds ratio; OC = oral contraceptive.

TABLE 16.3 Results of a Case-Control Study Using Only Fertile Women From the Hypothetical Study Base

	Total	
	MI$^+$	*MI*$^-$
OC$^+$	90	48
OC$^-$	70	112
N =	160	160
OR =	3.0 (2.0–4.4)	
χ^2 =	21.42	

NOTE: MI = myocardial infarction; OR = odds ratio; OC = oral contraceptive.

Table 16.3 does not change, although the loss of precision (as shown by χ^2) is less than Poole simulated.

Thus, Poole was technically correct in saying that the exclusion of sterile women decreases the power of the analysis. And Schlesselman was technically in error saying that such inclusion would dilute the apparent effect. However, the methodological problems of the case-control study are primarily ones of bias, not efficiency. We argue that the inclusion of cases and controls who would not normally have been exposed to the etiological factor under study can be a potential source of bias. Gaining a slight degree of statistical efficiency by exposing the resulting estimate to potential threats to validity is not a good research policy.

THE EXPERIMENTAL PARADIGM

Treating the randomized trial as an appropriate ideal in deciding details of the case-control study (Feinstein, 1985), we would ask, does a given practice imitate appropriately a random assignment experimental design procedure? Poole's simulated data on the associations between fertility status, recent exposure to oral contraceptives, and myocardial infarction as displayed in Table 16.1 can be used to simulate a randomized trial. In a hypothetical experiment that simulates a threefold increase of the relative risk of myocardial infarction as a result of exposure to oral contraceptives and that randomizes the intake of oral contraceptives, the estimated effect would be dependent on two possible hypotheses regarding the compliance of women with the assigned treatment. First, there is no differential compliance with the treatment within the group of sterile women. Second, there is much of an effect, meaning that sterile women assigned to oral contraception do not comply as often as do fertile women assigned to oral contraception. The latter seems most plausible. These two hypotheses are obviously extreme cases, and one would expect that a proportion of fertile women would not comply with the treatment, although this proportion should be larger for sterile women.

In Table 16.4, the relative risk is estimated for each of these two extreme hypotheses of the randomized trial. Any differential compliance between the control and experimental groups, as well as between fertile and sterile women, increases the variation within the experimental group and decreases the variation between the two groups, thus diluting the effect of oral contraceptives. Within-group heterogeneity increases the risk of differential attrition, and unless one is interested in studying possible interactions, one should try to keep the participant pool as homogeneous as possible. In the case in which the interaction is implausible in reality (known sterile women taking oral contraceptives), one would never recommend the inclusion of participants whose conditions prevent them from being exposed to the factor under study in their normal environment; there is no real benefit of doing so and one could jeopardize the internal validity of the study, not to mention the health of the participants.

In the case-control study, this requirement for homogeneity must be interpreted differently. Cases and controls should be as homogeneous as pos-

TABLE 16.4 Hypothetical Randomized Trial Using Poole's Simulated Data

			Group OC$^+$		*Group OC$^-$*	*RR*
Person-years	FS$^-$		250,000		250,000	
	FS$^+$		250,000		250,000	
	Total		500,000		500,000	
Hypothesis 1: All FS$^-$ are compliant						
Number of MI						
	FS$^-$	(.0006 × 250000)	150	(.0002 × 250000)	50	
	FS$^+$	(.0006 × 250000)	150	(.0002 × 250000)	50	
	Total		300		100	3.0
Hypothesis 2: No FS$^-$ is compliant						
Number of MI						
	FS$^-$	(.0002 × 250000)	50	(.0002 × 250000)	50	
	FS$^+$	(.0006 × 250000)	150	(.0002 × 250000)	50	
	Total		200		100	2.0

NOTE: FS = fertility status; RR = relative risk; OC = oral contraceptive; – = sterile; + = fertile.

sible at the time of the possible exposure, not at the time of selection for the study. The latter would bias the estimate toward the null hypothesis by removing sources of variation between the groups. However, keeping the groups homogeneous at the time of exposure is a safe means of ruling out a possible susceptibility bias (Feinstein & Horwitz, 1981). One might argue that a susceptibility bias is impossible when the exposure-opportunity variable is not causally related to the outcome, as is the case with oral contraceptives and myocardial infarction. Our second argument deals with the fact that one cannot rule out that possibility a priori.

INABILITY TO ASCERTAIN LACK OF CONFOUNDING

In Poole's simulation, fertility status is assumed to be not directly related to myocardial infarction (even though it is a direct cause of exposure to oral contraceptives). Therefore, it is not a confounding variable and one does not need to worry about its effect on the exposure–disease causal relation (Miettinen & Cook, 1981). In terms of causal paths, variable z is an exposure-opportunity variable in the three-variable model in case 1A of Figure 16.1, and a

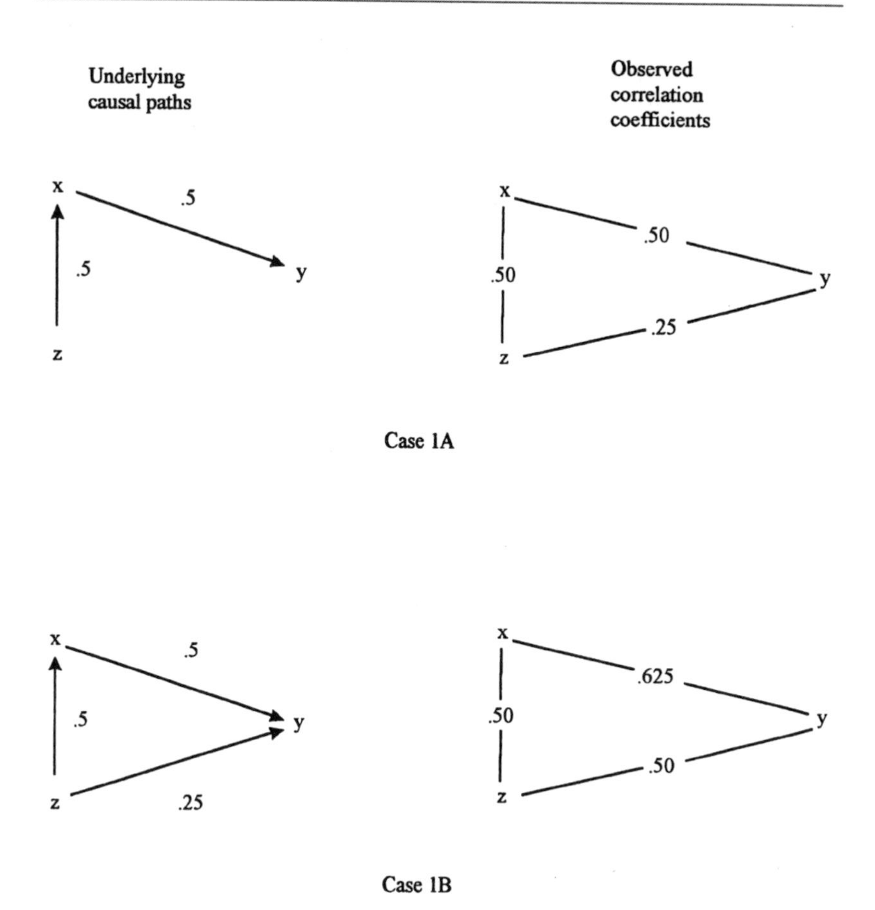

Figure 16.1. Possible causal models induced by hypothetical observed correlations: z = fertility status; x = oral contraception; y = myocardial infarction.
NOTE: In the tradition of path analysis, the causal path coefficients on the left generate the observed correlations on the right.

confounder in case 1B. Poole assumed that one can distinguish, a priori, between models similar to 1A and lB: If the situation is like the one in 1A, one does not need to control for z, and if it is as in 1B, one needs to control for z. For possible 1B cases, restricting the study to persons "identical" on z is commonly thought to be a means of controlling for confounding by z.

In the tradition of causal path analysis (Kenny, 1979; Kupper, 1984), causal models such as Figure 16.1 produce the observed correlations shown. Thus,

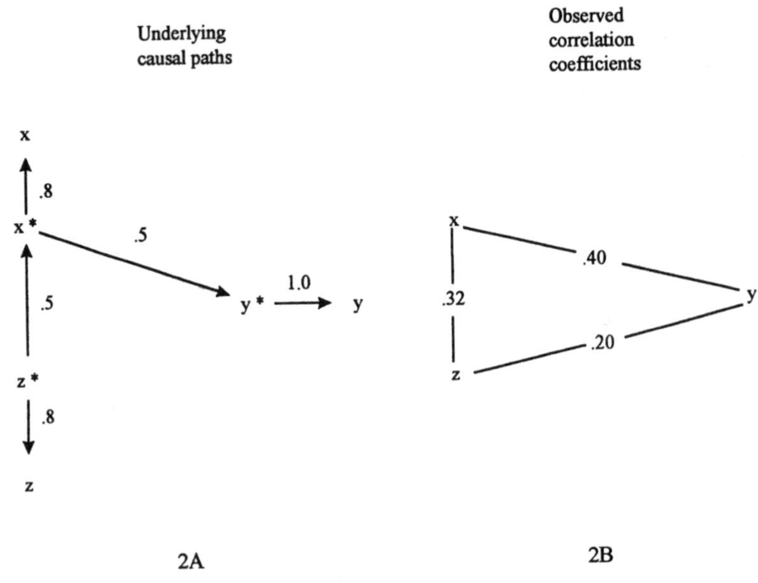

Figure 16.2. Causal model including errors in variables, induced by hypothetical correlations.

even in case 1A, in which there is no direct cause between *z* and *y*, the observed correlation between an exposure opportunity variable and the disease does not equal zero. Were the relationships in a case-control study to be expressed as correlation coefficients, and were one to assume the causal model of case 1A, then the estimate of the coefficient of the path from *z–y* would be zero for the correlation set of 1A, supporting Poole's condition.

But our simple illustration of Figure 16.1 should be made much more complex by the inclusion of the errors in variables, recognizing that an observed variable *x* is only a proxy measure for a latent construct x^*. In the model illustrated in Figure 16.2, 2A differs from the previous model, because 1A assumes a perfect measurement model for all three variables, whereas 2A assumes small errors for both the exposure opportunity and the exposure variables (i.e., the paths from z^* to *z* and from x^* to *x* are not 1.0). Taking into account errors in variables, the same causal model implies a set of lower correlations. This phenomenon of attenuation and its effect on the odds ratio has already been introduced in the epidemiological literature (Kupper, 1984). More important, lacking knowledge of the reliability and validity of x, y and z, and therefore assuming a model such as 1A or lB, one obtains a spurious nonzero path from y to z.

Using Poole's simulated data, it can be shown that in a case-control study one cannot establish a priori that an exposure opportunity variable is not a confounder. In Table 16.5 we present the risk ratios and confidence intervals between fertile and sterile women of being exposed to oral contraceptives and of having a myocardial infarction. As expected, the ratio of exposure to oral contraceptives cannot be estimated, involving a denominator equal to zero. However, the odds of having a myocardial infarction for fertile women is 1.5 times greater than for sterile women.

According to these results (which could have been obtained by an observational study prior to the real study), fertility status, related to both oral contraceptives and myocardial infarction, turns out to be a confounder of the relationship under study. It is only when using a post hoc stratified analysis (as Poole has done in his simulation) that one can show that the observed weak but significant relationship between fertility status and myocardial infarction is completely "explained" by the indirect link: fertility → oral contraceptives → myocardial infarction. Thus, if different subgroups have differential probabilities of being exposed, there is no a priori ground for ruling out the possibility that they also have a differential risk for the outcome condition.

This situation exemplifies the problems associated with Schlesselman's illustrations for the prescription of controlling for confounders (Schlesselman, 1982, p. 108). He has given as a basis for judgement causal links that can never be observed, only estimated from collected data. The use of path diagrams and path analysis in observational studies is helpful for both specifying possibilities and (when errors in variables problems are considered) illustrating the difficulties of ascertaining the underlying causal situation.

It might be, other things being equal, that the risk of myocardial infarction is the same for sterile and for fertile women not exposed to oral contraceptives, as it is assumed in Poole's simulation. However, in any epidemiological study this relationship would be confounded by age, general health condition, and other factors as well. Thus, in a cohort study of exposure to oral contraceptives

TABLE 16.5 Relationship Between Fertility and Oral Contraceptive and Myocardial Infarction Using Poole's Simulated Data

	Fertile		*Sterile*
Person-years OC+	150,000		0
Person-years total	500,000		500,000
Prevalence OC+	0.30		0
Risk ratio		30/0	
Person-years MI+	160		100
Person-years total	500,000		500,000
Incidence	0.00032		0.0002
Risk ratio		1.5	
95% confidence interval		(1.17–1.93)	

NOTE: OC = oral contraceptive; MI = myocardial infarction.

that includes women known to be sterile, one would certainly observe a difference with respect to age, health status, and other variables between the two cohorts. This is because sterile women are more likely to be older and to have different health problems than fertile women, whether or not they were exposed to oral contraceptives. Controlling for true confounders would practically result in the exclusion of sterile women from the analysis. It would be very difficult to find, in the cohort of exposed women, a match for each sterile woman, because there would be an excess of zero cells in a stratified analysis.

Poole's presentation of data for his simulation does not consider the lack of equilibrium brought in the design by the complete absence of sterile women in the group of exposed women. Presented as they are in Table 16.6, these data strongly suggest that the exposed cohort is homogeneous, whereas the unexposed cohort is made of two distinct populations. (Note that in the exposed cohort, the fertile women tend to be younger, and no sterile women are included; in the unexposed cohort, the fertile women are younger, but the sterile women tend to be older.) The only way that Poole's design can lead to unbiased results is through the assumption that sterile women and fertile women unexposed to oral contraceptives have the same risk of myocardial infarction. As we have shown, this assumption is both implausible and cannot be ascertained from the cases and controls available.

TABLE 16.6 Cohort Display of Poole's Simulated Study Base

	OC^+		OC^-	
	Fertile	*Sterile*	*Fertile*	*Sterile*
Person-years	150,000	0	350,000	500,000

NOTE: OC = oral contraceptive.

SHARPENING THE EXPERIMENTAL CONTRAST

Our third argument is that the inclusion of sterile women in the study base adds an irrelevant difference between the exposure distinctions, and needlessly expands the treatment contrast from a more precise *exposure to oral contraceptives* versus *nonexposure to oral contraceptives* to the less precise *exposure to oral contraceptives plus nonsterility* versus *nonexposure to oral contraceptives plus sterility.* Oral contraceptives and sterility thus become confounded variables on one side of the experimental contrast. A significant odds ratio resulting from such a comparison can logically be a result of any element of the complex treatment.

This way of stating the issue produces a methodological agreement with that of McFarlane, Feinstein, and Horwitz (1986) in discussing the case-control studies of D.E.S. exposure in mothers and increased risk of vaginal cancer in daughters. However, our support of their methodological principle does not in this case lead us to concur with their doubts about the D.E.S. causal link. In agreeing with their methodological principle, but disagreeing with the overall conclusion, we introduce a caveat borrowed from the quasi-experimental tradition. Although it is the duty of the methodologist to be alert to as many threats to validity as possible, these threats are then to be evaluated for their greater plausibility as rival hypotheses in comparison with the causal hypothesis under study. We ourselves regard D.E.S. treatment in mothers as causing vaginal cancer in daughters as the most plausible effective causal inference. We do not find McFarlane's "complex treatment package" threat to internal validity a plausible explanation of the results in this instance, but we do find it an important threat to be regularly examined. Their point can be stated as follows (McFarlane et al., 1986). D.E.S. was given to women who were at risk of miscarriage because of bleeding. Thus a precondition for exposure to D.E.S.

was a risky pregnancy. In the original case-control study, the experimental contrast was between pregnant controls versus the complex package: *D.E.S. plus risk of miscarriage.* Twenty years later, when case-control studies were done, the controls were not screened for their mother's risk during pregnancy. Had the control women been limited to daughters of mothers pregnant at the same time as the cases' mothers, and who also showed bleeding during pregnancy or for other reasons were judged to be in danger of miscarriage but who were *not* given D.E.S., then the experimental contrast would have been more nearly limited to D.E.S. alone. It is scientifically conceivable (although we judge not plausible overall) that the cause of vaginal cancer in the daughters was the threatened miscarriage and bleeding of the mothers. A more precise selection of controls would have ruled this out. But in spite of this weakness, we agree with Stolley (1985), and we think that the uncontrolled threat to validity is relatively implausible, given the many lines of evidence for a genuine D.E.S.–vaginal cancer association, including animal studies using randomized trials.

CONCLUSION

With regard to the Schlesselman–Poole debate, we end up agreeing with Schlesselman's recommendation that individuals not eligible for exposure to the etiological agent be excluded as cases and controls. We recommend this on several grounds: (a) loyalty to the experimental paradigm seeing the case-control study as an effort to approximate as nearly as possible the conditions of a randomized trial; (b) removing as many confounded variables as possible from the experimental contrast; (c) the unlikelihood of Poole's preconditions for his simulation (that known sterile women, ineligible for oral contraceptives, would have the same observed risk of myocardial infarction as would fertile women); and (d) the general inability of the epidemiologist to ascertain from observed relationships Poole's precondition of no indirect causal path through a third variable, linking exposure to the etiological agent and likelihood of disease.

On the less important issue of statistical power (the number of cases needed to determine an etiological relationship), we find Schlesselman technically wrong. Including ineligibles does not dilute the effect. Rather it produces a slight increase in power, as Poole claimed. However, we judge that this source of increase in power is, in practice, unavailable.

REFERENCES

Adams, J. S. (1965). Inequity in social exchange. In L. Berkowitz (Ed.), *Advances in experimental social psychology* (Vol. 2). New York: Academic Press.

Adorno, T. W., Frenkel-Brunswick, E., Levinson, D. J., & Sanford, R. N. (1950). *The authoritarian personality.* New York: Harper & Row.

Aronson, E., & Carlsmith, J. M. (1962). Performance expectancy as a determinant of actual performance. *Journal of Abnormal and Social Psychology, 65,* 178-182.

Aronson, E., & Linder, D. (1965). Gain and loss of esteem as determinants of interpersonal attraction. *Journal of Experimental Social Psychology, 1,* 156-171.

Arrow, K. J. (1951). *Social choice and individual values.* New York: Wiley.

Ashenfelter, O. (1978). Estimating the effect of training programs on earnings. *Review of Economics and Statistics, 60*(1), 47-57.

Aubert, V. (1959). Chance in social affairs. *Inquiry, 2,* 1-24.

Barnes, B. (1974). *Scientific knowledge and sociology theory.* London: Routledge & Kegan Paul.

Barnes, B. (1977). *Interests and the growth of knowledge.* London: Routledge & Kegan Paul.

Barnes, B. (1982). *T. S. Kuhn and social science.* New York: Columbia University Press.

Barnes, B., & Bloor, D. (1982). Relativism, rationality, and the sociology of knowledge. In M. Hollis & L. Lukes (Eds.), *Rationality and relativism.* Oxford: Blackwell.

Barnes, B., & Law, J. (1976). Whatever should be done with indexical expressions. *Theory and Society, 3,* 223-237.

Barnow, B. S. (1973). *The effects of Head Start and socioeconomic status on cognitive development of disadvantaged children.* Unpublished doctoral dissertation, University of Wisconsin.

Barnow, B. S., Cain, G. C., & Goldberger, A. S. (1980). Issues in the analysis of selectivity bias. In E. W. Stormsdorfer & G. Farkas (Eds.), *Evaluation Studies Review Annual* (Vol. 5, pp. 42-59). Beverly Hills, CA: Sage.

Beck, B. (1967). Welfare as a moral category. *Social Problems, 14,* 258-277.

Becker, H. S. (1979). Do photographers tell the truth? In T. D. Cook & C. S. Reichardt (Eds.), *Qualitative and quantitative methods in evaluation research* (Vol. 1, pp. 99-117). Beverly Hills, CA: Sage.

Beebe-Center, J. G. (1932). *Pleasantness and unpleasantness.* Princeton, NJ: Van Nostrand.

Bennett, J. W. (1946). The interpretation of pueblo culture. *Southwestern Journal of Anthropology, 2,* 361-374.

Bentler, P. M., & Woodward, J. A. (1978). A Head Start reevaluation: Positive effects are not yet demonstrable. *Evaluation Quarterly, 2,* 493-510.

Berliner, J. S. (1957). *Factory and manager in the U.S.S.R.* Cambridge, MA: Harvard University Press.

Bevan, W. (1963). The pooling mechanism and the phenomena of reinforcement. In O. J. Harvey (Ed.), *Motivation and social interaction, cognitive determinants.* New York: Ronald Press.

Bhaskar, R. (1975). *A realist theory of science.* Leeds, England: Leeds Books.

Blau, P. (1955). *The dynamics of bureaucracy.* Chicago: University of Chicago Press.

Blau, P. (1956). *Bureaucracy in modern society.* New York: Random House.

Blau, P. M. (1963). *The dynamics of bureaucracy.* (Rev. ed.). Chicago: University of Chicago Press.

Bloor, D. (1976). *Knowledge and social imagery.* London: Routledge & Kegan Paul.

Bloor, D. (1983). *Wittgenstein: A social theory of knowledge.* London: Macmillan.

Bloor, D. (1984). Reply to J. W. Smith. *Studies in History and Philosophy of Science, 15*(3), 245-249.

Bloor, D. (1989). Professor Campbell on models of language learning and the sociology of science: A reply. In S. L. Fuller, M. DeMey, T. Shinn, & S. Woolgar (Eds.), *The cognitive turn.* Boston: Kluwer Academic.

Boeckmann, M. E. (1981). Rethinking the results of a negative income tax experiment. In R. F. Boruch, P. M. Wortman, & D. S. Cordray (Eds.), *Reanalyzing Program Evaluations* (pp. 341-363). San Francisco: Jossey-Bass.

Boring, E. G. (1923). Intelligence as the tests test it. *The New Republic, 34,* 33-36.

Boruch, R. F. (1973). *Regression-discontinuity designs revisited.* Unpublished manuscript, Northwestern University.

Boruch, R. F. (1974, May). *Regression-discontinuity designs: A summary.* Paper presented at the Annual Meeting of the American Educational Research Association, Chicago.

Boruch, R. F. (1975a). Coupling randomized experiments and approximations to experiments in social program evaluation. *Sociological Methods and Research, 4*(1), 31-53.

Boruch, R. F. (1975b). *Executive summary: Issues in the analysis of regression discontinuity data.* Psychology Department, Northwestern University, Evanston, IL.

Boruch, R. F. (1975c). *Regression-discontinuity evaluation of the Mesa Reading Program: Background and technical report.* Unpublished report, NIE Project on Secondary Analysis, Psychology Department, Northwestern University.

Boruch, R. F. (1978a). *Secondary analysis* (Vol. 4). San Francisco: Jossey-Bass.

Boruch, R. F. (1978b). *Double pretests for checking certain threats to the validity of some conventional evaluation designs or, stalking the null hypothesis.* Unpublished manuscript, Northwestern University.

Boruch, R. F., & Cecil, J. S. (1979). *Assuring privacy and confidentiality in social research.* Philadelphia: University of Pennsylvania Press.

Boruch, R. F., & Cecil, J. S. (Eds.). (1982). *Solutions to ethical and legal problems in social research.* New York: Academic Press.

Boruch, R. F., & DeGracie, J. S. (1975). *Regression-discontinuity evaluation of the Mesa reading program: Background and technical report.* Unpublished manuscript, Northwestern University.

Boruch, R. F., & DeGracie, J. S. (1977, April). *The use of regression-discontinuity models with criterion referenced testing in the evaluation of compensatory education.* Paper presented at the Annual Meeting of the American Educational Research Association, New York.

Boruch, R. F., Wortman, D. M., & Cordray, D. S., et al. (1981). *Reanalyzing program evaluations.* San Francisco: Jossey-Bass.

Box, G. E. P., & Tiao, G. C. (1965). A change in level of a non-stationary time series. *Biometrika, 52,* 181-192.

Box, G. E. P., & Tiao, G. C. (1975). Intervention analysis with applications to economic and environmental problems. *Journal of the American Statistical Association, 70,* 70-92.

Bradburn, N. M., & Caplovitz, D. (1965). *Reports on happiness.* Chicago: Aldine.

Brady, I. (Ed.). (1983). Speaking in the name of the real: Freeman and Mead on Samoa. Contributions by A. B. Weiner, T. Schwartz, L. Holmes, and B. Shore. *American Anthropologist, 85,* 908-947.

Brewer, M. B., Crano, W. D., & Campbell, D. T. (1970). Testing a single-factor model as an alternative to the misuse of partial correlations in hypothesis-testing research. *Sociometry, 33*(1), 1-11.

Brickman, P., & Campbell, D. T. (1971). Hedonic relativism and planning the good society. In M. H. Appley (Ed.), *Adaptation-level theory: A symposium* (pp. 287-302). New York: Academic Press.

Bridgman, P. W. (1927). *The logic of modern physics.* New York: Macmillan.

Brinton, C. C. (1965). *The anatomy of revolution.* New York: Random House.

Brown, D. R. (1953). Stimulus-similarity and the anchoring of subjective scales. *American Journal of Psychology, 66,* 199-214.

Brown, J. S. (1961). *The motivation of behavior.* New York: McGraw-Hill.

Brown, T. M. (1952). Habit persistence and lags in consumer behavior. *Econometrica, 20,* 355-371.

Bryk, A. (Ed.). (1983). *Stakeholder-based evaluation* (Vol. 17). San Francisco: Jossey-Bass.

Bultmann, R. K. (1956). *History and eschatology.* Edinburgh: Edinburgh University Press.

Burks, A. W. (1977). *Chance, cause, reason: An inquiry into the nature of scientific evidence.* Chicago: University of Chicago Press.

Cain, G. G. (1975). Regression and selection models to improve nonexperimental comparisons. In C. A. Bennett & A. A. Lumsdaine (Eds.), *Evaluation and experiment* (pp. 297-317). New York: Academic Press.

Campbell, D. T. (1956). Perception as substitute trial and error. *Psychological Review, 63*(5), 330-342.

Campbell, D. T. (1957). Factors relevant to the validity of experiments in social settings. *Psychological Bulletin, 54*(4), 297-312.

Campbell, D. T. (1959). Methodological suggestions from a comparative psychology of knowledge processes. *Inquiry, 2,* 152-182.

Campbell, D. T. (1960). Recommendations for APA test standards regarding construct, trait, or discriminant validity. *American Psychologist, 15,* 546-553.

Campbell, D. T. (1963). From description to experimentation: Interpreting trends as quasi-experiments. In C. W. Harris (Ed.), *Problems in measuring change.* Madison: University of Wisconsin Press.

Campbell, D. T. (1964). Distinguishing differences of perception from failures of communication in cross-cultural studies. In F. S. C. Northrop & H. H. Livingston (Eds.), *Cross-cultural understanding: Epistemology in anthropology.* New York: Harper & Row.

Campbell, D. T. (1966a). Pattern matching as an essential in distal knowing. In K. R. Hammond (Ed.), *The psychology of Egon Brunswik.* New York: Holt, Rinehart & Winston.

Campbell, D. T. (1966b). *The principle of proximal similarity in the application of science.* Unpublished manuscript.

Campbell, D. T. (1967a). Administrative experimentation, institutional records, and nonreactive measures. In J. C. Stanley (Ed.), *Improving experimental design and statistical analysis.* Chicago: Rand McNally.

Campbell, D. T. (1967b). *The effects of college on students: Proposing a quasi-experimental approach.* Unpublished manuscript.

Campbell, D. T. (1968). Quasi-experimental design. In D. L. Sills (Ed.), *International Encyclopedia of the Social Sciences* (Vol. 5, pp. 259-263). New York: Macmillan and Free Press.

Campbell, D. T. (1969a). Reforms as experiments. *American Psychologist, 24*(4), 409-429.

Campbell, D. T. (1969b). Prospective: Artifact and control. In R. Rosenthal & R. Rosnow (Eds.), *Artifact in behavior research* (pp. 351-382). New York: Academic Press.

Campbell, D. T. (1969c). Ethnocentrism of disciplines and the fish-scale model of omniscience. In M. Sherif & C. W. Sherif (Eds.), *Interdisciplinary relationships in the social sciences* (pp. 328-348). Chicago: Aldine.

Campbell, D. T. (1969d). A phenomenology of the other one: Corrigible, hypothetical and critical. In T. Mischel (Ed.), *Human action: Conceptual and empirical issues* (pp. 41-69). New York: Academic Press.

Campbell, D. T. (1971a). Comments on the comments by Shaver and Staines. *Urban Affairs Quarterly, 7*(2), 187-192.

Campbell, D. T. (1971b, September 5). *Methods for the experimenting society.* Paper presented at the annual meeting of the American Psychological Association, Washington, DC.

Campbell, D. T. (1971c). Temporal changes in treatment-effect correlations: A quasi-experimental model for institutional records and longitudinal studies. In G. V. Glass, (Ed.), *Proceedings of the 1970 Invitational Conference on Testing Problems.* Princeton, NJ: Educational Testing Service.

Campbell, D. T. (1972). Herskovits, cultural relativism, and metascience. In M. J. Herskovits (Ed.), *Cultural relativism* (pp. v-xxiii). New York: Random House.

Campbell, D. T. (1973). The social scientist as methodological servant of the experimenting society. *Policy Studies Journal, 2,* 72-75.

Campbell, D. T. (1974a). Evolutionary epistemology. In P. A. Schilpp (Ed.), *The philosophy of Karl Popper* (pp. 413-463). LaSalle, IL: Open Court.

Campbell, D. T. (1974b). Unjustified variation and selective retention in scientific discovery. In F. J. Ayala & T. Dobzhansky (Eds.), *Studies in the philosophy of biology.* London: Macmillan.

Campbell, D. T. (1975a). "Degrees of freedom" and the case study. *Comparative Political Studies, 8*(2), 178-193.

Campbell, D. T. (1975b). On the conflicts between biological and social evolution and between psychology and moral tradition. *American Psychologist, 30,* 1103-1126.

Campbell, D. T. (1976). Focal local indicators for social program evaluation. *Social Indicators Research, 3,* 237-256.

Campbell, D. T. (1977a). *Descriptive epistemology: Psychological, sociological, and evolutionary.* Unpublished manuscript.

Campbell, D. T. (1977b). Reforms as experiments. In F. G. Caro (Ed.), *Readings in evaluation research* (2nd ed.). New York: Russell Sage Foundation.

Campbell, D. T. (1978). Qualitative knowing in action research. In M. Brenner, P. Marsh, & M. Brenner (Eds.), *The social contexts of method* (pp. 184-209). London: Croom Helm.

Campbell, D. T. (1979a). Assessing the impact of planned social change. *Evaluation and Program Planning, 2,* 67-90.

Campbell, D. T. (1979b). Comments on the sociobiology of ethics and moralizing. *Behavioral Science, 24,* 37-45.

Campbell, D. T. (1979c). A tribal model of the social system vehicle carrying scientific knowledge. *Knowledge: Creation, Diffusion, Utilization, 2,* 181-201.

Campbell, D. T. (1981). ERISS Conference. *4S Newsletter (Society for the Social Studies of Science), 6*(3), 24-25.

Campbell, D. T. (1982). Experiments as arguments. *Knowledge: Creation, Diffusion, Utilization, 3*(3), 327-337.

Campbell, D. T. (1983). The two distinct routes beyond kin selection to ultrasociality: Implications for the humanities and social sciences. In D. L. Bridgeman (Ed.), *The Nature of Prosocial Development: Theories and Strategies* (pp. 11-41). New York: Academic Press.

Campbell, D. T. (1984a). Can we be scientific in applied social science? In R. F. Conner, D. G. Altman, & C. Jackson (Eds.), *Evaluation studies: Review annual* (Vol. 9, pp. 26-48). Beverly Hills, CA: Sage.

Campbell, D. T. (1984b). Foreword to Robert K. Yin Case Study Research (pp. 7-9). Beverly Hills, CA: Sage.

Campbell, D. T. (1984c). Foreword to Trochim, W. M. K. *Research design for program evaluation: The regression-discontinuity approach* (pp. 15-43). Beverly Hills, CA: Sage.

Campbell, D. T. (1984d). Hospital and landsting as continuously monitoring social polygrams: Advocacy and warning. In B. Cronholm & L. Von Knorring (Eds.), *Evaluation of mental health services programs* (pp. 13-39). Stockholm, Sweden: Forskningsraadet Medicinska.

Campbell, D. T. (1984e, November 11-13). *Types of evolutionary epistemology extended to a sociology of scientific belief.* Paper presented at the International Conference on Evolutionary Epistemology, University of Ghent.

Campbell, D. T. (1985). Toward an epistemologically-relevant sociology of science. *Science, Technology, and Human Values, 10,* 38-48.

Campbell, D. T. (1986a). Relabeling internal and external validity for applied social scientists. In W. M. K. Trochim (Ed.), *Advances in quasi-experimental design and analysis: New directions for program evaluation.* San Francisco: Jossey-Bass.

Campbell, D. T. (1986b). Science policy from a naturalistic sociological epistemology. *Philosophy of Science Association, 1984,* 2.

Campbell, D. T. (1986c). Science's social system of validity-enhancing collective belief change and the problems of the social sciences. In D. W. Fiske & R. A. Schweder (Eds.), *Metatheory in social science: Pluralisms and subjectivities* (pp. 108-135). Chicago: University of Chicago.

Campbell, D. T. (1987a). Guidelines for monitoring the scientific competence of preventive intervention research centers. *Knowledge: Creation, Diffusion, Utilization, 8*(3), 389-430.

Campbell, D. T. (1987b). Selection theory and the sociology of scientific validity. In W. Callebaut & R. Pinxten (Eds.), *Evolutionary epistemology: A multiparadigm program* (pp. 139-158). Dordrecht: Reidel.

Campbell, D. T. (1988a). A general 'selection theory' as implemented in biological evolution and in social belief-transmission-with-modifications in science. *Biology and Philosophy, 3,* 171-177.

Campbell, D. T. (1988b). *Methodology and epistemology for social science: Selected papers.* Chicago, IL: University of Chicago Press.

Campbell, D. T. (1988c). Provocation on reproducing perspectives, Part V. *Social Epistemology, 2,* 189-192.

Campbell, D. T. (1988d). The experimenting society. In E. S. Overman (Ed.), *Methodology and epistemology for social science: Selected papers* (pp. 290-314). Chicago: University of Chicago Press.

Campbell, D. T. (1989). Models of language learning and their implications for social constructionist analyses of scientific belief. In S. L. Fuller, M. DeMey, T. Shinn, & S. Woolgar (Eds.), *The cognitive turn.* Boston: Kluwer Academic.

Campbell, D. T. (1990). Epistemological roles for selection theory. In N. Rescher (Ed.), *Evolution, cognition, realism* (pp. 1-19). Lanham, MD: University Press of America.

Campbell, D. T. (1991). Quasi-experimental research designs in compensatory education. In E. M. Scott (Ed.), *Evaluating intervention strategies for children and youth at risk: Proceedings of the O.E.C.D. Conference in Washington, DC.* Washington, DC: U.S. Department of Education, Office of Planning, Budget and Evaluation and the U.S. Government Printing Office.

Campbell, D. T. (1993). Plausible coselection of belief by referent: All the "objectivity" that is possible. *Perspectives on Science, 1*(1), 88-108.

Campbell, D. T. (1994). How individual and face-to-face group selection undermine firm selection in organizational evolution. In J. A. C. Baum & J. V. Singh (Eds.), *Evolutionary dynamics of organizations.* New York: Oxford University Press.

Campbell, D. T., & Boruch, R. F. (1975). Making the case for randomized assignment to treatments by considering the alternatives: Six ways in which quasi-experimental evaluations in compensatory education tend to underestimate effects. In C. A. Bennett & A. Lumsdaine (Eds.), *Evaluation and experimentation: Some critical issues in assessing social programs* (pp. 195-296). New York: Academic Press.

Campbell, D. T., Boruch, R. F., Schwartz, R. D., & Steinberg, J. (1977). Confidentiality-preserving modes of access to files and to interfile exchange for useful statistical analysis. *Evaluation Quarterly, 1*(2), 269-299.

Campbell, D. T., & Cecil, J. S. (1982). A proposed system of regulation for the protection of participants in low-risk areas of applied social research. In J. E. Sieber (Ed.), *The ethics of social research: Fieldwork, regulation, and publication* (pp. 97-121). New York: Springer-Verlag.

Campbell, D. T., & Clayton, K. N. (1961). Avoiding regression effects in panel studies of communication impact. *Studies in Public Communication, 3,* 99-118.

Campbell, D. T., & Clayton, K. N. (1964). Avoiding regression effects in panel studies of communication impact. *Studies in public communication. Department of Sociology, University of Chicago, 1961, No. 3, 99-118.* (Vol. reprint series in the social sciences, S-353). Indianapolis: Bobbs-Merrill.

Campbell, D. T., & Erlebacher, A. E. (1970). How regression artifacts in quasi-experimental evaluations can mistakenly make compensatory education look harmful. In J. Hellmuth (Ed.), *Compensatory education: A national debate* (Vol. 3, pp. 185-210). New York: Brunner/Mazel.

Campbell, D. T., & Fiske, D. W. (1959). Convergent and discriminant validation by the multitrait-multimethod matrix. *Psychological Bulletin, 65,* 81-105.

Campbell, D. T., & Frey, P. W. (1970). The implications of learning theory for the fade-out of gains from compensatory education. In J. Hellmuth (Ed.), *Compensatory education: A national debate* (Vol. 3). New York: Brunner/Mazel.

Campbell, D. T., & Kimmel, A. J. (1985). *Guiding preventive intervention research centers for research validity* (Order no. 83MO54244901D). Rockville, MD: Department of Health and Human Services.

Campbell, D. T., & LeVine, R. A. (1970). Field-manual anthropology. In R. Naroll & R. Cohen (Eds.), *A handbook of method in cultural anthropology.* New York: Natural History Press/Doubleday.

Campbell, D. T., & O'Connell, E. (1982). Methods as diluting trait relationships rather than adding irrelevant systematic variance. In D. Brinberg & L. Kidder (Eds.), *Forms of validity in research. (New directions for methodology of social and behavioral science, Number 12)* (pp. 93-111). San Francisco: Jossey-Bass.

Campbell, D. T., & Reichardt, C. S. (1983, June 5-8). *Quasi-experimental forecasting of the impact of planned interventions.* Paper presented at the Third International Symposium on Forecasting, Philadelphia, PA.

Campbell, D. T., & Reichardt, C. S. (1991). Problems in assuming the comparability of pretest and posttest in autoregressive and growth models. In R. E. Snow & D. E. Wiley (Eds.), *Improving inquiry in social science: A volume in honor of Lee J. Cronbach* (pp. 201-219). Hillsdale, NJ: Erlbaum.

Campbell, D. T., Reichardt, C. S., & Trochim, W. (1979). *The analysis of the "fuzzy" regression-discontinuity design: Pilot simulations.* Unpublished manuscript, Northwestern University.

Campbell, D. T., & Ross, H. L. (1968). The Connecticut crackdown on speeding: Time-series data in quasi-experimental analysis. *Law and Society Review, 3*(1), 33-53.

Campbell, D. T., & Stanley, J. C. (1966). *Experimental and quasi-experimental designs for research.* Boston: Houghton Mifflin.

Caplan, N. (1968). Treatment intervention and reciprocal interaction effects. *Journal of Social Issues, 24*(1), 63-88.

Cappelleri, J. C., Trochim, W., Stanley, T. D., & Reichardt, C. S. (1991). Random measurement error doesn't bias the treatment effect estimate in the regression-discontinuity design: I. The case of no interaction. *Evaluation Review, 15*(4), 395-419.

Chapin, F. S. (1947). *Experimental design in sociological research.* New York: Harper.

Cicirelli, V., et al. (1969). *The impact of Head Start: An evaluation of the effects of Head Start on children's cognitive and affective development* (Contract B89-4536): Office of Economic Opportunity.

Clausner, J. I., & Shimony, A. (1978). Bell's theorem: Experimental tests and implications. *Reports on Progress in Physics, 41,* 1881-1927.

Coleman, J. S., Campbell, E. Q., Hobson, C. J., McPortland, J., Mood, A. M., Weinfeld, F. D., & York, R. L. (1966). *Equality of educational opportunity.* Washington, DC: Office of Education, U. S. Department of Health, Education, and Welfare.

Collins, H. (1985). *Changing order: Replication and induction in scientific practice.* Beverly Hills, CA: Sage.

Collins, H. M. (1975). The seven sexes: A study in the sociology of a phenomenon, or the replication of experiments in physics. *Sociology, 9,* 205-224.

Collins, H. M. (1981a). Son of seven sexes: The social destruction of a physical phenomenon. *Social Studies of Science, 11,* 33-62.

Collins, H. M. (1981b). Stages in the empirical programme of relativism. *Social Studies of Science, 11,* 3-10.

Consortium for Longitudinal Studies. (1983). *As the twig is bent . . . : Lasting effects of preschool programs.* Hillsdale, NJ: Erlbaum.

Cook, T. D. (1985, May 3). *Recent attacks on well-known validity distinctions: An appreciative rejoinder.* Paper presented at the Annual Convention of the Midwestern Psychological Association, Chicago.

Cook, T. D., Appleton, H., Conner, R., Shaffer, A., Tamkin, G., & Weber, S. J. (1975). *Sesame Street revisited: A case study in evaluation research.* New York: Russell Sage Foundation.

Cook, T. D., & Campbell, D. T. (1976). The design and conduct of quasi-experiments and true experiments in field settings. In M. D. Dunnette (Ed.), *Handbook of industrial and organizational psychology* (pp. 223-326). Chicago: Rand McNally.

Cook, T. D., & Campbell, D. T. (1979). *Quasi-experimentation: Design and analysis for field settings.* Boston: Houghton Mifflin.

Cook, T. D., & Reichardt, C. S. (1979). *Qualitative and quantitative methods in evaluation research.* Beverly Hills: Sage.

Crain, R. L., Hawes, J. A., Miller, R. L., & Peichert, J. R. (1984). *A longitudinal study of a metropolitan voluntary school desegregation plan.* Washington, DC: RAND.

Cronbach, J., & Furby, L. (1970). How we should measure "change"—or should we? *Psychological Bulletin, 74*(1), 68-80.

Cronbach, L. J., Rogosa, D. R., Floden, R. E., & Price, G. G. (1977). *Analysis of covariance in nonrandomized experiments: Parameters affecting bias.* Occasional paper, Stanford University, Stanford Evaluation Consortium.

Cummings, N. A., & Follette, W. T. (1968). Brief psychotherapy and medical utilization in a prepaid health plan setting: Part II. *Medical Care, 6,* 31-41.

Cziko, G. A., & Campbell, D. T. (1990). Comprehensive evolutionary epistemology bibliography. *Journal of Social and Biological Structures, 13*(1), 41-82.

Dahrendorf, R. (1958). Out of Utopia: Toward a reorientation of sociological analysis. *American Journal of Sociology, 64,* 115-127.

Darlington, R. (1990). *Regression and linear models.* New York: McGraw Hill.

Davies, J. C. (1962). Toward a theory of revolution. *American Sociological Review, 27,* 5-18.

DeGracie, J. S., & Boruch, R. F. (1977). *Regression-discontinuity analysis of the Mesa reading compensatory programs.* Paper presented at the Annual Meeting of the American Educational Research Association, New York.

Director, S. M. (1979). Underadjustment bias in the evaluation of manpower training. *Evaluation Quarterly, 3*(2), 190-218.

Dirsmith, M. W., & Jablonsky, S. F. (1978). The pattern of PPB rejection: Something about organizations, something about PPB. *Accounting, Organizations and Society, 3,* 3-4.

Dittes, J. E. (1959). Attractiveness of group as a function of self-esteem and acceptance by group. *Journal of Abnormal and Social Psychology, 54,* 77-82.

Divgi, D. R. (1979). *Choosing polynomials in the regression-discontinuity design.* Unpublished manuscript.

Duesenberry, J. S. (1949). *Income, saving, and the theory of consumer behaviour.* Cambridge, MA: Harvard University Press.

Dunn, E. S. (1971). *Economic and social development: A process of social learning.* Baltimore: Johns Hopkins Press.

Dunn, O. J. (1961). Multiple comparisons among means. *Journal of the American Statistical Association, 56,* 52-64.

Editors of Sport. (1970). Four general managers sound off: How to make money, win, and keep the peace. *Sport, 50,* 46.

Ehrenberg, A. S. C. (1968). The elements of lawlike relationships. *Journal of the Royal Statistical Society, Series A, 131,* 280-302.

Ellis, B. (1979). *Rational belief systems.* Totowa, NJ: Rowman & Littlefield.

Etzioni, A. (1968). *The active society.* New York: Free Press.

Exline, R. V., & Ziller, R. C. (1959). Status congruency and interpersonal conflict in decision-making groups. *Human Relations, 12,* 147-162.

Feinstein, A. R. (1985). *Clinical epidemiology: The architecture of clinical research.* New York: Saunders.

Feinstein, A. R., & Horwitz, R. I. (1981). An algebraic analysis of biases due to exclusion, susceptibility, and protopathic prescription in case-control research. *Journal of Chronic Diseases, 34,* 393-403.

Festinger, L. (1954). A theory of social comparison processes. *Human Relations, 7,* 117-140.

Feyerabend, P. K. (1975). *Against method.* London: NLB Press.

Firth, R. (1983). Review of Derek Freeman's *Margaret Mead and Samoa. RAIN, 57,* 11-12.

Fishbein, M., Raven, B. H., & Hunter, R. (1963). Social comparison and dissonance reduction in self evaluation. *Journal of Abnormal Social Psychology, 67,* 491-501.

Fiske, D. W. (1982). Convergent-discriminant validation in measurements and research strategies. In D. Brinberg & L. Kidder (Eds.), *Forms of validity in research. (New directions for methodology of social and behavioral science, Number 12)* (pp. 77-92). San Francisco: Jossey-Bass.

Fiske, D. W., & Schweder, R. A. (Eds.). (1986). *Metatheory in social science: Pluralism and subjectivity.* Chicago: University of Chicago Press.

Fleck, L. (1979). *Genesis and development of a scientific fact.* Chicago: University of Chicago Press.

Fleiss, J. L., & Tanur, J. M. (1972). The analysis of covariance in psychopathology. In M. Hammer, K. Salzinger, & S. Sutton (Eds.), *Psychopathology* (pp. 509-527). New York: Wiley.

Follette, W. T., & Cummings, N. A. (1967). Psychiatric services and medical utilization in a prepaid health plan setting: Part I. *Medical Care, 5,* 25-35.

Franklin, A. (1990). *Experiment, right or wrong.* Cambridge: Cambridge University Press.

Franklin, A. D. (1981). Millikan's published and unpublished data on oil drops. *Historical Studies in the Physical Sciences, 11*(2), 185-201.

Franks, L., & Powers, T. (1970, September 14-18). Dianna: The making of a terrorist. *Chicago Daily News.*

Freeman, D. (1983). *Margaret Mead and Samoa.* Cambridge: Harvard University Press.

Fuller, S. (1988). Provocation on reproducing perspectives, Part III. *Social Epistemology, 2,* 99-101.

Galbraith, J. K. (1958). *The affluent society.* Boston: Houghton.

Galison, P. (1987). *How experiments end.* Chicago: University of Chicago Press.

Gardiner, J. A. (1969). *Traffic and the police: Variations in law-enforcement policy.* Cambridge, MA: Harvard University Press.

Garfinkel, H. (1967). "Good" organizational reasons for "bad" clinic records. In H. Garfinkel (Ed.), *Studies in ethnomethodology* (pp. 186-207). Englewood Cliffs, NJ: Prentice Hall.

Gergen, K. J. (1982). *Toward transformation in social knowledge.* New York: Springer-Verlag.

Giere, R. N. (1984). Toward a unified theory of science. In J. T. Cushing, C. F. Delaney, & G. Gutting (Eds.), *Science and reality* (pp. 5-31). Notre Dame, IN: University of Notre Dame Press.

Giere, R. N. (1985a). Constructive realism. In P. M. Churchland & C. A. Hooker (Eds.), *Images of science* (pp. 75-98). Chicago, IL: University of Chicago Press.

Giere, R. N. (1985b). Philosophy of science naturalized. *Philosophy of Science, 52,* 331-356.

Giere, R. N. (1988). *Explaining science: A cognitive approach.* Chicago: University of Chicago Press.

Gieryn, T. F. (1982). Relativist/constructionist programmes in the sociology of science redundance and retreat. *Social Studies of Science, 12,* 279-297.

Gieryn, T. F. (1983). Boundary-work and the demarcation of science from non-science: Strains and interests in professional ideologies of scientists. *American Sociological Review, 48,* 781-795.

Ginsberg, P. E. (1984). The dysfunctional side effects of quantitative indicator production: Illustrations from mental health care. *Evaluation and Program Planning, 7,* 1-12.

Glass, G. V. (1968). Analysis of data on the Connecticut speeding crackdown as a time-series quasi-experiment. *Law and Society Review, 3*(1), 55-76.

Glass, G. V., Tiao, G. C., & Maguire, T. O. (1971). Analysis of data on the 1900 revision of the German divorce laws as a quasi-experiment. *Law and Society Review, 6,* 539-562.

Goffman, E. (1952). On cooling the mark out: Some aspects of adaptation to failure. *Psychiatry, 15,* 451-463.

Goldberger, A. S. (1972). Selection bias in evaluation treatment effects: Some formal illustrations, *Discussion Papers* (pp. 123-172). Madison: Institute for Research on Poverty, University of Wisconsin.

Goldman, A. I. (1986). *Epistemology and cognition.* Cambridge, MA: Harvard University Press.

Gooding, D. (1990). *Experiment and the making of meaning.* Dordrecht: Kluwer Academic.

Goodman, N. (1972). Likeness. In N. Goodman (Ed.), *Problems and projects.* Indianapolis, IN: Bobbs-Merrill.

Granick, D. (1954). *Management of the industrial firm in the U.S.S.R.* New York: Columbia University Press.

Greenwood, E. (1945). *Experimental sociology: A study in method.* New York: King's Crown Press.

Guttman, L. (1955). A generalized simplex for factor analysis. *Psychometrika, 20,* 173-192.

Habermas, J. (1970a). On systematically distorted communication. *Inquiry, 13,* 205-218.

Habermas, J. (1970b). Toward a theory of communicative competence. *Inquiry, 13,* 360-375.

Habermas, J. (1983). Interpretive social science vs. hermeneuticism. In N. Haan, R. N. Bellah, P. Rabinow, & W. M. Sullivan (Eds.), *Social science as moral inquiry.* New York: Columbia University Press.

Haner, C. F., & Brown, P. A. (1955). Clarification of the instigation to action concept in the frustration-aggression hypothesis. *Journal of Abnormal and Social Psychology, 51,* 204-206.

Hanson, N. R. (1958). *Patterns of discovery.* Cambridge, MA: Harvard University Press.

Harrod, R. F. (1956). *Foundations of inductive logic.* London: Macmillan.

Haworth, L. (1960). The experimental society: Dewey and Jordan. *Ethics, 71*(1), 27-40.

Helson, H. (1964). *Adaption-level theory: An experimental and systematic approach to behavior.* New York: Harper.

Hoffman, P. J., Festinger, L., & Lawrence, D. H. (1954). Tendencies toward group comparison in competitive bargaining. *Human Relations, 7,* 141-159.

Hollis, M. (1967). Reason and ritual. *Philosophy, 43,* 231-247. In W. B. R. (Ed.), *Reprinted in Rationality.* New York: Harper & Row, Harper Torchbooks.

Holmes, D. S. (1970). Differential change in affective intensity and forgetting of unpleasant personal experience. *Journal of Personality and Social Psychology, 15,* 234-239.

Holmes, L. D. (1957). *The restudy of Manu'an culture: A problem in methodology.* Unpublished doctoral dissertation, Northwestern University.

Holton, G. (1978). Subelectrons, presuppositions, and the Millikan-Ehrenhaft dispute. In *The scientific imagination: Case studies* (pp. 25-83). Cambridge: Cambridge University Press.

Homans, G. C. (1961). *Social behavior: Its elementary forms.* New York: Harcourt, Brace & World.

Hooker, R. W. (1901). Correlation of the marriage-rate with trade. *Journal of the Royal Statistical Society, 64,* 485-492.

Hull, D. L. (1978). Altruism in science: A sociobiological model of cooperative behaviour among scientists. *Animal Behaviour, 26,* 685-697.

Hull, D. L. (1988a). A mechanism and its metaphysics: An evolutionary account of the social and conceptual development of science. *Biology and Philosophy, 3,* 123-156.

Hull, D. L. (1988b). *Science as process.* Chicago: University of Chicago Press.

Humphreys, L. G. (1960). Investigations of the simplex. *Psychometrika, 25,* 313-323.

Hyman, H. H., & Wright, C. R. (1967). Evaluating social action programs. In P. F. Lazarsfeld, W. H. Sewell, & H. L. Wilensky (Eds.), *The uses of sociology* (pp. 741-782). New York: Basic Books.

Janousek, J. (1970). Comments on Campbell's "Reforms as experiments." *American Psychologist, 25*(2), 191-193.

Jones, E. E., & Gerard, H. B. (1967). *Foundations of social psychology.* New York: Wiley.

Jones, E. E., Gergen, K. J., & Jones, R. G. (1963). Tactics of ingratiation among leaders and subordinates in a status hierarchy. *Psychological Monographs, 77* (Whole No. 566).

Jones, K., & Vischi, T. (1979). Impact of alcohol, drug abuse, and mental health treatment on medical care utilization: A review of the research literature. *Medical Care, 17*(12, Suppl.), i-vi, 1-82.

Keat, R., & Urry, J. (1975). *Social theory as science.* London: Routledge & Kegan Paul.

Kenny, D. A. (1970a). *A model for temporal erosion and common factor effects in cross-lagged panel correlation.* Unpublished master's thesis, Northwestern University.

Kenny, D. A. (1970b, November 12-16). *Testing a model of dynamic causation.* Paper presented at the Conference on Structural Equations, University of Wisconsin.

Kenny, D. A. (1979). *Correlation and causality.* New York: Wiley.

Kershaw, D., & Fair, J. (1976a). *The New Jersey income-maintenance experiment.* New York: Academic Press.

Kershaw, D., & Fair, J. (1976b). *The New Jersey income maintenance experiment, Vol. 1: Operations, surveys, and administration.* New York: Academic Press.

Kim, K. M. (1989). *Explaining scientific consensus: Toward a social system grounding of scientific validity.* Unpublished doctoral dissertation, University of Chicago, Chicago.

Kitsuse, J. K., & Circourel, A. V. (1963). A note on the uses of official statistics. *Social Problems, 11,* 131-139.

Knafl, G. J. (1978). *Implementing approximately linear models.* Unpublished manuscript.

Knorr-Cetina, K. D. (1981). *The manufacture of knowledge: An essay on the constructivist and contextual nature of science.* Oxford: Pergamon.

Knorr-Cetina, K. D. (1987). Evolutionary epistemology and the sociology of science. In W. Callebaut & R. Pinxten (Eds.), *Evolutionary epistemology: A multiparadigm program* (pp. 179-201). Dordrecht: Reidel.

Knorr-Cetina, K. D., & Mulkay, M. (1983). Introduction: Emerging principles in social studies of science. In K. D. Knorr-Cetina & M. Mulkay (Eds.), *Science observed* (pp. 1-17). Beverly Hills, CA: Sage.

Kogan, W. S., Thompson, D. J., Brown, J. R., & Newman, H. F. (1975). Impact of integration of mental health service and comprehensive medical care. *Medical Care, 13,* 934.

Kornblith, H. (Ed.). (1985). *Naturalizing epistemology.* Cambridge: MIT Press.

Krause, M. S., & Howard, K. I. (1976). Program evaluation in the public interest. *Community Mental Health Journal, 5,* 291-300.

Kuhn, T. S. (1970). *The structure of scientific revolutions* (2nd ed.). Chicago: University of Chicago Press.

Kupper, L. L. (1984). Effects of the use of unreliable surrogates on the validity of epidemiologic research studies. *American Journal of Epidemiology, 120,* 643-648.

Latane, B. (1966). Studies in social comparison—Introduction and overview. *Journal of Experimental Social Psychology, Supplement 1,* 1-5.

Latour, B. (1987). *Science in action.* Cambridge, MA: Harvard University Press.

Latour, B., & Woolgar, S. (1979). *Laboratory life: The social construction of scientific facts.* Beverly Hills, CA: Sage.

Laudan, L. (1981a). A confutation of convergent realism. *Philosophy of Science, 48,* 19-49.

Laudan, L. (1981b). *Science and hypothesis.* Dordrecht: Reidel.

Laudan, L. (1984). *Science and values.* Berkeley: University of California Press.

Lawler, E. E., III. (1965, April). *Managerial perceptions of compensation.* Paper presented at the Midwestern Psychological Association Convention, Chicago.

Lazar, I., & Darlington, R. (1982). Lasting effects of early education: A report from the Consortium for Longitudinal Studies. *Monographs of the Society for Research in Child Development, 47,* 2-3.

Lehrer, K. (1974). *Knowledge.* Oxford: Clarendon Press.

Lehrer, K. (1989). Knowledge reconsidered. In M. Clay & K. Lehrer (Eds.), *Knowledge and skepticism* (pp. 131-154). Boulder, CO: Westview.

Lenski, G. (1967). Status inconsistency and the vote: A four-nation test. *American Sociological Review, 32,* 298-301.

Levine, J. (1978). The autonomy of history: R. G. Collingwood and Agatha Christie. *Clio, 7,* 253-264.

Lewin, K. (1946). Action research and minority problems. *Journal of Social Issues, 2,* 34-46.

Lewin, K. (1948). *Resolving social conflicts.* New York: Harper.

Lewin, K., Dembo, T., Festinger, L., & Sears, P. (1944). Level of aspiration. In J. M. Hunt (Ed.), *Personality and the behavior disorders.* New York: Ronald Press.

Lewis, O. (1951). *Life in a Mexican village: Tepoztlan restudied.* Urbana: University of Illinois Press.

Lindblom, C. E., & Cohen, D. K. (1979). *Usable knowledge.* New Haven, CT: Yale University Press.

Lohr, B. W. (1972, January). *An historical view of the research on the factors related to the utilization of health services.* Duplicated Research Report, National Center for Health Services Research and Development, Social and Economic Analysis Division. Rockville, MD: U.S. Government Printing Office.

Lord, F. M. (1960). Large-scale covariance analysis when the control variable is fallible. *Journal of the American Statistical Association, 55,* 307-321.

Lord, F. M. (1967). A paradox in the interpretation of group comparisons. *Psychological Bulletin, 68,* 304-405.

Lubin, A. (1969). *Time series analysis of repeated measures.* Colloquium at Northwestern University, October 1969.

MacKenzie, D. A., & Barnes, B. (1975). Biometrician versus Mendelian: A controversy and its explanation (in German). *Kölner Zeitschrift für Soziologie und Sozialpsychologie, 18,* 165-196.

MacKenzie, D. A., & Barnes, B. (1979). Scientific judgment: The biometry-Mendelism controversy. In B. Barnes & S. Shapin (Eds.), *The natural order: Historical studies of scientific culture* (pp. 191-210). Beverly Hills, CA: Sage.

Magidson, J. (1977). Toward a causal model approach for adjusting for preexisting differences in the nonequivalent control group situation: A general alternative to ANCOVA. *Evaluation Quarterly, 1*(3), 399-420.

Maguire, T. O., & Glass, G. V. (1967). A program for the analysis of certain time-series quasi-experiments. *Educational and Psychological Measurement, 27,* 743-750.

Maynard-Smith, J. (1988). Mechanisms of advance. *Science, 242*(25), 1182-1183.

McCall, W. A. (1923). *How to experiment in education.* New York: Macmillan.

McCarthy, T. (1973). A theory of communicative competence. *Philosophy of the Social Sciences, 3,* 135-156.

McClelland, D. C., Atkinson, J. W., Clark, R. A., & Lowell, E. L. (1953). *The achievement motive.* New York: Appleton.

McCord, J. (1978). A thirty-year follow-up of treatment effects. *American Psychologist, 33,* 284-289.

McCord, J. (1981). Consideration of some effects of a counseling program. In S. E. Martin, L. B. Sechrest, & R. Redner (Eds.), *New directions in the rehabilitation of criminal offenders* (pp. 394-405). Washington, DC: National Academy Press.

McFarlane, M. J., Feinstein, A. R., & Horwitz, R. I. (1986). Diethylstilbestrol and clear cell vaginal carcinoma: Reappraisal of the epidemiologic evidence. *American Journal of Medicine, 81,* 855-863.

McHose, J. H. (1970). Relative reinforcement effects: S_1/S_2 and S_1/S_1 paradigms in instrumental conditioning. *Psychological Review, 77,* 135-146.

McNemar, Q. (1940). A critical examination of the University of Iowa studies of environmental influences upon the I.Q. *Psychological Bulletin, 37,* 63-92.

Meehl, P. E. (1970). Nuisance variables and the ex post facto design. In M. Radner & S. Winokur (Eds.), *Analyses of theories and methods of physics and psychology, Vol. IV. Minnesota studies in the philosophy of science* (pp. 373-402). Minneapolis: University of Minnesota Press.

Merton, R. K. (1973). *The sociology of science.* Chicago: University of Chicago Press.

Merton, R. K., & Gieryn, T. F. (1982). Institutionalized altruism: The case of the professions. In R. K. Merton (Ed.), *Social research and the practicing professions* (pp. 109-134). Cambridge, MA: Abt Books.

Merton, R. K., & Kitt, A. S. (1950). Contributions to the theory of reference group behavior. In R. K. Merton & P. F. Lazarsfeld (Eds.), *Continuities in social research: Studies in the scope and method of "The American Soldier."* Glencoe, IL: Free Press of Glencoe.

Miettinen, O. S., & Cook, E. F. (1981). Confounding: Essence and detection. *American Journal of Epidemiology, 114,* 593-603.

Miller, S. M., Roby, T., & Steenwijk, A. A. V. (1970). Creaming the poor. *Trans-action, 7*(8), 38-45.

Mishan, E. J. (1960). A survey of welfare economies, 1939-59. *The Economic Journal, 70,* 197-265.

Mood, A. M. (1950). *Introduction to the theory of statistics.* New York: McGraw-Hill.

Morrissey, W. R. (1972). Nixon anti-crime plan undermines crime statistics. *Justice Magazine, 1*(5/6), 8-11, 14.

Mosteller, F., & Moynihan, D. P. (Eds.). (1972). *On equality of educational opportunity.* New York: Vintage Books.

Moyer, D. F. (1979). Revolution in science: The 1919 eclipse test of general relativity. In B. Kursunoglu, A. Perlmutter, & L. F. Scott (Eds.), *On the path of Albert Einstein* (pp. 55-101). New York: Plenum Press.

Mulkay, M. (1979). *Science and the sociology of knowledge.* London: Allen & Unwin.

Olbrisch, M. E. (1977). Psychotherapeutic interventions in physical health: Effectiveness and economic efficiency. *American Psychologist, 32*(9), 761-777.

Olbrisch, M. E. (1980). Psychological intervention and reduced medical care utilization: A modest interpretation. *American Psychologist, 35,* 760-761.

Olby, R. (1974). *The path to the double-helix.* Seattle: University of Washington Press.

Orne, M. T. (1969). Demand characteristics and the concept of quasi-controls. In R. Rosenthal & R. L. Rosnow (Eds.), *Artifact in behavioral research.* New York: Academic.

Paller, B. T., & Campbell, D. T. (1989). Maxwell and van Fraassen on observability, reality, and justification. In M. L. Maxwell & C. W. Savage (Eds.), *Science, mind and psychology: Essays in honor of Grover Maxwell.* Lanham, MD: University Press of America.

Pelz, D. C., & Andrews, F. M. (1964). Detecting causal priorities in panel study data. *American Sociological Review, 29,* 836-848.

Pelz, D. C., & Andrews, F. M. (1966). *Scientists in organizations.* New York: Wiley.

Pettigrew, T. F. (1964). *A profile of the Negro American.* Princeton, NJ: Van Nostrand.

Pickering, A. (1984). *Constructing Quarks: A sociological history of particle physics.* Edinburgh: Edinburgh University Press.

Pickering, A. (1989). Living in the material world: On realism and experimental practice. In D. Gooding, T. J. Pinch, & S. Schaffer (Eds.), *The uses of experiment: Studies of experimentation in the natural sciences* (pp. 275-297). Cambridge: Cambridge University Press.

Polanyi, M. (1958). *Personal knowledge: Toward a post-critical philosophy.* London: Routledge & Kegan Paul.

Polanyi, M. (1966a). The message of the Hungarian revolution. *American Scholar, 35,* 261-276.

Polanyi, M. (1966b). A society of explorers. In *The tacit dimension* (pp. 53-92). Garden City, NY: Doubleday.

Polanyi, M. (1967). The growth of science in society. *Minerva, 5,* 533-545.

Polanyi, M. (1969). *Knowing and being.* London: Routledge & Kegan Paul.

Pollock, J. (1974). *Knowledge and justification.* Princeton, NJ: Princeton University Press.

Poole, C. (1986). Exposure opportunity in case-control studies. *American Journal of Epidemiology, 123,* 352-358.

Poole, C. (1987). Critical appraisal of the exposure-potential restriction rule. *American Journal of Epidemiology, 125,* 179-183.

Popper, K. R. (1935). *Logik der Forschung.* Vienna: Springer.

Popper, K. R. (1944). *The poverty of historicism.* London: Routledge & Kegan Paul.

Popper, K. R. (1952). *The open society and its enemies* (2nd ed.). London: Routledge & Kegan Paul.

Popper, K. R. (1959). *The logic of scientific discovery.* New York: Basic Books.

Popper, K. R. (1963). *Conjectures and refutations.* New York: Basic Books.

Popper, K. R. (1970). Normal science and its dangers. In I. Lakatos & A. Musgrave (Eds.), *Criticism and the growth of knowledge* (pp. 51-58): Cambridge University Press.

Porter, A. C. (1967). *The effects of using fallible variables in the analysis of covariance.* Unpublished doctoral dissertation, University of Wisconsin (University Microfilms, Ann Arbor, Michigan, 1968).

Potvin, L., & Campbell, D. T. (1996). *Exposure opportunity, the experimental paradigm and the case-control study.* Unpublished manuscript.

Prakhash, P., & Rappaport, A. (1977). Information inductance and its significance for accounting. *Accounting, Organizations and Society, 2,* 29-38.

Provine, W. (1971). *Origins of theoretical population genetics.* Chicago: University of Chicago Press.

Putnam, H. (1975). *Mind language and reality.* Cambridge: Cambridge University Press.

Quine, W. V. (1951). Two dogmas of empiricism. *Philosophical Review, 60,* 20-43.

Quine, W. V. (1969a). Epistemology naturalized. In W. V. Quine (Ed.), *Ontological relativity and other essays* (pp. 69-90). New York: Columbia University Press.

Quine, W. V. (1969b). *Ontological relativity and other essays.* New York: Columbia University Press.

Quine, W. V. (1974). *The roots of reference.* LaSalle, IL: Open Court.

Quine, W. V. (1975). The nature of natural knowledge. In S. Guttenplan (Ed.), *Mind and language* (pp. 67-81). Oxford: Clarendon Press.

Quinsey, V. L. (1970). Some applications of adaptation-level theory to aversive behavior. *Psychological Bulletin, 73,* 441-450.

Raser, J. R., Campbell, D. T., & Chadwick, R. W. (1970). Gaming and simulation for developing theory relevant to international relations. In *General systems research.* Ann Arbor, MI: Society for General Systems Research.

Ravetz, J. (1971). *Scientific knowledge and its social problems.* Oxford: Clarendon Press.

Reichardt, C. S. (1979). The statistical analysis of data from nonequivalent group designs. In T. D. Cook & D. T. Campbell (Eds.), *Quasi-experimentation: Design and analysis issues for field settings.* Boston: Houghton Mifflin.

Rescher, N. (1977). *Methodological pragmatism.* Oxford: Basil Blackwell.

Richards, R. J. (1981). Natural selection and other models in the historiography of science. In M. B. Brewer & B. E. Collins (Eds.), *Scientific inquiry and the social sciences.* San Francisco: Jossey-Bass.

Richards, R. J. (1987). The natural selection and other models in the historiography of science. *Darwin and the emergence of evolutionary theories of mind and behavior* (pp. 559-594). Chicago: Chicago University Press.

Rickard, S. (1972). The assumptions of causal analysis for incomplete causal sets of two multilevel variables. *Multivariate Behavioral Research, 7,* 317-359.

Ridgeway, V. (1956). Dysfunctional consequences of performance measures. *Administrative Science Quarterly, 1*(2), 240-247.

Riecken, H. W., Boruch, R. F., Campbell, D. T., Caplan, N., Glennan, T. K., Pratt, J., Rees, A., & Williams, W. (1974). *Social experimentation: A method for planning and evaluating social intervention.* New York: Academic Press.

Roll-Hansen, N. (1983). The death of spontaneous generation and the birth of the gene: Two case studies of relativism. *Social Studies of Science, 13,* 481-519.

Rose, A. M. (1952). Needed research on the mediation of labor disputes. *Personnel Psychology, 5,* 187-200.

Ross, H. L., Campbell, D. T., & Glass, G. V. (1970). Determining the social effects of a legal reform: The British "breathalyser" crackdown of 1967. *American Behavioral Scientist, 13,* 493-509.

Rossi, P. H. (1969). Practice, method, and theory in evaluating social-action programs. In J. L. Sundquist (Ed.), *On fighting poverty* (pp. 217-235). New York: Basic Books.

Rothman, K. J. (1986). *Modern epidemiology.* Boston: Little-Brown.

Rozelle, R. M., & Campbell, D. T. (1969). More plausible rival hypotheses in the cross-lagged panel correlation technique. *Psychological Bulletin, 71,* 74-80.

Runciman, W. G. (1961). Problems of research on relative deprivation. *European Journal of Sociology, 2,* 315-323.

Ryan, T. A. (1959). Multiple comparisons in psychological research. *Psychological Bulletin, 56,* 26-47.

Ryan, T. A. (1960). Significance tests for multiple comparisons of proportions, variances, and other statistics. *Psychological Bulletin, 57,* 318-328.

Sackett, D. L. (1979). Bias in analytic research. *Journal of Chronic Diseases, 32,* 51-63.

Sacks, J., & Ylvisaker, D. (1978). Linear estimates for approximately linear models. *Annals of Statistics, 6,* 1122-1138.

Sampson, E. E. (1963). Status congruence and cognitive consistency. *Sociometry, 26,* 146-162.

Sanford, N. (1970). Whatever happened to action research? *Journal of Social Issues, 26,* 2-23.

SAS Institute Inc. (1982). *SAS User's Guide: Statistics, 1982 Edition.* Cary, NC: Author.

Sawyer, J. (1971). Relative deprivation: A politically-based concept. *Psychiatry, 34,* 97-99.

Schanck, R. L., & Goodman, C. (1939). Reactions to propaganda on both sides of a controversial issue. *Public Opinion Quarterly, 3,* 107-112.

Schlesselman, J. J. (1982). *Case-control studies: Design, conduct, analysis.* New York: Oxford University Press.

Schlesselman, J. J., & Stadel, B. V. (1987). Exposure opportunity in epidemiologic studies. *American Journal of Epidemiology, 125,* 174-178.

Schmookler, J. (1966). *Invention and economic growth.* Cambridge, MA: Harvard University Press.

Schwartz, R. D. (1961). Field experimentation in sociolegal research. *Journal of Legal Education, 13*(401-410).

Schwarzlander, H. (1978). The process utopia. *Alternative Futures, 1*(2), 99-102.

Sears, D. O., & McConahay, J. B. (1970). Racial socialization, comparison levels, and the Watts riot. *Journal of Social Issues, 26,* 121-140.

Seaver, L. B. (1970). *College impact on student personality as reflected by increases in treatment-outcome correlation.* Unpublished master's thesis, Northwestern University.

Seaver, W. B., & Quarton, R. J. (1976). Regression-discontinuity analysis of Dean's List effects. *Journal of Educational Psychology, 68,* 459-465.

Segal, D. R., Segal, M. W., & Knoke, D. (1970). Status inconsistency and self-evaluation. *Sociometry, 33,* 347-357.

Segall, M. H., Campbell, D. T., & Herskovits, M. J. (1966). *The influence of culture on visual perception.* Indianapolis, IN: Bobbs-Merrill.

Seidman, D., & Couzens, M. (1974). Getting the crime rate down: Political pressure and crime reporting. *Law & Social Review, 8*(3), 457-493.

Selltiz, C., Jahoda, M., Deutsch, M., & Cook, S. W. (1959). *Research methods in social relations. Published for the Society for the Psychological Study of Social Issues.* (Rev. ed.). New York: Holt.

Shapin, S. (1982). History of science and its sociological reconstructions. *History of Science, 20,* 157-211.

Shaver, P., & Staines, G. (1971). Problems facing Campbell's "Experimenting Society." *Urban Affairs Quarterly, 7*(2), 173-186.

Siegel, S. (1957). Level of aspiration and decision making. *Psychological Review, 64,* 253-262.

Simmel, G. (1895). Über eine Beziehung der Selectionslehre zur Erkenntnistheorie. *Archiv für systematische Philosophie, 1,* 34-45.

Simon, J. L. (1966). The price elasticity of liquor in the U. S. and a simple method of determination. *Econometrica, 34,* 193-205.

Sivin, N. (1980). Science in China's past. In L. A. Orleans (Ed.), *Science in contemporary China* (pp. 1-29). Stanford, CA: Stanford University Press.

Skolnick, J. H. (1966). *Justice without trial: Law enforcement in democratic society.* New York: Wiley.

Smith, H. F. (1957). Interpretation of adjusted treatment means and regressions in analysis of covariance. *Biometrics, 13,* 282-308.

Smith, M. S., & Bissell, J. S. (1970). Report analysis: The impact of Head Start. *Harvard Educational Review, 40,* 51-104.

Sober, E. (1981). The evolution of rationality. *Synthese, 46,* 95-120.

Solomon, R. W. (1949). An extension of control group design. *Psychological Bulletin, 46,* 137-150.

Spiegelman, C. H. (1976). *Two methods of analyzing a nonrandomized experiment "adaptive" regression and a solution to Reiersol's problem.* Unpublished doctoral dissertation, Northwestern University, Chicago.

Spiegelman, C. H. (1977). *A technique for analyzing a pretest-posttest nonrandomized field experiment.* Florida State University, Statistics Report M 435.

Spiegelman, C. H. (1979a). *A technique for analyzing an educational program with Monte-Carlo results.* Unpublished manuscript, Northwestern University, Chicago.

Spiegelman, C. H. (1979b, August). *Estimating the effect of a large scale pretest-posttest social program.* Paper presented at the Proceedings of the Social Statistics Section, American Statistical Association, Washington, DC.

Spiegelman, C. H. (1979c). On estimating the slope of a straight line when both variables are subject to error. *The Annals of Statistics, 7*(1), 201-206.

Stake, R. E. (1971). Testing hazards in performance contracting. *Phi Delta Kappan, 52*(10), 583-588.

Stegmuller, W. (1976). *The structure and dynamics of theories.* New York: Springer-Verlag.

Stern, E., & Keller, S. (1953). Spontaneous group reference in France. *Public Opinion Quarterly, 17,* 208-217.

Stieber, J. W. (1949). *Ten years of the Minnesota Labor Relations Act.* Minneapolis: Industrial Relations Center, University of Minnesota.

Stolley, P. D. (1985). Faith, evidence and the epidemiologist. *Journal of Public Health Policy, 6,* 37-42.

Struening, E. L., & Brewer, M. B. (Eds.). (1983). *Handbook of evaluation research: University edition.* Beverly Hills, CA: Sage.

Sween, J. A. (1971). *The experimental regression design: An inquiry into the feasibility of nonrandom treatment allocation.* Unpublished doctoral dissertation, Northwestern University, Chicago.

Sween, J. A. (1977). *Regression discontinuity: Statistical tests of significance when units are allocated to treatments on the basis of quantitative eligibility.* Unpublished manuscript, Department of Sociology, DePaul University.

Sween, J., & Campbell, D. T. (1965). *A study of the effect of proximally autocorrelated error on tests of significance for the Interrupted Time Series Quasi-Experimental Design.* Evanston, IL: Northwestern University.

Tallmadge, G. K., & Horst, D. P. (1976). *A procedural guide for validating achievement gains in educational projects.* Washington, DC: U.S. Dept. of Health, Education and Welfare (Education Division), Office of Education.

Tallmadge, G. K., & Wood, C. T. (1977). *User's guide: ESEA Title I evaluation and reporting system.* Mountain View, CA: RMC Research Corporation, State Board of Education, Illinois Office of Education.

Thibaut, J. W., & Kelley, H. H. (1959). *The social psychology of groups.* New York: Wiley.

Thistlethwaite, D. L., & Campbell, D. T. (1960). Regression-discontinuity analysis: An alternative to the ex post facto experiment. *Journal of Educational Psychology, 51,* 309-317.

Thorndike, R. L. (1942). Regression fallacies in the matched groups experiment. *Psychometrika, 7,* 85-102.

Toulmin, S. E. (1961). *Foresight and understanding: An inquiry into the aims of science.* Bloomington: Indiana University Press.

Toulmin, S. E. (1967). The evolutionary development of natural science. *American Scientist, 55,* 456-471.

Toulmin, S. E. (1972). *Human understanding: The evolution of collective understanding* (Vol. 1). Princeton, NJ: Princeton University Press.

Toulmin, S. E. (1981). Evolution, adaption, and human understanding. In M. B. Brewer & B. E. Collins (Eds.), *Scientific inquiry and the social sciences* (pp. 18-36). San Francisco: Jossey-Bass.

Trochim, W. (1982). Methodologically based discrepancies in compensatory education evaluations. *Evaluation Review, 6,* 443-480.

Trochim, W. (1984). *Research design for program evaluation: The regression-discontinuity approach.* Beverly Hills, CA: Sage.

Trochim, W. M. K., & Campbell, D. T. (1996). *The regression point displacement design for evaluating community-based pilot programs and demonstration projects.* Unpublished manuscript.

Trochim, W., Cappelleri, J. C., & Reichardt, C. S. (1991). Random measurement error doesn't bias the treatment effect estimate in the regression-discontinuity design: II. When an interaction effect is present. *Evaluation Review, 15*(5), 571-604.

Trochim, W., & Spiegelman, C. H. (1980, August). *The relative assignment variable approach to selection bias in pretest-posttest group designs.* Paper presented at the Proceedings of the Social Statistics Section, American Statistical Association, Houston.

Twigg, R. (1972). Downgrading of crimes verified in Baltimore. *Justice Magazine, 1*(5/6).

van Fraassen, B. (1980). *The scientific image.* Oxford: Clarendon Press.

Verinis, J. S., Brandsma, J. M., & Cofer, C. N. (1968). Discrepancy from expectation in relation to affect and motivation: Tests of McClelland's hypothesis. *Journal of Personality and Social Psychology, 9,* 47-58.

Vertinsky, A. (1969). Research note: The use of aspiration-level behavior models in political science. *American Behavioral Scientists New Studies* (May-June), NS9-NS12.

Watson, J. (1968). *The double helix.* New York: Signet.

Watts, H. W., & Rees, A. (Eds.). (1977a). *The New Jersey income maintenance experiment. Vol. 2: Labor-supply responses.* New York: Academic Press.

Watts, H. W., & Rees, A. (Eds.). (1977b). *The New Jersey income maintenance experiment. Vol. 3: Expenditures, health and social behavior and the quality of the evidence.* New York: Academic Press.

Webb, E. J., Campbell, D. T., Schwartz, R. D., & Sechrest, L. B. (1966). *Unobtrusive measures: Nonreactive research in the social sciences.* Chicago: Rand McNally.

Weikart, D. P., Berrueta-Clement, J. R., Schweinhart, L. J., Barnett, W. S., & Epstein, A. S. (1984). *Changed lives: The effects of the Perry Preschool Program on youths through age 19.* Ypsilanti, MI: High/Scope Press.

White, R. W. (1959). Motivation reconsidered: The concept of competence. *Psychological Review, 66,* 297-333.

Wilder, C. S. (1972). *Physician visits, volume, and interval since last visit, U. S., 1969.* National Center for Health Statistics, July (Series 10, No. 75; DHEW Pub. No. (HSM)72-1064).

Wimsatt, W. C. (1981). Robustness, reliability and overdetermination. In M. B. Brewer & B. E. Collins (Eds.), *Scientific inquiry and the social sciences* (pp. 124-163). San Francisco: Jossey-Bass.

Wright, G. H. von. (1971). *Explanations and understanding.* Ithaca, NY: Cornell University Press.

Yin, R. (1984). *Case study research.* Beverly Hills, CA: Sage.

Zeisel, H. (1971). *The future of law enforcement statistics: A summary view.* Washington, DC: Federal Statistics, Report of the President's Commission.

Zuckerman, H. (1977). Deviant behavior and social control in science. In E. Sagarin (Ed.), *Deviance and social change* (pp. 87-138). Beverly Hills: Sage.

AUTHOR INDEX

SUBJECT INDEX

ABOUT THE AUTHORS

Donald T. Campbell spent the major part of his academic career at Northwestern University. Before retiring, he was Professor at Lehigh University. He received his AB and PhD from the University of California at Berkeley. He held teaching positions at Syracuse University, University of Chicago, and Ohio State University. During his career, he also lectured at Oxford, Harvard, and Yale Universities. He served as president of the American Psychological Association and was a member of the National Academy of Sciences; he received numerous honorary degrees and awards. He wrote more than 235 articles in the areas of social psychology, sociology, anthropology, education, and philosophy, covering a broad scope of topics from social science methodology to philosophy of science.

M. Jean Russo is a research scientist at the Center for Social Research and the Center for Innovation Management Studies at Lehigh University. She holds academic degrees from Moravian College (BA) and Lehigh University (MA, PhD). She was a student of Donald T. Campbell's while pursuing her master's degree in Social Relations and her doctoral degree in Applied Social Research. She has applied her research expertise in a number of diverse areas, including studies of severe childhood discipline, technology transfer, the sources and allocation of R&D funds in industry, and program evaluation.